Developments in Valves and Actuators for Fluid Control

BHRA

Proceedings of the

2nd International Conference on

Developments in Valves and Actuators for Fluid Control

Manchester, England: 28-30 March 1988

SPRINGER-VERLAG BERLIN HEIDELBERG GMBH

Editors: Dr. P. Wood
Mrs. L. Grove

EDITORIAL NOTES

The organisers are not responsible for statements or opinions made in the papers. The papers have been reproduced by offset printing from the authors' original typescript to minimise delay.

When citing papers from the volume, the following reference should be used:- Title, Author(s), Paper No., Pages, Proceedings of the 2nd International Conference on Developments in Valves and Actuators for Fluid Control, Manchester, England. Organised and sponsored by BHRA, The Fluid Engineering Centre, Cranfield, Bedford, England. (28-30 March, 1988).

ISBN 978-3-662-11465-0 **ISBN 978-3-662-11463-6 (eBook)**
DOI 10.1007/978-3-662-11463-6

Originally published by Springer-Verlag Berlin Heidelberg New York in 1988

ACKNOWLEDGEMENTS

The valuable assistance of the Organising Committee, Corresponding Members and Panel of Referees is gratefully acknowledged.

ORGANISING COMMITTEE

A. Henderson (Chairman)	BVMA
Dr. D.R. Airey	CEGB (MEL)
J. Franklin	British Gas PLC
M. Greenhalgh	Shaw Son & Greenhalgh Ltd.
W.W. Mott	Copes-Vulcan
A. Salisbury	Girdleston Pumps Ltd.
A. Scriven	Crane Ltd.
Dr. P. Wood	BHRA, The Fluid Engineering Centre
Lorraine Grove Conference Organiser	BHRA, The Fluid Engineering Centre

OVERSEAS CORRESPONDING MEMBERS

J.B. Arant	E.I. Du Pont De Nemours & Co. Inc., U.S.A.
C. Davila	TRW Mission Drilling Products Division, U.S.A.
L. Driskell	Consultant, U.S.A.

2nd International Conference on

Developments in Valves and Actuators for Fluid Control

Manchester, England: 28-30 Manchester 1988

CONTENTS

2nd International Conference on

Developments in Valves and Actuators for Fluid Control

Manchester, England: 28-30 March 1988

PAPER A1

A VARIABLE ORIFICE ROTARY CONTROL VALVE

Wayne Broadway
Daniel Industries, Inc.
P.O. Box 19097
Houston, Texas 77224
U.S.A.

Summary

A variable orifice rotary control valve with precise control capabilities has been developed for high pressure drop, compressible or noncompressible flow control. The valve is similar to a conventional ball valve except that its stem has been tilted back off vertical and identical variable orifice slots have been incorporated around and through the ball on both the upstream and downstream sides. The valve achieves its precise control by stretching the variable orifice slots out over a 270-degree rotation from fully open to fully closed. A special full bore or reduced bore trim is available for noise and cavitation control. The trim swings out of the flow path when the valve is in its full open position, giving the control valve an exceptionally high flow coefficient.

1. Introduction

Over the past several years, a tight economy and increasing energy costs have resulted in industry turning to the use of high speed digital computers to improve the control of fluid flows. To take full advantage of computer control, a new, more accurate control valve has been developed. The valve achieves its operating characteristic from a pair of specially-designed and contoured orifices built into a rotating ball control element. Precise control is obtained by increasing the rotation of the control element to 270 degrees. This means that for each degree of rotation, the controlled variable will change a very small amount. A cross-sectional view of the valve assembly is shown in Figure 1.

Held in Manchester, England. Organised and sponsored by BHRA, The Fluid Engineering Centre and co-sponsored by the British Valve Manufacturers Association and the Institution of Mechanical Engineers Process Industries Division

2. Valve construction

The valve design makes extensive use of castings to produce a precision manufactured valve at a competitive price. Straight through flow assures low wear of internal valve parts while throttling. This has been proven by examination of valves after six months on high pressure drop liquid service. In an aerated abrasive powder service, the valve proved to be a non-clogging design as well.

2.1 Center split body

The simplicity of the center split body produces an economical and rugged body design which will withstand the pipe stresses encountered in severe service flow control. The identical body halves are joined by a rigid flange to facilitate easy field maintenance. Flanged end connections are machined to ANSI standards but can be furnished to special customer specifications.

2.2 Ball control element

One piece stem and ball insures hysteresis free operation while controlling fluid flows. Oversized stems with rigid trunion-mounted body supports assure the ball control elements' integrity while throttling pressure drops equal to 0.97 of the inlet pressure. Torque requirements for positioning the ball control element while throttling are very low due to the pressure balanced ball and straight through flow.

Stems are supported in the trunions by self-lubricated, metal-backed, PTFE-lined sleeve bearings capable of supporting compressive loads as high as 310 MN/m^2 (44,000 psi). They are resistant to shock loading and can operate at temperatures from cryogenic to 280 degrees C (536 degrees F).

2.3 Tight shut-off

Body and trunion connections are sealed with totally confined o-rings. The rotary stems are sealed with metallic spring-energized, pressure-responsive lip seals. These seals feature ultra-low breakout and running friction and are self-compensating for wear. The seals are capable of handling temperatures from cryogenic to 315 degrees C (600 degrees F) and are available in a variety of elastomers and PTFE with a choice of fillers.

Upstream and downstream seats are spring-loaded to achieve an initial low pressure seal and are pressure-responsive so that seal integrity increases as line pressure increases. The dual seat arrangement gives the variable orifice valve the capability of bi-directional operation and tight shut-off.

3. Flow characteristic

The valve's inherent flow characteristic is determined by the shape of the orifice slot. A linear or equal percentage characteristic is standard, but other curves are attainable by reshaping the orifice slot. The full bore capability gives the valve a very high flow coefficient (Cv) in comparison to conventional control valves as shown in Table 1. Variable Orifice Valve Cv's were established per ISA procedures (1).

4. Low noise/anti-cavitating trim

Since most applications for this valve are for high pressure drop service, the valve will contain either low noise or anti-cavitating trim. The trim is located directly in the orifice slots and leaves an unobstructed, full bore through the ball control element. Fluid entering the valve will have an unobstructed flow path in the full open position, but will flow through the trim while the valve is throttling. The trim is a series of staged pressure drops, each stage being constructed from perforated plate with a wire mesh backing. Fluid flow is broken into multiple flow channels in each stage and is then mixed and, in the case of a gas, is expanded immediately after the major restriction in each stage. The restrictive design of the trim dissipates most of the energy of the fluid entering the valve without creating choked flow conditions at any point in the trim. The energy of velocity is dissipated at the same rate the pressure energy is being converted into velocity. The inherent flow curve for the trim may be varied anywhere from linear to hyperbolic by varying the number of stages built into the orifice slot.

4.1 Liquid testing trim

Water tests with the anti-cavitating trim have been completed on a DN100 (4") variable orifice valve. At 7-1/2% travel from the closed position the valve controlled a flow of 51 m^3/h (225 GPM) with an inlet pressure of 98 bar (1425 psig) and an outlet pressure of 2.3 bar (40 psig). That is a 95 bar (1385 psi) pressure drop, or $0.97P_1$, pressure drop, without noise, vibration or cavitation.

A DN50 (2") valve on Diathanolamine level control, with an inlet pressure of 55-62 bar (800-900 psig) and an outlet pressure of 6.9 bar (100 psig), was dismantled after six months of service. The valve exhibited no wear of the internal parts. Previous control valves in this application exhibited significant wear during the same service period and required a back pressure valve to be partially closed to control cavitation. The variable orifice valve controlled this severe service without the use of a back pressure valve and had no cavitation, noise or vibration.

4.2 Gas testing trim

Aerodynamic noise tests were completed in the Noise Control Laboratory of a leading university in the USA. The valve was tested at each 10% of rotation from 10% to 90%. The accoustic environment was provided by an anechoic chamber in each of these tests. The electrical connection for all transducers and instrumentation is represented in Figure 2. The total standard error for this chain was found to be 0.7dB.

Figure 3 is a graph of the measured A-weighted sound pressure levels versus the pressure ratio P1/P2 for the variable orifice valve at 10, 50, and 90% of rotation. Added to these graphs are plots of predicated levels for a conventional control valve having the same estimated Cv. The predictions for the conventional control valve were obtained from the procedure developed by Ward and Reethof(2). The plots indicate that a variable orifice valve can perform the same throttling function as a conventional control valve with roughly 10 to 20 dB less noise.

Figure 4 is a graph of the valve's correction factor for use in the graphical method of aerodynamic noise prediction as presented in the ISA Handbook of Control Valves, Second Edition. Curves of commonly used control valve types are shown for comparison.

5. Conclusion

Laboratory tests and field applications have proven the variable orifice rotary control valve to be an excellent final control element on high pressure drop liquid flow control. Field testing has proven the valve suitable as a non-clogging valve in the flow control of semi-solids. Laboratory tests indicate the valve has potential as a low-noise valve on gas service. Continued testing in the laboratory and field and further development of the low noise trim is anticipated.

6. Acknowledgement

Aerodynamic test data was taken from a report compiled by William C. Ward, Consulting in Acoustics and Noise Control, State College, PA 16801, USA. His tireless efforts in compiling the report in time for this paper are very much appreciated.

7. References

1. "Control Valve Capacity Test Procedure", ANSI/ISA S75.02-1982, Instrument Society of America, Research Triangle Park, NC, USA.

2. Ward, W. C. and Reethof, G., "Graphical Implementation of Fundamental-Based Method for the Prediction of Aerodynamic Control Valve Noise", presented at the ASME Conference on Pressure Vessels and Piping Systems, June 1985, New Orleans, LA, USA.

3. Fagerlund, Allen: "Predicting Aerodynamic Noise from Control Valves". Chemical Engineering, May 14, 1984, pp.65-67.

TABLE 1: Comparison of Flow Coefficients (C_V).

VALVE SIZE mm (in)	CONVENTIONAL SINGLE PORT	CONVENTIONAL DOUBLE PORT	CONVENTIONAL CAGED	VARIABLE ORIFICE ROTARY CONTROL
DN25 (1")	10.1	12.5	18.0	52.7
DN50 (2")	43.4	51.6	64.9	220.0
DN80 (3")	99.0	116.0	135.0	480.0
DN100 (4")	153.0	196.0	274.0	860.0

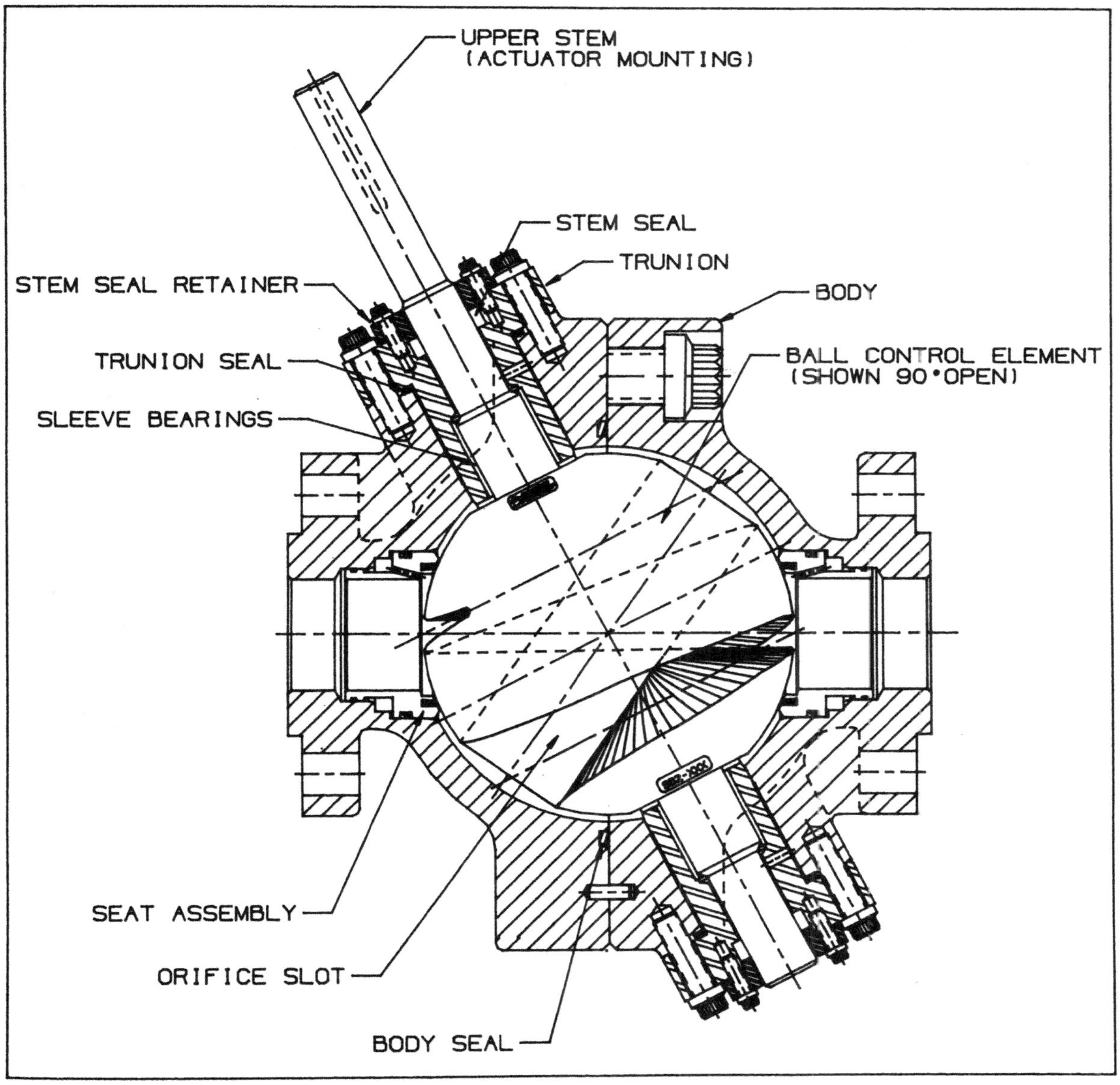

FIGURE 1: Cross-sectional view of valve assembly

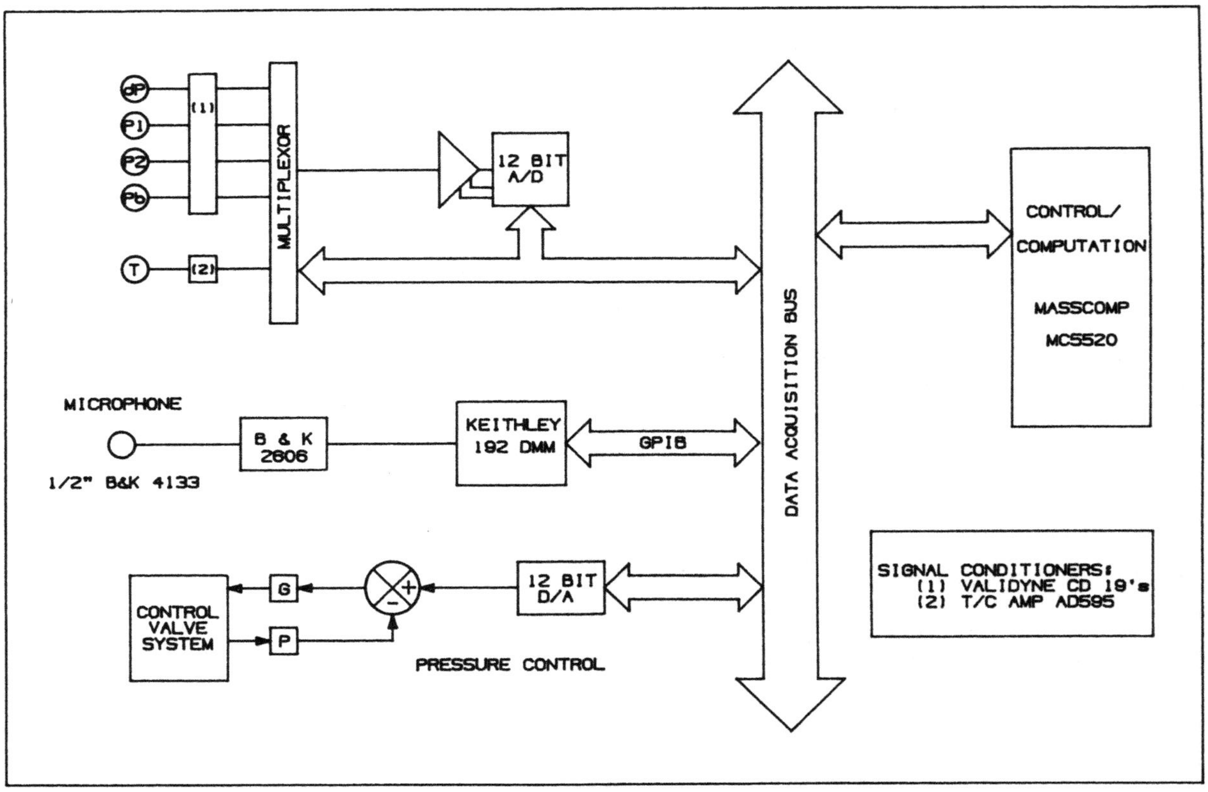

FIGURE 2: Instrumentation schematic

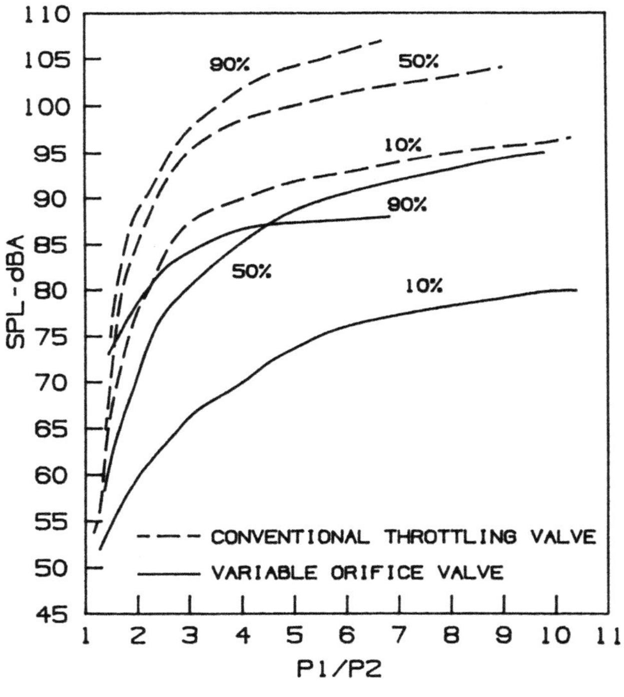

FIGURE 3: SPL for different valve positions

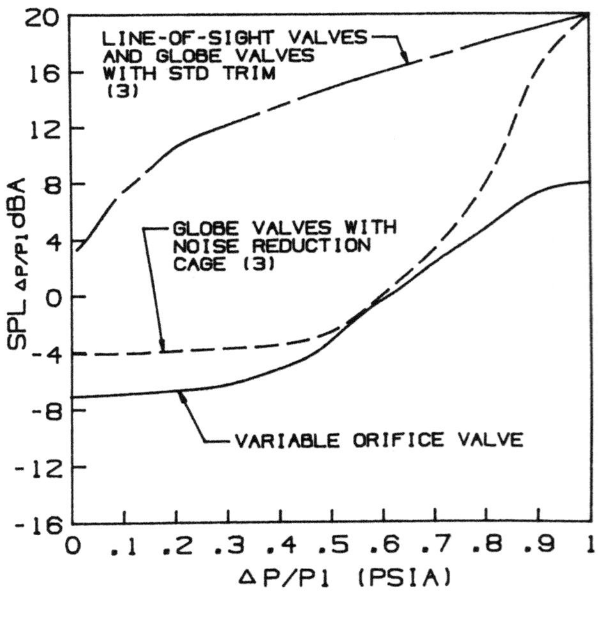

FIGURE 4: SPL for different valve styles

2nd International Conference on

Developments in Valves and Actuators for Fluid Control

Manchester, England: 28-30 March 1988

PAPER A3

CONTIGUOUS SEALING OF METAL SEATED BUTTERFLY VALVES

Robert C. Sansone
C&S VALVE COMPANY, TRICENTRIC DIVISION, USA

1 SUMMARY

Metal seated butterfly valves are being used more and more throughout the world. The selection and design of any metal seated butterfly valve for use as a replacement for resilient seated valves or in applications well above the physical and mechanical limits of pressure and temperature associated with resilient materials can be difficult and confusing. An understanding of the different sealing types and sealing concepts being used today will benefit not only the users, but the engineers as well.

2 NOMENCLATURE

Contiguous - All around, sharing an edge or boundary, touching on all sides.

Resilient - Rubber (various types), TFE, PTFE, Fluorocarbons, etc.

R_N - Normal reactions

R_T - Tangential reactions

3 INTRODUCTION

As butterfly valves gain acceptance in industry, manufacturers are responding with designs capable of meeting the "new found" uses for butterfly valves. The butterfly valve was once thought to be only useful in low pressure and low temperature applications. The original resilient seated "high performance" valves were the initial step in expanding the application limits of the butterfly valve. These valves did succeed in expanding the uses for butterfly valves up to the point at which the

Held in Manchester, England. Organised and sponsored by BHRA, The Fluid Engineering Centre and co-sponsored by the British Valve Manufacturers Association and the Institution of Mechanical Engineers Process Industries Division

temperature limits of the seal materials could not be overcome. The next logical step in the advanced development of the butterfly valve would be to expand the temperature/pressure limits of the seal materials; hence the era of the metal seated butterfly valve.

Metal seated butterfly valves are currently available which can satisfy not only the relatively moderate applications, but also the severe applications ranging from vacuum to high pressure, and at temperatures ranging from cryogenic to extremely elevated. With the current emphasis on plant safety regarding fires, some metal butterfly valves are now being designed which meet the variety of different "Fire Tested" criteria being specified. The butterfly valve long known for its light weight, compact design, quarter turn ease of operation, control capabilities, and now the ability to seal under severe conditions, has allowed the metal seated butterfly valve to be used in applications where gate, globe, plug, and ball valves were once used exclusively.

4 DISCUSSION

All metal seated butterfly valves can be broken down into two basic categories based on the sealing concept used.

4.1 Type 1 - Laminated Seal Butterfly Valves

The first category can be defined as those valves which incorporate a relatively rigid seal which contacts an extremely rigid seat. Typically these seals are made up of multiple metal laminations retained in the periphery of the valve disc. The use of multiple laminations results in a surface contact as opposed to a line contact. A typical design concept of this type of seal is shown in Figure 1.

During closure of this type of valve, no deflection of the seat, and only a minimal flexure of the seal takes place. The contact of the two mating surfaces must occur virtually instantaneously at all points around the 360° of interface. For the instantaneous mating to occur and to allow the valve disc to rotate freely during opening, a unique geometry is required. Essential to the design of such a sealing system are three eccentricities (Fig. 2). The first eccentricity is that the valve shaft centerline is displaced a positive distance from the sealing surface so that an uninterrupted 360° of seat/seal interface is available. The second eccentricity shall be defined as the positive distance between the seal axis and the shaft centerline. The third eccentricity is the distance between the cone axis of the seat and the centerline of the seal.

The relationship of these three eccentricities is critical to the operation of this type of sealing system. When the proper geometric relationship of the eccentricities exists, a very effective seal is possible. Although Figures 1 and 2 are only two-dimensional, it should be noted that the mating surfaces between the seat and the seal actually represent the surface described by the frustrum of a cone. To this point only the geometry necessary to allow the rotation of one continuous seal element into a mating stationary seat has been discussed. Now that the two surfaces can be mated, a closer look at the actual sealing is required. The metal laminations used in this type of seal can be thought of as multiple independent sealing members acting in unison to effect a seal. An additional benefit of multiple laminations is that a labyrinth sealing effect occurs in which pressure drops across each sealing member until all pressure is dissipated resulting in a tight seal.

Since the forces and reactions between the seal and seat are acting on rigid members, the elastic properties of the members cannot be expected to generate the forces in sufficient magnitude to effect a seal once closure has occurred. It is for this reason that some external force(s) be maintained during all times at which shut-off is to occur. Because the valve seat in this type of valve is actually the closed position stop, these valves are considered to be "Torque Seated". What this means is that the valve actuator must never absorb the torque or limit the travel in the closed direction. When this type seal is installed in a specific orientation, that being with the higher line pressure applied to the shaft side of the seat/seal interface, a torque results that is independent of the torque produced by the actuator. The pressure on the shaft side of the valve acts on a lever arm of length "eccentricity 2" producing a torque which in turn is reacted to by the seat (Fig.3).

As can be seen in Fig. 3, the torque "T" can be generated by the actuator alone or with the addition of the torque produced by the pressure "P". When an equal pressure is applied to the opposite side of the valve, the torque component of "T" caused by the pressure will have the same magnitude as before, but will now have a negative value and must therefore be subtracted from the total torque. A valve installed in this orientation must have an actuator with sufficient torque output to not only overcome the negative torque produced by pressure, but also maintain a minimum value of R_N to effect a seal for the two materials in contact.

Just as the Normal Reaction (R_N) of a gasket to the forces applied by two bolted flanges effects a seal in a pipeline, the normal reaction (R_N) to the closing force caused by torque between the seat/seal causes a seal to occur in a valve. Shown in Fig. 3 is the fact that the tangential reaction (R_T) not only varies in magnitude, but also completely changes direction as forces and reactions are considered around the full 360° of interface.

Although R_N does change in magnitude around the interface, its directions always remain the same, causing a compressive load to be exerted radially inward on the seal. The loading between the seat/seal may actually be high enough to cause macroscopic yielding or "bedding in" of the materials during the first several loadings resulting in a pseudo lapping action which allows the two surfaces to form an even better seal. Due to the fact that these valves are torque seated, they are also repeatable in effective sealing since the seal will always rotate closed until the reactions equal the applied closing force. Another point to be considered is the fact that since normal and tangential reactions occur instantaneously, the factor causing wear or rubbing (R_T) as the valve closes is minimized in this type of seal.

By using this theory, sealing between such metals as Stainless Steel/Stainless Steel, Stellite/Stellite,[1] Inconel/Inconel,[2] Monel/Monel,[3] and Hasteloy/Hasteloy[4] have been achieved. With the selection of metals available and the high sealing forces generated using this type of seal arrangement the application and design possibilities using this type of seal are almost unlimited.

4.2 Type 2 - Flexible Metal Seal Butterfly Valves

The second category of metal seated butterfly valves shall be defined as those valves which have an extremely rigid disc edge which mates with a relatively flexible seal. In general, these metal seals are retained in the body of the valve and act similar to their non-metallic counterparts. To obtain the flexure required of this type of seal, they usually have a very thin cross-section and are usually pressure assisted in sealing. A general design concept of this type of valve is shown in Fig. 4.

The geometry of this type of butterfly valve may appear to be similar to the first type of valve discussed but actually differs radically. In this type of design the first eccentricity described previously is always present, the second eccentricity may or may not be incorporated, and the third is never present, (Fig. 5).

Since eccentricity 1 is always present, this type of valve also has a 360° continuous seal that is uninterrupted by the shaft. Some designs incorporate eccentricity 2 into the design but it is usually a very minimal displacement when compared to the other type metal seated butterfly valve. The lack of the third eccentricity is the major difference between the two sealing concepts. Without the third eccentricity, the second eccentricity becomes irrelevant whether it is present or not. Some designs may claim that with the second eccentricity, the disc rotates "cleanly" out of the seal. Although this may slightly reduce the contact time (drag)' as the disc rotates in or out of the seal, it cannot minimize or eliminate it as can be shown by plotting a path of constant chords for a specific valve design.

With the knowledge that instantaneous contact will not occur, an alternative approach must be considered. When the valve disc rotates to the closed position, a given outside diameter of a disc is trying to engage a smaller inside diameter of a seal. Most typically the disc edge will have a spherical radius which helps the engagement take place.

The way complete engagement occurs is when the seal acting as a resilient member is required to compensate for the interference between the two mating surfaces. Seals in this type of valve have the small cross-section mentioned earlier to allow them to expand with a relatively small seating torque. Since the sealing members have one position of engagement in which the mating diameters and flexure of the seal are optimized, these valves are position-seated. Further, rotation of the disc into the seal will not yield an increase in seal flexure all points around circumference.

Hoop stresses developed in the seal can be shown to cause normal and tangential reactions similar to the Type One valves except that the tangential reactions do not change in direction. The initial value of R_N may be high enough to effect a partial or complete seal depending on the materials selected. As pressure increases, R_N will undoubtedly need to be increased to effect a seal. Also R_N will also have to increase to prevent the pressure from blowing the seal away from the disc, causing leakage.

The only source left to provide extra force which is translated to increases in R_N and R_T is the differential pressure across the disc. Seals are designed to take advantage of this pressure and are therefore considered to be pressure activated.

A derivation of the Type Two valve may incorporate one or more resilient sealing components in addition to the metal seal. The value of R_N required to seal these resilient materials is much lower than that required for metals. By combining the two types of materials, a primary seal using the resilient member and a secondary metal seal, an attempt to satisfy all application requirements is made. This type of combination seal has been used to meet certain fire tested requirements. An industry concern exists relative to valve leakage, if the resilient portion of the seal is partially or completely destroyed. This mutant type valve should not be considered a metal seated butterfly valve since it still has the same material limitations present in pure resilient seated valves, and is therefore incapable of operating beyond the material limits for any length of time.

5 APPLICATIONS

Water Intake - Crustaceans, such as clams, can accumulate on the seals while the valves are open. When the valve is required to close, these crustaceans are dragged across the resilient seals and may tear or cut them and cause sufficient damage to prevent sealing. The metal seated butterfly valve will not be susceptible to such damage as the metal is harder than the crustaceans and can literally destroy them so a seal can be effected.

Blast Furnace Gas - The temperature at which the valves operate is usually at or above the operating limits of resilient materials. Particulate matter may also be present which can be very abrasive to resilient seals and result in a service life that is less than desirable due to costs associated with valve change outs.

Sulphur - The solidification of sulphur in valves has been a continual problem for both valve designers and users. The metal seated butterfly valve has proven to be a solution by using the higher seating torque required to clear the solidified sulphur away from the seat.

Cryogenic - Some resilient material may get brittle at extremely low temperatures. As the disc enters the seal, the initial interference between the seal and disc can cause destruction of the seal. Another point to consider is the difference in coefficient of expansion between metals and resilient materials which can result in a lack of seat/seal engagement causing leakage.

Bottom Ash - Resilient seated valves may be unable to close due to ash build-up and the lack of torque to break through the ash built-up. Erosion in this service can also be a problem causing a reduced service life of resilient seated valves.

Steam - The real world problems with resilient seated valves in steam service continue to frustrate the industry. Certain steam applications can present valving headaches due to temperature, pressure, velocity, or erosion. The metal seated butterfly valve can resolve the problems on many steam applications.

General - The metal seated butterfly valve should be considered when pressure and temperature concerns prohibits the use of resilient materials. Where high velocities are present resilient seals can actually be completely or partially sucked out of the valve and pushed down stream leaving the valve inoperable. With the use of a rigid metal seal, this problem can be avoided as the metal cannot be deformed. Erosion and abrasion can be extremely detrimental to resilient materials. The proper selection of available metals can reduce the problems of erosion and abrasion to acceptable levels.

6 CONCLUSION

For reasons of patent infringement and disclosure, this paper does not go into the complete engineering and mathematical solution to each valve design, instead a basic insight has been presented. With a better understanding and evaluation of different concepts used to effect seals in a metal seated butterfly valve it is hoped that the job of the designer and user will be made easier. As with any technology, the current metal seated butterfly valve will only be state of the art until the next major advancements are made.

FOOTNOTES

[1] (Haynes) Stellite is a registered Trademark of the Cabot Corp.

2 Inconel is a registered Trademark of International Nickel Co.

[3] Monel is a registered Trademark of International Nickel Co.

[4] Hasteloy is a registered Trademark of the Cabot Corp.

REFERENCES

Litvin, F.L., "Synthesis of Conjugated Surfaces for Sealing Systems," ASME Journal of Mechanisms, Transmissions, and Automation in Design, Vol. 105, 1983.

Lyons, J.L. and Askland, C.L., Jr., Lyons' Encyclopedia of Valves, Van Nostrand Reinhold Co. 1975.

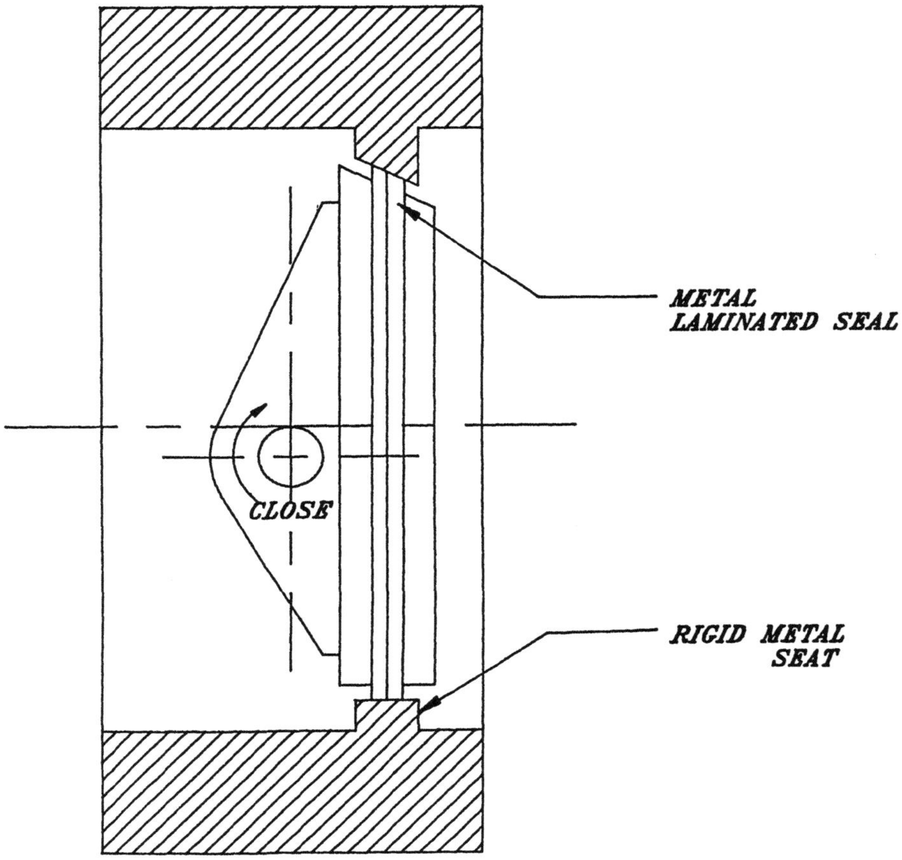

Figure 1 Typical Laminated Seal Butterfly Valve

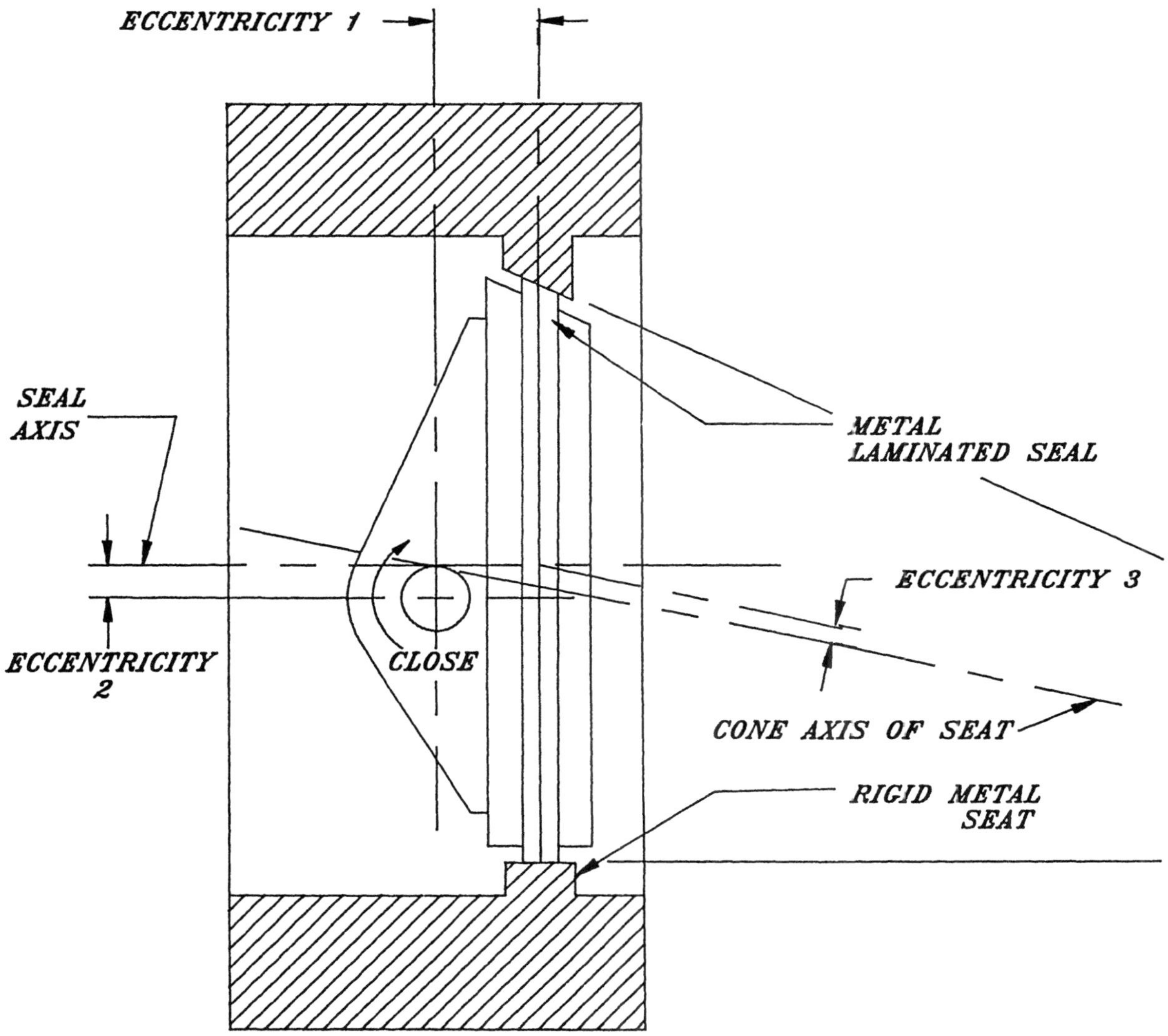

Figure 2 Three Eccentricities

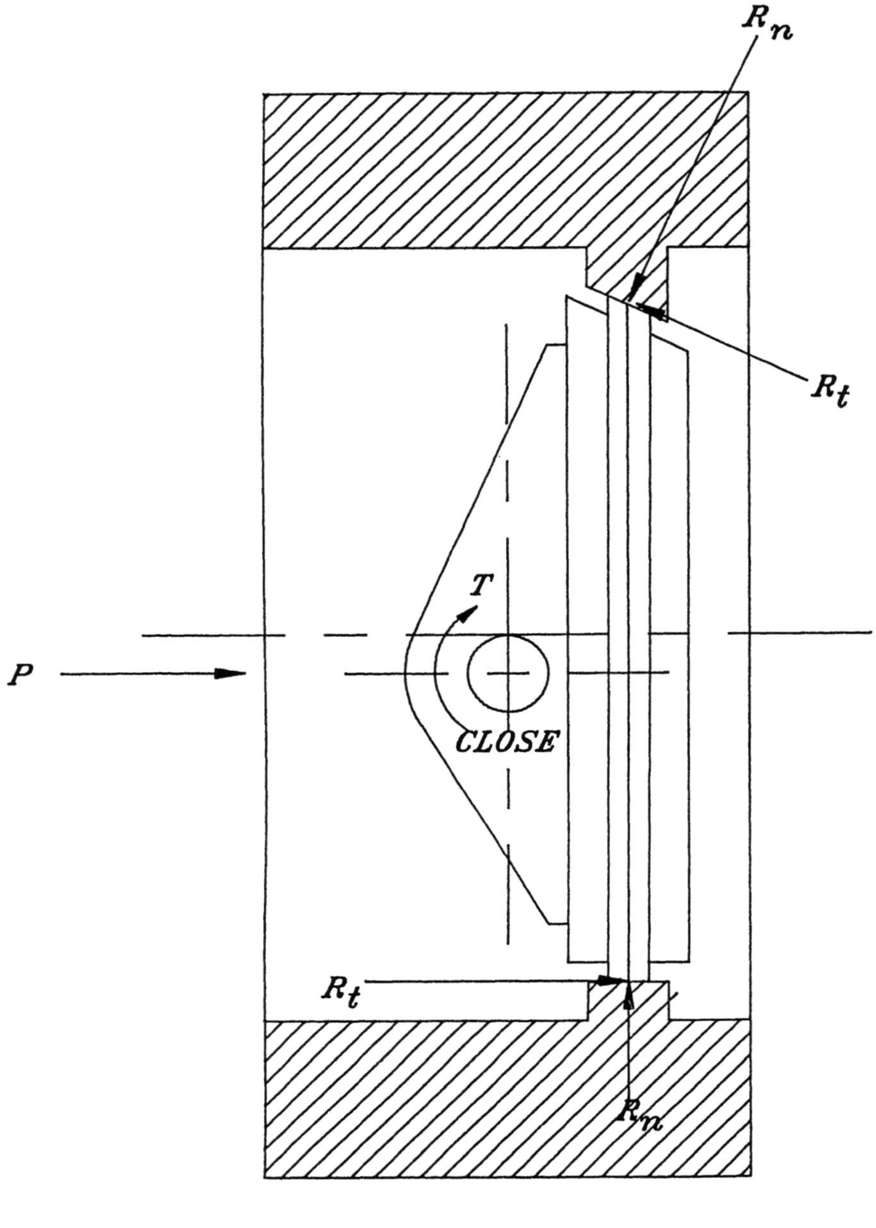

Figure 3 Torques, Forces, and Reactions

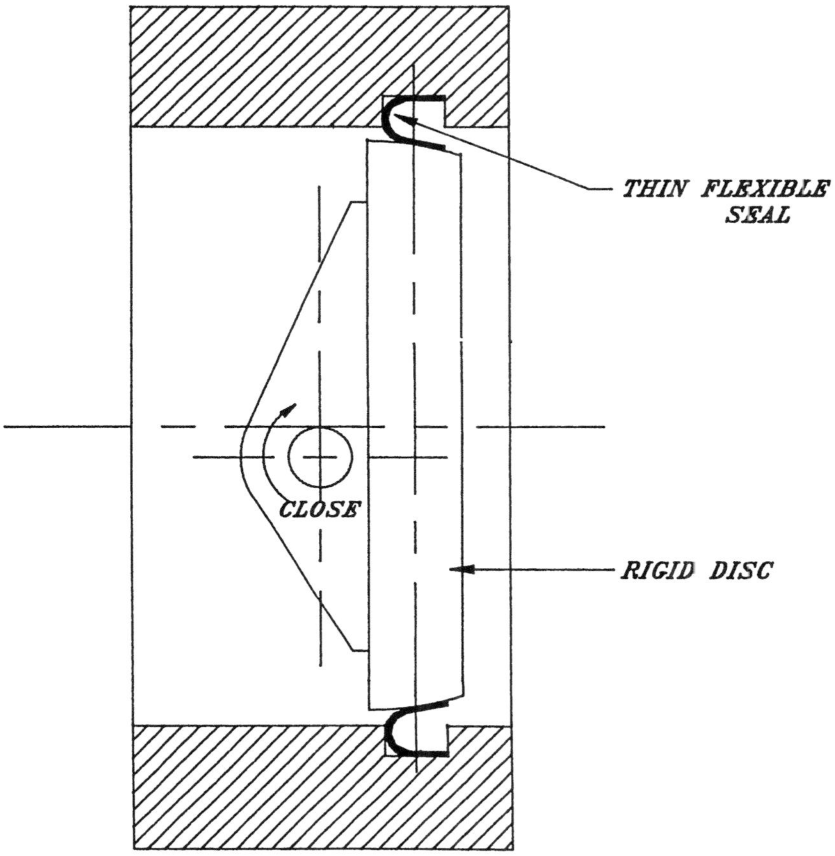

Figure 4 Typical Flexible Metal Seated Butterfly Valve

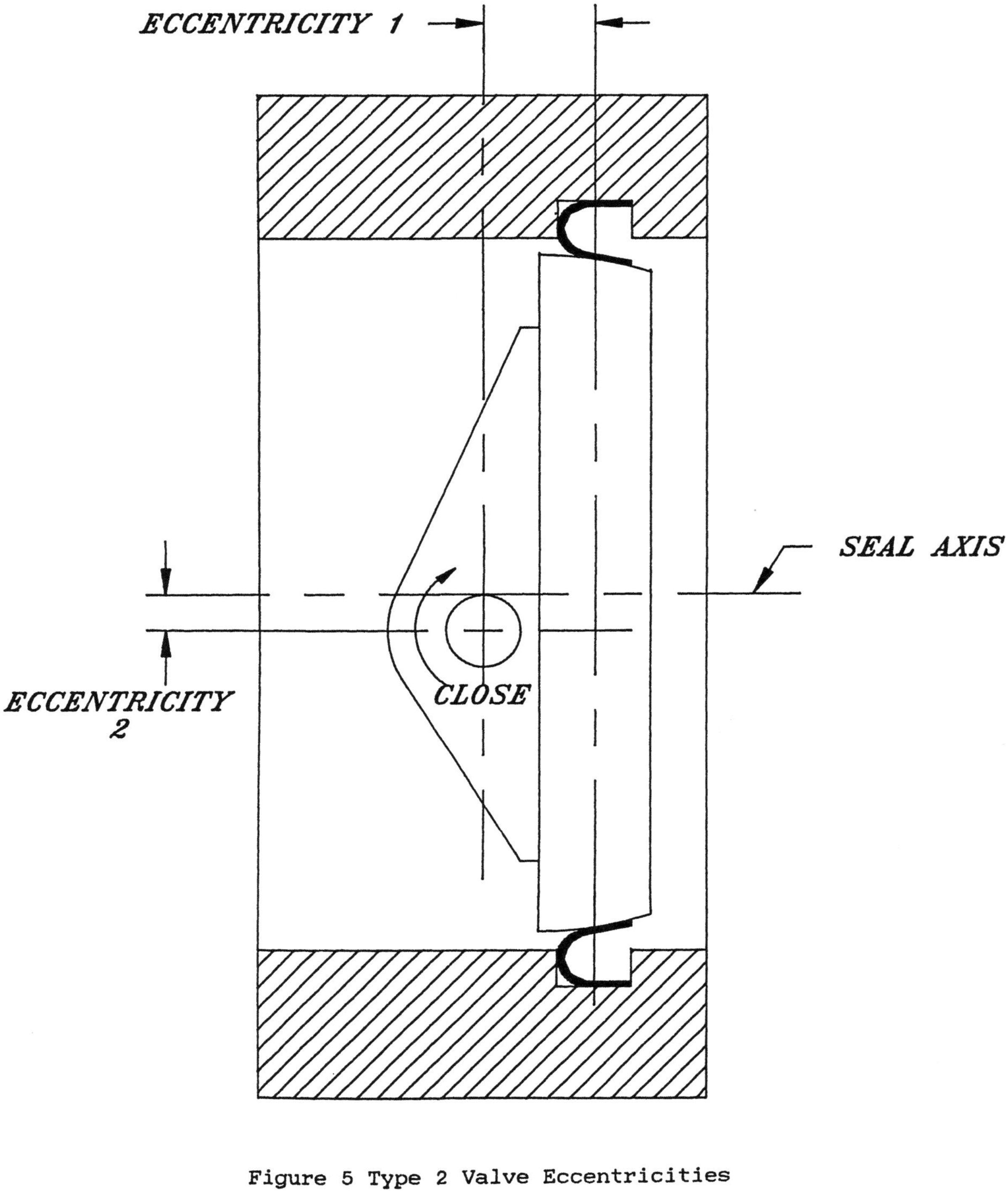

Figure 5 Type 2 Valve Eccentricities

2nd International Conference on

Developments in Valves and Actuators for Fluid Control

Manchester, England: 28-30 March 1988

PAPER B1

TESTING OF VALVES AND ACTUATORS FOR FIRE ENVIRONMENTS

Joseph B. Wright, Vice-President Technical Marketing
Jamesbury Corp.

The period of the mid-1950s saw a fundamental change in the selection of valves for chemical plants and refineries. That change was brought about with the commercialization of soft-seated, quarter turn ball valves with their attendant ease of operation, position indication and tight shut-off. These characteristic have been extended to other types of quarter-turn valves resulting in these types of devices gaining a significant share of the valve market in the ensuing three decades.

While process demands have brought about many design evolutions of the soft-seated quarter turn valves, perhaps none has had greater impact than the need for fire resistance. Historically, those operating chemical plants and refineries saw the soft seated ball valve as a desireable alternate to gate valves with their tendencies toward leakage or freezing in position.

Held in Manchester, England. Organised and sponsored by BHRA, The Fluid Engineering Centre and co-sponsored by the British Valve Manufacturers Association and the Institution of Mechanical Engineers Process Industries Division

Since the gate valves generally depended on metal-to-metal contact to achieve sealing, exposure to industrial fires had little affect on their leakage characteristics. Users concern with this new generation of soft-seated valves was a valid one that fire exposure could destroy the soft seats resulting in valve leakage.

The problem of exposure of soft-seated ball valves in the fire environment was studied by several manufacturers in the time period of 1955 through 1957. Within a short time, several designs were marketed which incorporated secondary metal seats to minimize leakage in the event of soft seat destruction by fire as well as ancillary stem and body seals of fire resistant materials.

The question of the hour was how to prove out those designs and rate their effectiveness relative to fire resistance. Two problems existed with that evaluation; defining an industrial fire that these valves should be expected to resist and devising a test be devised to simulate that fire environment. Both deserve examination.

The issue of defining the industrial fire is not trivial. One can postulate fires that range from smoking oxygen-poor low temperature fires with their resultant low heat flux to the extreme of a hydrogen jet fire with flame temperatures exceeding 2800°C. Additionally, no valve is an entity by itself in a fire. Instead, one must look at an industrial fire in its entirety and include effects on pipe supports, pressure-retaining bolting, tanks, and even concrete structures. In fact, this area has been the thorniest one relative to the issue of fire-testing of valves for over two and one-half decades.

After several years of debate, it became apparent that the definition of a fire was not going to be resolved quickly. Groups on both sides of the Atlantic iniated their own fire tests in an effort to do "something" rather than wait for the perfect definition of a fire.

In the U.K., the Oil Companies Materials Association, OCMA, postulated a fire test later to be known as OCMA FSV-1. This test procedure was founded on the thought that since 95% of all valves in service were in the open position, they should be evaluated in the open position. After much debate, the issues of fire exposure time and temperatures were resolved by requiring the valve to be burned for sufficient time at temperature so to totally decompose the software.

During the same period of time in the U.S., a major ball valve manufacturer collaborated with the Factory Mutual Research Labs to develop a fire test. Fundamentally, this test (FM-6033) revolved around burning the valve also until soft parts decomposition occurred. This test differed from the British standard in that the valve was in the closed position during the burn position.

In the early 1960s, the American Petroleum Institute's Division of Refining began to study the issue of fire effects on piping. The initial portion of this study on complete piping systems led the work group to realize that the task of looking at all piping components was entirely too enormous for any accomplishment. After review with the Committee on Refinery Equipment, the work group began to focus solely on the issue of soft-seated ball valves.

The period of the late 1960s and 1970s saw actual fire tests run by a very few valve companies to research exactly what happened to valves during a fire. Test results from these few companies were freely shared in an effort to increase the state of knowledge relative to valve exposure. During this time period, increased liason of petroleum refinery engineers on both sides of the Atlantic took place through host efforts on the part of both the British Standards Institute and the American Petroleum Institute's Division of Refining. Not suprisingly, the research in fire test developments drew a great deal of interest during the 1970s.

The valve companies performing tests developed a host of information which pointed to fundamental flaws in both the OCMA FSV-1 and the FM 6033 fire tests. Most significant of these flaws was the issue of burning in the open position. This issue was followed by the duration of fire exposure. Both issues deserve a closer look.

Soft-seated valves which employ seats on both the upstream and downstream side of the obturator will trap fluid in the cavity formed by these seats and the pressure shell. If this fluid is an incompressible liquid, increases in temperature will cause the pressure of the trapped fluid to rise dramaticlly. Calculations and rather simple experiments will both show that a trapped hydrocarbon liquid will increase in pressure by 12 Bar for every degree C of temperature rise. This is a known problem which has been so often overlooked that paragraph 2.3.3 of the American National Standard, B16.34 "Valves-Flanged and Buttwelding End" warns about the effects of thermal expansion of fluids trapped in double-seated valves.

While some ball valve designs are capable of automatically relieving this cavity pressure when in the closed position to the upstream side, others may not. The fatal flow in the original OCMA FSV-1 was found to be that the test did not evaluate the valve's resistance to cavity pressure rise as conditions with the valve open and a vent hole in the ball stem slot did not represent those of a closed valve.

Research of this problem led API to insist on a cavity filled, valve closed test. After seeing cavity pressure increases of 200 Bar within three minutes of flame ignition, members of the API Task Force on fire testing recommended a change in test fluid from kerosene to water in the interests of test personnel safety. Naturally, these results were shared with the work group in the U.K.

The thorny issue of test duration caused years of debate and testing. Happily, this issue led to what is really, in my opinion, the future area of study of fire testing; the piping system. Two problems in test duration became evident; how to standardize the heat flux to the valve, and for what period of time the valve should be exposed to this fire.

The issue of fire duration came easiest. The early U.S. and U.K. fire tests required total destruction of valve soft components. In attempting this with large valves, such as size 8, that were liquid filled, flame exposure times ran from 3 to 8 hours. In discussing this with refinery engineers with actual industrial fire experience, it became evident that fires of that duration caused problems far and above valve performance. A strategy of fire fighting seemed to be universal throughout knowledgeable users; namely if the fire isn't beaten in one-half hour, a withdrawal and containment policy is instituted. This, of course, comes from structural component failures such as pipe rack collapse, flange bolt failures, and even concerete eruptions due to water of hydration changing to steam. Based on the limiting factors of ancillary equipment, a test duration of one-half hour was established.

Heat flux to the test valve has been a major issue for years. Early fire test standards including API 607 (Division of Refining) and API RP6F (Division of Production) attempted to deal with this by establishing valve temperatures during the fire. Unfortunately, tests of ANSI pressure class 2500 where the contained test fluid (water) was above its critical pressure showed this to be an unworkable solution. Additionally the issue of valve mass came into question. After many years of work, a step back to college thermodynamics was taken by placing calorimeter blocks around the test valve and subjecting the assembly to a fire environment causing in a standardized thermal response of these calorimeters.

The mid-80s became an exciting time for those of us who have devoted a substantial portion of our careers to the issue of developing valves for the environments of industrial fires. The excitement came from the fact that the three major fire tests for valves became aligned; today we have the British Standards Institute and the two Divisions of the American Petroleum Institute all having the same test procedures. Perhaps it speaks well to the sharing of technical information through conferences like this one that achievements such as standardizing on fire tests can happen.

Today we have common international standards for fire tests which evaluate the following three critical elements of a valve design;

1. Ability to either withstand or relieve any internal pressure buildup

2. Operability while hot

and 3. Leakage behavior during a rapidly controlled fire

Evaluation of these three issues to conform to end users needs will advance the design of soft-seated quater turn products in a manner which should improve them in even normal daily services. The obvious question is where to go from here.

As a valve engineer whose involvement in fire testing spans twenty years and who has been fortunate to participate in these new standards, I do see three areas for future work. These involve the following issues and each issue deserves a brief comment.

1. API/BSI fire test standard rationalization

2. Adoption of these standards by users who currently have unique corporate standards

and 3. Extension of these valve fire test standards to cover automated valves

Looking at rationalization of the existing API and BSI fire tests, step one of this is in motion in that the American Petroleum Institute steering group has directed both the Production and Refining Divisions to rationalize to a single common standard. The obvious question is whether it would be logical for an ISO standard to replace both the BSI and API documents. From my perspective, this would be a serious error in that ISO may be too large and bureacratic to effect changes in the standard based on either technology or additional research. More to the point would be a shared API/BSI standard stewarded by a joint task force or steering group.

The issue of corporate standards being replaced by either API or BSI documents is a difficult and emotional one. Not only does the NIH problem exist; I suspect there are some cases where no one knows why a particular fire test was adopted by their corporation and no one will take the responsibility to change their documents. Perhaps these corporate standards' groups should study the fundamental changes in the API and BSI approaches and the reasons why these changes were accepted. Clearly though, it would seem in the best interest of all to switch to a common standard and to advance the state of the art through that standard rather than maintain a series of unique and, in some cases, obsolete standards. Fundamentally, the largest obstacle to replacing unique corporate standards with a common standard lies in education relative to valve behavior. It never ceases to amaze me that years of test experience and theoretical studies can be overlooked by a large refiner who defines artificial and unrealistic conditions such as partially-burned valves without any consideration for either the conditions which might cause this or their effects on the balance of the piping components. Clearly the top operating engineering management of those companies who maintain unique fire tests should review the state of today's art and then act.

The remaining area for future fire test development is that of automated valves. With the trend to taking workers out of the plant through increased automation, there is a need for automated valve product evaluations in fire environments. While the fire testing of automated valves at first seems trivial, this is distinctly not the case. Once again, those engineering groups that undertake study of this will have difficulty in the area of problem definition.

Perhaps looking at three topics for automated valves in the environment will present a starting point. The cases worth examining are:

1. An automated valve with a built-in device to allow the assembly to automatically go to either valve open or closed and remain there in a fire

2. A protection scheme to allow multiple operations of an automated valve for some duration of fire exposure

and 3. The study of the effects of automation devices on valve leakage during fire exposure

While these topics represent ideas for a starting point, they are not absolute in any sense of the word. Clearly, the time has come for bringing together knowledgable engineers to discuss the subject of automated valves in the fire environment and to chart a course of action. Based on my past experience, I would recommend a joint effort between the API and the BSI to study this subject.

In summary, the past two decades have been exciting ones in the testing and design of quarter-turn valves engineered for fire resistance. Not only has the technology been of interest, the cooperation between end user and valve manufacturers to solve a problem has been refreshing.

2nd International Conference on

Developments in Valves and Actuators for Fluid Control

Manchester, England: 28-30 March 1988

PAPER B2

COLD WATER LEAK DETECTION IN HIGH PRESSURE STAINLESS STEEL ISOLATION VALVES USING ULTRASONIC NOISE EMISSION

By D R Airey and O Lloyd
Marchwood Engineering Laboratories, CEGB, Marchwood,
Southampton, SO4 4ZB.

SUMMARY

This Paper investigates the ultrasonic seat/disc leakage noise, in the 10 to 100 kHz frequency range, emitted by 3 different stainless steel globe isolation valves subjected to differential pressures up to 170 bar in cold demineralised water. The measurements were made using a piezo-electric accelerometer and a commercially available leak analyser linked to a micro-computer to facilitate data storage and analysis. The dynamic range and accuracy of the noise measurements were enhanced by computer integration of the area under the whole frequency spectrum following subtraction of the background noise.

Noise was detected at flow rates as low as 150 ml/hr which is of the same order as calculations of turbulence/cavitation inception in a hypothetical leak channel. The noise integral varied linearly with differential pressure, with flow rate to the power 1.4, and was substantially independent of the three valve designs studied. Repeatable measurements with a small scatter were obtained at leak rates ≥ 1000 ml/hr suggesting the possibility of the development of

Held in Manchester, England. Organised and sponsored by BHRA, The Fluid Engineering Centre and co-sponsored by the British Valve Manufacturers Association and the Institution of Mechanical Engineers Process Industries Division

a reliable quantitative instrument for power station valve leakage measurement in this flow range.

1. INTRODUCTION

There are a number of applications in power stations where the internal leakage rate of fluid across a valve seat can have important operational or safety related consequences. However, establishing which valve is leaking, or the leakage rate, can be a time consuming and laborious exercise if the suspected faulty valve is one of many connected to a high pressure pipework system. Thus the development of a method of leak detection which does not involve the uncoupling of high pressure pipework is particularly attractive.

Such diagnostic/maintenance tasks are a persistent problem on existing CEGB power stations. In PWR nuclear power stations however, the very limited access to primary circuit valves inside the containment building during power generation means that valve investigations and repairs are confined to reactor refuelling outages. Therefore, investigations of suspected leaking valves together with regular leak testing of critical valves inside the containment building, can require a large number of personnel over a short period of time and contribute significantly to the radiation exposure burden for the station.

A potentially unobtrusive approach is to detect the high frequency acoustic noise emitted in the region of the fluid leak path and transmitted through the body of the valve to the surroundings. On the basis that the intensity of the noise should increase with leak rate for a given set of fluid, valve and measurement conditions, this method offers potential for the development of a working power station instrument capable of calibration to yield leak rate data from any suspect valve without mechanical interference with the valve or it's associated pipework. Conversion of the noise signal into a frequency spectrum might help the operator, or preferably the instrument, to distinguish between low frequency background plant noise and the high frequency leak signal.

Techniques of this type have been in use for some years in the USA, France and Japan mainly to distinguish between badly leaking and non-leaking valves, References 1 to 4. Attempts to quantify the leak rate in terms of the noise level have been carried out, Reference 5, and leaks of the order of 500 ml/hr have been detected in cold liquid by monitoring noise emission at 20 kHz under laboratory conditions. However, power station measurements require a large element of experience on the part of the operator in interpreting the noise signal and an awareness of the potential pitfalls; no unique relationship exists between noise and leakage for all valves under all possible different operating conditions of fluid, temperature, pressure and background noise.

Therefore, the present paper is concerned with investigating the various relationships between noise emission and some of the parameters which might affect it. A commercially available leak analyser was used and, in an attempt to overcome data interpretation problems and to facilitate data analysis, the instrument was linked to a micro computer with data storage and hard copy facilities. A novel approach was used to enhance the dynamic range and accuracy of the noise measurement; namely integration of the area under the high frequency spectrum curve following subtraction of the background noise.

2. SOURCES OF COLD WATER NOISE IN LEAKING VALVES

A brief assessment of noise sources in valves under the present cold-water leakage conditions suggests three main mechanisms: turbulence, cavitation and mechanical vibration, see for example the brief review and example calculations in Reference 6.

Turbulence gives rise to random pressure fluctuations over a broad frequency band. For a typical seat/disc radial seal dimension of 1 mm and assuming a single leak path, simple calculations suggest that the Reynolds number would exceed 2,000 and the flow become turbulent for a leak rate of about 100 ml/hr as shown in Figure 1. Theoretically the intensity of the noise generated by turbulence in incompressible flow varies as the 8th power of the fluid velocity in the leak. However, once the flow is turbulent the distribution of velocities is likely to depend only on the pressure gradient and leak geometry. This will tend to set an upper limit to the turbulent velocity for a given valve and pressure differential.

Cavitation is the rapid formation of bubbles when the local fluid pressure falls below the vapour pressure. Subsequent increase of the fluid pressure further downstream causes bubble collapse and the generation of continuum noise emission with an inverse law dependence frequency. Similar calculations to those above for a single leak channel and for a fluid temperature of 20 deg C (water vapour pressure = 0.02 bar) suggest a cavitation inception leak rate of about 600 ml/hr. This figure would be modified, and the noise intensity changed, if a significant quantity of gas is dissolved in the fluid, Reference 5, or if irregularities in the leak channel walls caused low pressure "pockets". The relationship between cavitation sound level and fluid velocity is not clear. However, we note from simple theory that the local pressure falls as the square of the velocity and we might anticipate the number of bubbles proportional to the flow rate. References 1 & 5 suggest that the acoustic power varies as the square of the differential pressure. Simple calculations of the time to collapse of a 1 mm radius bubble under the above conditions suggest sound wave oscillation frequencies of 20 kHz and greater for smaller bubbles.

Mechanical noises may be excited by turbulent flow and cavitation. Calculations on globe valves similar to those used in the present investigations, Reference 7, indicate natural frequencies of internal components in the range 1 to 10 kHz. Extraneous background noise arises from pumps, (motor, bearing and blade noise), and valve actuator noise. This should be in the range < 1 kHz. Most pipework operates with fully turbulent flow giving some contribution, but air actuator noise, although of high frequency, is poorly coupled into the valve pipework.

The total fluid power available for noise production by some unspecified mechanism varies linearly with both the pressure drop across the leak and the leak flow rate. However, in the case of small leaks, much of this power goes into overcoming fluid viscous forces at the walls of the leak channel.Such a viscous power dissipation tends to be non-turbulent, might inhibit cavitation, and might not therefore produce noise. An alternative is to consider the fraction of the total power available in the form of the kinetic energy of the leak stream, Figure 1. This fraction may be more useful in considering the power available for driving acoustic emission processes. In this case the power available varies linearly with the pressure drop but as the square of the flow rate. We note that this power is released not at the walls of the leak but in the efflux of the jet into the downstream fluid.

3. VALVES EXAMINED AND FLUID CONDITIONS

The present investigations are confined to stainless steel globe isolation valves which comprise some of the 3,500 or so valves carrying primary circuit fluids in a PWR. The valve seats and discs are hard-faced with Stellite 6 this being the established material from USA PWR experience capable of withstanding differential pressures up to 172 bar with borated, 2000 ppm, sub-cooled water normally at a temperature of 326 degrees C.

All of the measurements reported here were carried out using cold demineralised water but in the PWR pressure range of interest. The aim was to carry out the investigations under the more difficult measurement conditions; high temperature sub-cooled water flashes to steam when crossing a pressure gradient which transits the saturation curve and this is a strong noise generating process. Noise detection at condensate flow levels as low as 20 ml/hr has been reported, Reference 1. However, a low temperature test fluid might be used when checking the leak rate of containment isolation valves before a PWR was put back into service. The acceptance cold water leak rate for newly manufactured 1" and 1.5" valves at 172 bar DP is 3.0 and 4.5 ml/hr respectively. A further aim of the work was to establish whether leakage noise could be detected above the background at this very low order of flow rate.

In order to obtain calibration curves of noise versus leak rate, the valves were "cracked" open using the actuator handwheel to reduce the torque, and hence the seating stress, until the desired flow was obtained. This may not be representative of real leak paths in valves and consequently the noise intensity, or it's frequency distribution, may also be different to that from a real leaking valve. However, the work reported in Reference 3 found no significant difference in noise levels from damaged valves, cracked open valves or from valves with simulated leak paths operating with sub-cooled water. The valves examined were:-

A - 1.5" ANSI # 2,500 lb packed gland globe valve.
B - 1" ANSI # 1,500 lb packed gland globe valve.
C - 1" ANSI # 1,620 lb bellows sealed globe valve.

Of these valves, A was a new valve but both B and C had completed 725 cycles of blowdown testing in the Marchwood Engineering Laboratory high pressure valve endurance test loop, Reference 8. The background noise whilst making the measurements consisted mainly of pump and pneumatic actuator noise from adjacent test rigs together with noises associated with test valve changing and rig maintenance. Some of these noises, like the use of a hammer some 20 m distant, clearly affected the readings and these data were rejected.

4. MEASUREMENT EQUIPMENT, TECHNIQUES AND SOFTWARE

4.1 Leak Rate

The Leak Test Rig is shown in Figure 2. It is normally used to evaluate seat/disc and atmospheric leakage of isolation valves prior to and following life endurance tests in the main facilities. Pressurisation of the upstream side of the test valve is achieved using a pneumatic/ hydraulic reciprocating pressure amplifier pump with a bourdon tube calibrated pressure gauge. A nitrogen filled accumulator was used to provide compressibility in the demineralised water system and to damp out pulsations from the pump with an impulse frequency of 1 to 10 Hz. The leak rate across the seat/disc of the valve was found by observing the rate of rise of water in a calibrated manometer tube, nominally at atmospheric pressure, piped to the downstream side of the valve. Thus the downstream side of the leak was always flooded with water.

4.2 Acoustic Analysis Equipment

The Acoustic Valve Leak Analyser (AVLA) was obtained from Leak Detection Services Inc, of Maryland, USA. It consists of a piezo- electric accelerometer whose output is fed to a pre-amplifier and a variable frequency "notch" filter with a bandwidth of about 3 kHz. The output is in the form of a graph with relative acoustic amplitude in dB's as the "Y" axis and frequency in the range 10 to 100 kHz as the "X" axis. The measured noise signal should ideally be continuous and steady during the frequency scan time which is about 30 seconds. Leakage is indicated by an increase in general signal level compared to background; either the difference between adjacent and remote transducers or a comparison of leak/no leak fluid pressure conditions in the suspect valve. The instrument is portable and can be driven either from mains or internal batteries.

The piezo-electric accelerometer supplied with the AVLA was the Endevco type 2272. This transducer has the attractive features of relatively high gain, 13 pC/g, and stable time and temperature characteristics. However, the manufacturers data shows that the frequency response is only linear up to 7 kHz and a significant increase in output is observed as the transducer's resonant frequency of about 37 kHz is approached. Attaching the transducer to the body of a valve may well lower the resonant frequency.

4.3 Computer Equipment

Two modifications were made to the AVLA to facilitate attachment to the microcomputer. The first alteration involved connecting the "reset/sweep" switch, which controls frequency scanning, in series with a programmable relay unit so that operation could be controlled at the computer keyboard. The second alteration involved connecting the AVLA to two channels of an A/D converter so that analogue signals proportional to the frequency being scanned, and to the amplitude at that frequency, could be fed to the computer's memory. A photograph of the complete equipment together with a sample valve spectrum on the computer screen is shown in Figure 3. The computer equipment consisted of a CBM 3032 computer fitted with a high resolution graphics board, 3040 disc drive for main programme/data storage, 4022 dot matrix printer, CIL PCI 2000 programmable relay unit and an Ortholog 181 A/D converter with 1 part in 256 resolution.

4.4 Computer Software

The main programme enables data from the A/D converter to be input via the user port and stored in random access memory (RAM). It also enables spectra to be stored on disc, printed, subtracted and mathematically compared, or displayed on the computer screen. Initially the programme presents a list of possible options on the screen:-

i) Input level test: enables the setting of a resistor attenuator on the AVLA before the measurement to yield maximum screen resolution.

ii) Input of spectrum to RAM: first asks for file header test data, eg. test number, valve number, differential pressure, closing torque and leak rate. Then commences data sampling and averaging over a number of spectra upon keyboard request. Each spectrum contains 210 pairs of amplitude/frequency measurements.

iii) Storage of spectrum on disc file.

iv) Display of spectrum from RAM: enables the user to observe the spectra as it is being recorded so that a decision can be taken if more averages are required.

v) Display of previously recorded spectra.

vi) Hard copy of spectra on screen.

vii) Subtraction of spectra: this subroutine was used to calculate the true noise integral by subtracting the background noise with zero leakage, from the leakage signal with background noise superimposed, at each common frequency interval. The resultant difference spectrum in the range 10 to 100 kHz corresponds to a signal to noise ratio. These calculations were carried out by first converting each signal to a common attenuation factor of zero dB which permitted subsequent direct comparison of each leakage result.

5. MEASUREMENTS PROCEDURE AND RESULTS

The measurement procedure was as follows:-

The pressure at the valve inlet was increased to the required value using the air pressure regulator and the leak rate set using the valve handwheel before the acoustic measurements were taken. Except at the highest flow rate, measurements were possible without the pump operating using the compressibility of the accumulator and the "dead band" of the air regulator valve. This significantly reduced the amount of background noise in the low flow rate acoustic measurements at the expense of a small reduction in pressure during the measurement.

Measurements of the noise generated by the leaking valves were made with the transducer clamped to a flat vertical plate on the valve casing, level with the supposed site of the acoustic emission, namely the valve seat. Holding the transducer on the valve body during the measurement was found to be unsatisfactory and instead a simple clamp was constructed using Dow Corning 340 heat sink compound for acoustic coupling between the casing and the transducer. Even so, varying the clamp tension and the amount of compound could alter the signal by as much as 50%. The cable coupling the transducer to the instrument was sensitive to background acoustic noise, movement and electrical noise. Wrapping the cable in heat-insulating foam was successful in isolating the recorded spectra from the former sources. Problems were also experienced with signal gain dependant on battery state. Consistent results were only obtained after 30 minutes of continuous operation. For each test valve and induced leak rate, noise spectra were taken and subtracted both with and without leakage flow. Typical spectra and the noise integrals are given in Figures 4 and 5.

Figures 6 and 7 are plots of noise integral, I, against leak rate, Q, for DP = 170 bar and for DP = 120 bar respectively. Figure 6 incorporates results from all three test valves whereas those in Figure 7 are from valves A and C only; no results were taken with valve B at 120 bar. The range of noise integrals covered is about 250 to 110,000 corresponding to leak rates in the range of about 150 to 8000 ml/hr. On average, valve C tends to produce more noise at low leak rate levels at

DP = 170 bar than does valve A or B. At leak rates above about 1000 ml/hr the data points tend to merge, even accounting for the logarithmic presentation, suggesting a power law relationship. Regression analysis gave a best fit power law curve for the 170 bar data of :-

$$I = 0.43\ Q^{1.38} \qquad (1)$$

with a correlation coefficient of 0.90.

Using a similar analysis for the more meagre data in Figure 7 at 120 bar gave:-

$$I = 1.05\ Q^{1.22} \qquad (2)$$

with a correlation coefficient of 0.83.

6. DISCUSSION AND CONCLUSIONS

The zero leakage, background noise, curves reproduced in Figures 4 & 5 are identical although different valves, B & C, were used. This observation; and the harmonic peaks in these curves, suggests that the transducer is primarily responsible. Little information can be gleaned from the position of the peaks in the case of leaking valve signals since they will also be dominated by the transducer response. It is suspected that the transducer may be stimulated by fast-rising shock pulses resulting from mechanical impacting elsewhere in the system. The observed background spectra strongly resemble the Fourier Transform of a pulse at about 15 kHz as might be expected from a shock-excited solid state transducer/mount system. More information on the nature of the leak would be obtained using a transducer having a higher frequency response, although probably at the expense of sensitivity. Despite these limitations, computer integration of the leak noise in the complete spectral range 10 to 100 kHz following background subtraction yields the reasonably monotonic data given in Figures 6 & 7.

The smallest leak rate detected is about 150 ml/hr and lies in the same order of magnitude range as the level predicted for turbulence inception. However, turbulence may provide the conditions for cavitation inception so that both phenomena may commence at the same low leak rate. The curve A.2 in Figure 4 shows that there is no detectable difference between the background transducer output and that with a superimposed leak rate of 0.25 ml/hr. Therefore, detection of cold water leak rates of 3.0 to 4.5 ml/hr is unlikely using the present equipment.

The slopes of the derived "best fit" power law curves for all valves tested at two pressures are very similar. Using the exponent derived at the higher pressure for the lower pressure data makes little difference to the degree of "fit", (see the dashed line in Figure 7), but leads to the interesting conclusion that the new coefficients vary linearly with the pressure difference DP. Thus, an equation describing the noise integral dependence on pressure difference and flow rate becomes:-

$$I = 2.55 \times 10^{-3} \, DP \times Q^{1.38} \qquad (3)$$

We note that the exponent of Q lies between the total power available exponent and that of the leak stream kinetic energy component discussed in Section 2. Since Q depends mainly on DP, there seems little justification on the basis of equation (3) to support turbulence alone as the dominant noise production mechanism which predicts noise level varying as the 8th power of the flow velocity. On the other hand there is no indication from the present data of a square law dependence of noise level on differential pressure as required by a cavitation mechanism. Cavitation damage has been found on the disc from an identical valve to valve B which showed characteristic indentations in the Stellite 6 surface following high temperature/ pressure sub-cooled water "blowdown" testing, Reference (6). Clearly, further fundamental investigations are required to unravel the noise production mechanism.

Considering firstly the potential errors due to transducer acoustic coupling, changes in background noise level and instrument gain variations etc, and secondly systematic variations in the three valve designs, surprisingly good correlation has been found between noise integral, leakage flow rate and differential pressure independent of these factors. This is particularly so at the higher leak rates above about 1000 ml/hr. Although much work remains to be done, it is suggested that the technique is capable of development to yield a working power station instrument for reliable leak rate evaluation at least in this leakage range. Following the investigations reported in Reference 1, the correlation may also be improved at smaller leakage rates by counting cavitation events rather than integrating the emitted noise. These investigations are continuing.

7. ACKNOWLEDGEMENTS

Most of the computer software was written and the experiments carried out by Miss Sarah Greasely of Jesus College, Cambridge University whilst carrying out a vacation project at MEL. Her contribution is gratefully acknowledged. This paper is published by permission of the Central Electricity Generating Board.

8. REFERENCES

1 Dimmick J G, Nicholas J R, Dickey J W, & Moore P M "Acoustic Valve Leak Detector for Fluid System Maintenance" Naval Engineers Journal, pp 71-83, April 1979.

2 Dumousseau P & Roget J "Application of Acoustic Emission for Leak Characterisation and Testing" CETIM Informations No 72, pp 85-91, 1983. CEGB Translation T 17046 (TRAN 18086502)

3 Nakamura T & Terada M "Development of Leak Monitoring System for Pressurizer Valves" Progress in Nuclear Energy, V 15, pp 175-179, 1985.

4 McElroy J W "Light Water Reactor Performance Surveys Utilizing Acoustic Techniques" EPRI Power Plant Valves Symposium, Kansas City, Missouri, USA, August 1987.

5 Dickey J W, Dimmick J G, & Moore P M "Acoustic Measurement of Valve Leak Rates" Materials Evaluation, pp 67-77, January 1978.

6 Greasely S J & Airey D R "Evaluation of an Acoustic Leak Analyser for use on PWR valves in Sub-Cooled Water" CEGB Research Report No TPRD/M/1621/R87, September 1987.

7 "Acoustic Monitoring of Power Plant Valves" EPRI report NP-2444, June 1982.

8 Airey D R, Richards D J W, Hilsley D E & Bryant S "Testing Primary Circuit Valves for the UK PWR" BHRA Conference on Valves and Actuators for Fluid Control, Paper E 1, Oxford, September 1985.

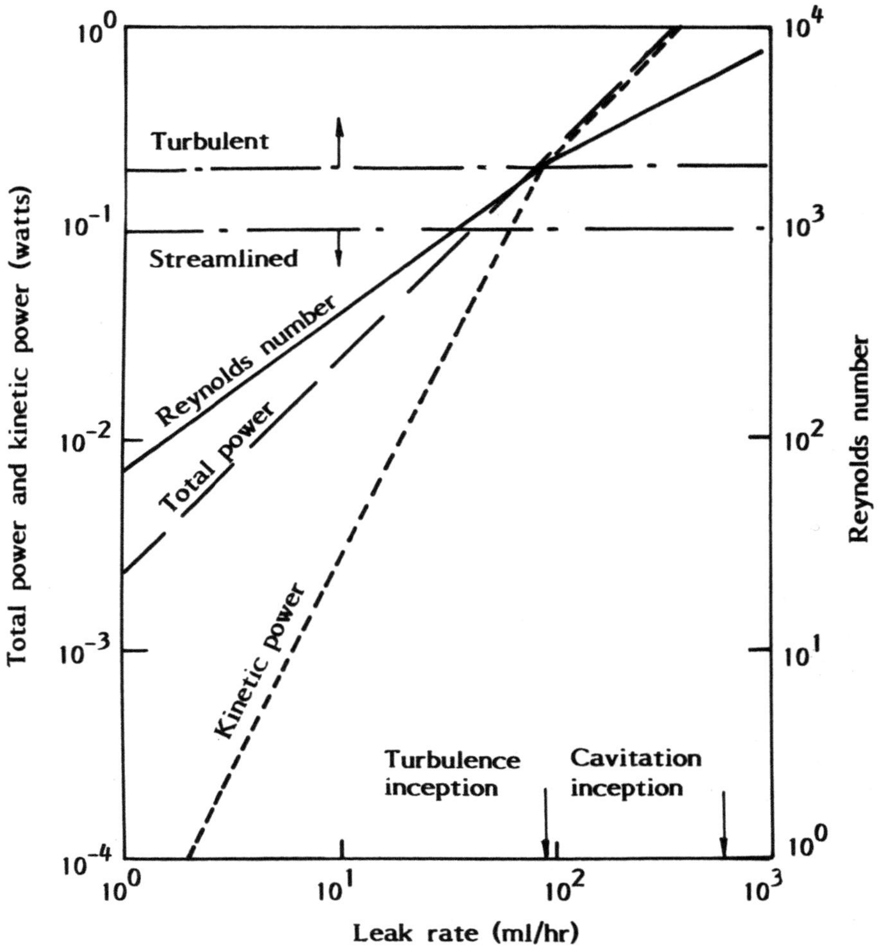

FIGURE 1. AVAILABLE POWER AND REYNOLDS NUMBER FOR A COLD WATER (20°C) LEAK, LENGTH 1mm, DIFFERENTIAL PRESSURE 170bar.

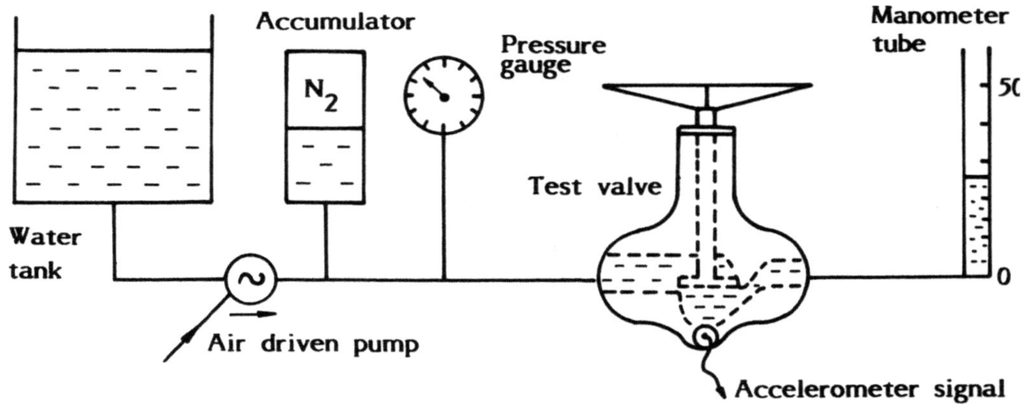

FIGURE 2. VALVE LEAKAGE TEST RIG

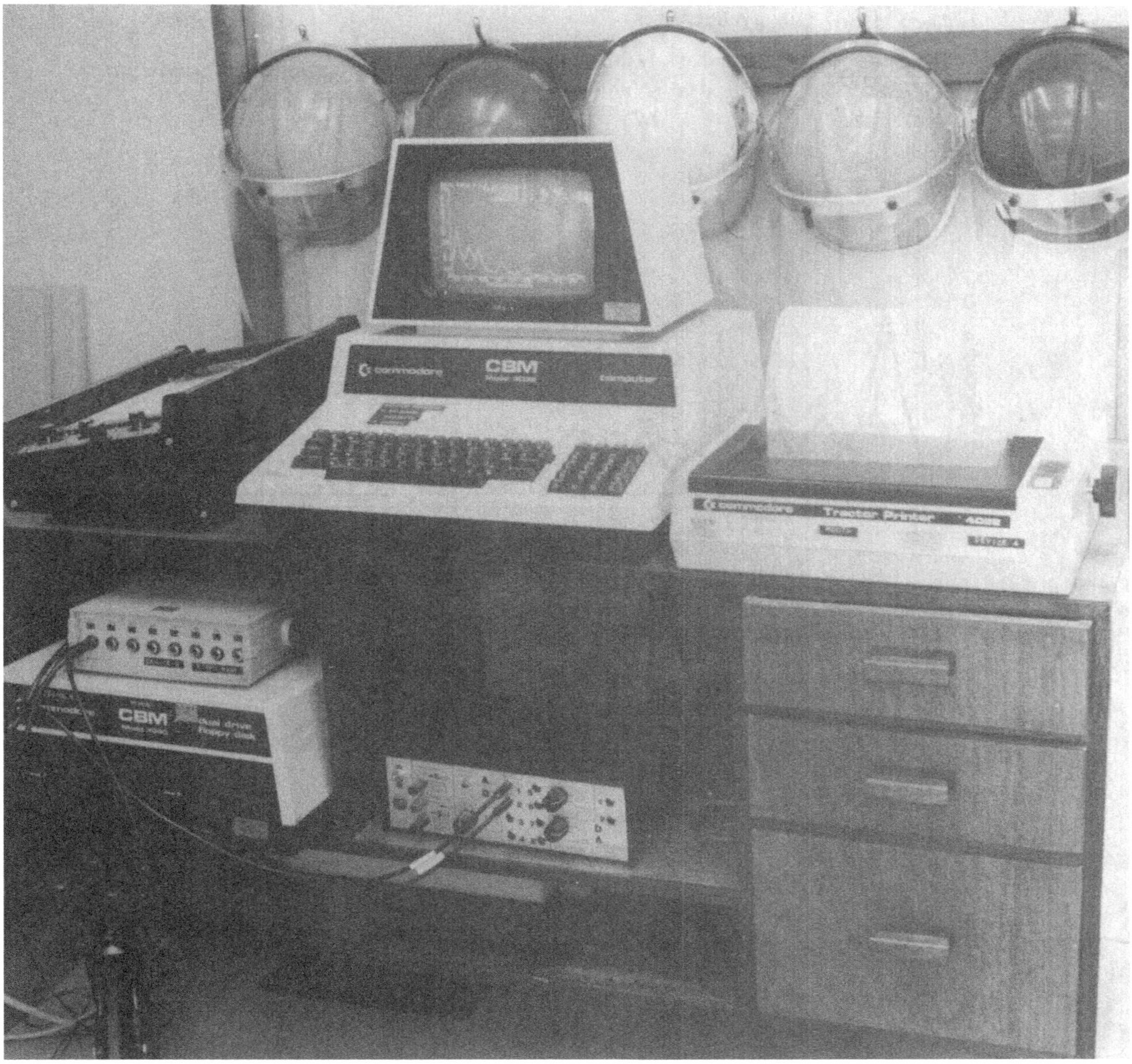

FIGURE 3. ACOUSTIC ANALYSER AND COMPUTER HARDWARE

Data from file 1A/002/A.6

Pressure difference 1 = 170bar
Torque 1 = 35 Nm
Leak rate 1 = 1300ml/hr

Acoustic spectrum with valve leaking. (Includes background noise)

Data from file 1A/002/A.2

Pressure difference 1 = 170bar
Torque 1 = 50 Nm
Leak rate 1 = 0.25ml/hr

Acoustic background spectrum with minimal leakage

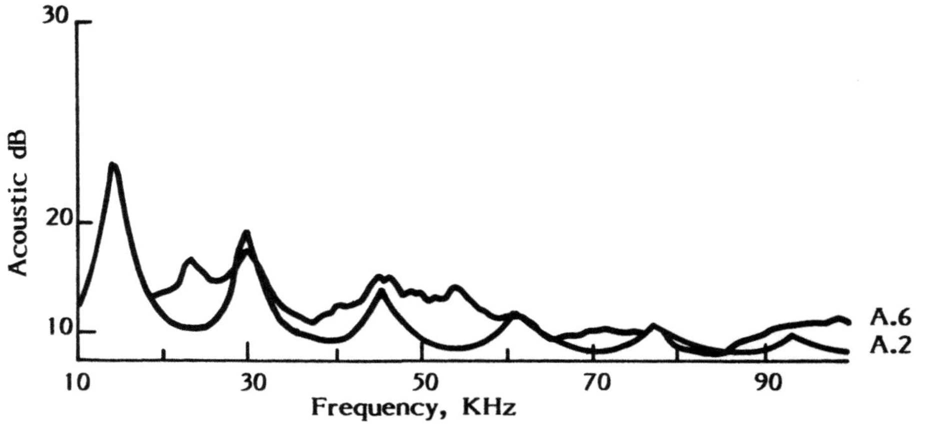

Data from file 1A/002/A.2

pressure difference 1 = 170bar
2 = 170bar
Torque 1 = 35Nm
2 = 50Nm
Leak rate 1 = 1300ml/hr
2 = 0.25ml/hr

Difference spectrum (true noise)

Acoustic noise

I = 10 448

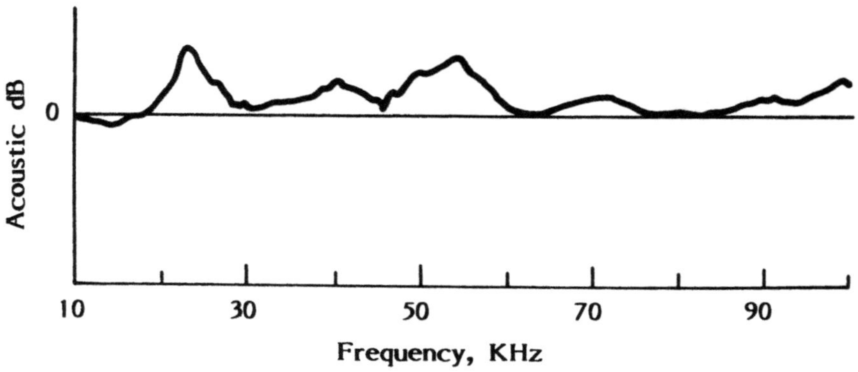

FIGURE 4. SPECTRA PRODUCED BY FLUID LEAKING THROUGH VALVE B

Data from file 1B/001/A.1

Pressure difference	1	= 0bar
Torque	1	= 70Nm
Leak rate	1	= 0ml/hr

Data from file 1B/001/A.5

Pressure difference	1	= 170bar
Torque	1	= 65Nm
Leak rate	1	= 750ml/hr

Data from file 1B/001/A.9

Pressure difference	1	= 160bar
Torque	1	= 65Nm
Leak rate	1	= 3600ml/hr

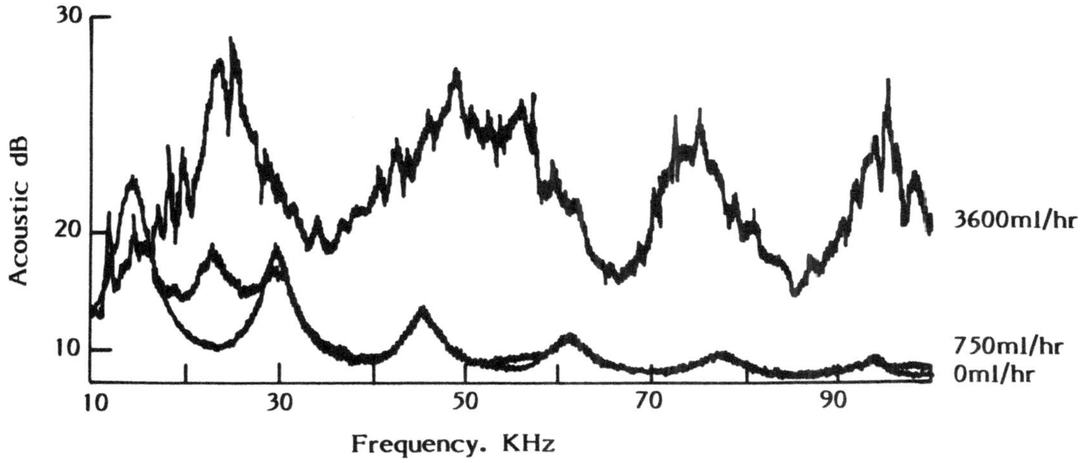

FIGURE 5. SPECTRA PRODUCED BY FLUID LEAKING THROUGH VALVE C

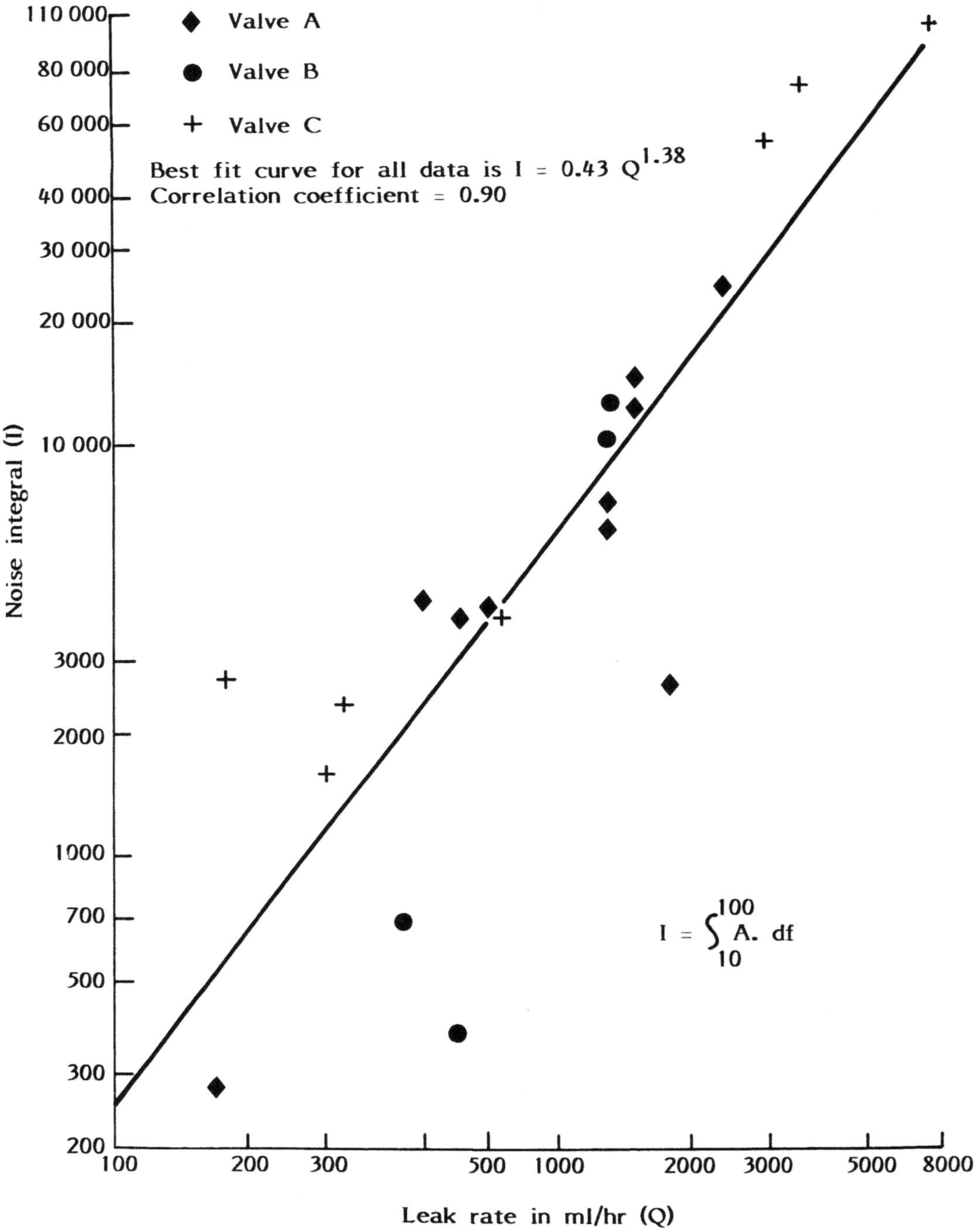

FIGURE 6. NOISE INTEGRAL AGAINST LEAK RATE AT 170bar PRESSURE DIFFERENCE FOR 3 VALVES

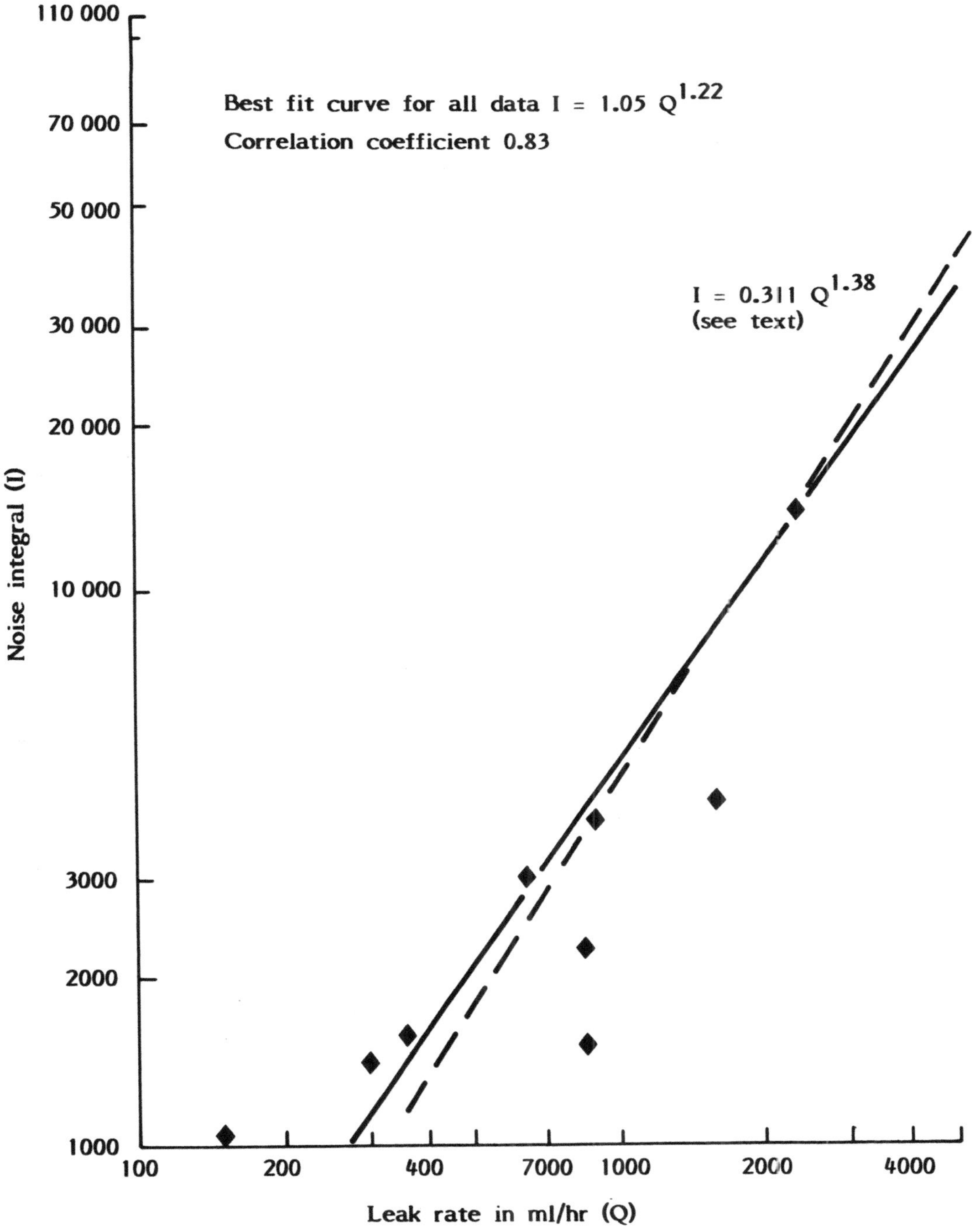

FIGURE 7. NOISE INTEGRAL AGAINST LEAK RATE AT 120bar PRESSURE DIFFERENCE

2nd International Conference on

Developments in Valves and Actuators for Fluid Control

Manchester, England: 28-30 March 1988

PAPER C1

ELECTRONIC ACTUATOR DESIGN

George M. Schwab
Milton Roy Company Electronic Actuators and Controls

Summary

A way has been found to engineer a true solid-state actuator which meets two major industrial requirements. First, this new solid-state actuator is competitive with traditional pneumatic, hydraulic, and electric designs. Second, it can take maximum advantage of the significant benefits offered by computer control.

1. Introduction

1.1 Synopsis

The science of optimal control systems devotes primary attention to such aspects of servomechanism control as sensitivity, stability, controllability, repeatability, and linearity.(1) These and other mathematically modelled system parameters are intrinsically understood in the algorithm of a P.I.D. controller. The controller provides the mathematical computation in a process, but the final element -- that is, the actuator -- must be able to follow closely the controller's commands if the system is to reap the rewards of computer control. The closer the actuator can follow the P.I.D. command signal, the better will be the quality of the final product and the greater the efficiency in materials consumption management. Such an approach is crucially needed today.

1.2 Product Quality

In the steel industry, in petrochemicals, and in many other markets, what can and does provide the competitive edge is product quality. An oil company, for example, may wish to control its mixing system to produce gasoline uniform within one-tenth of an octane. One company has chosen to reach this

Held in Manchester, England. Organised and sponsored by BHRA, The Fluid Engineering Centre and co-sponsored by the British Valve Manufacturers Association and the Institution of Mechanical Engineers Process Industries Division

objective with the electronic (or solid state) actuation method described in this manuscript.

1.3 Economy of Precision

As economists speak of the "economies of scale", so process engineers are recognizing the tremendous importance of the economies of precision. Improving the ability to control the flow of material by as little as 1% can represent savings -- or profit -- amounting to hundreds of thousands of dollars each year. An example of this is in chocolate production. The temperature of butterfat crystallization determines the shelf-life of the chocolate. A deviation of only ± one-tenth of one degree Fahrenheit can make the chocolate worth more or worth less.

1.4 Industrial Standards

With the advent of factory automation and robotics, it was merely a matter of time until process industries also began to automate -- and to require a fully computer-controllable "robot valve" based on microchip electronic technology. As this advance continues (with computer networking moving into production facilities) industrial standards have begun to reflect the technology of the late 1980's and the 1990's. (The ISA committee 50 is presently formulating such standards.) Part of this advance is an emphasis on digital control. Besides allowing discrete steps, an actuator that responds only to a defined digital signal will stay in place in the event of a brownout or a blackout. An analog unit, however, would follow a drifting command. Pure analog devices will simply be phased out in the face of digital requirements, and this has already occurred in many instances. We may expect "smart" actuators with the P.I.D. algorithm on board to become quite commonplace.(2)

1.5 Pneumatic

The pneumatic type of actuation relies on a balance of forces: valve force versus I/P transducer generated pressure. Obviously, if the line flow changes, the valve will move out of place until the transducer can modify its output. In many applications, the pneumatic actuator can be relied upon only to within ± 7.5% of the desired position -- despite the fact that, in laboratory tests, it can be measured to be much more precise. Pneumatic actuators are pure analog devices. Unless a pneumatic actuator is designed that will "clock" discrete steps with precise, short bursts of air, the pneumatic approach will remain outside the digital world.(3)

1.6 Hydraulic

Since it uses a noncompressible fluid, the hydraulic actuator cannot "hunt" for the same reason the pneumatic actuator does. Rather, the pressure generated by a continuously running motor produces the forces that must be balanced. The hydraulic cylinder and piston constitute a pure analog device.

1.7 Electric

With the electric actuator, an 1800 RPM motor is geared down and then controlled with relays, limit switches, and pushbuttons. These units tend to burn out with rigorous use because angular momentum is proportional to the square of motor speed. High angular momentum results from the electric actuator's high motor speed, thereby causing overshoots which in turn result in hunting. Most of the so-called "electronic" actuators in use today are actually only electric actuators with electronic embellishments. The electric motor is an analog device.

2. Theory

What is needed to overcome the limitations of the present conventional means of servomechanism control of valves may be best explained by starting with a brief review of some elementary points in control theory.

A valve actuator is a servomechanism, and as such it exhibits complex response to a control signal. The P.I.D. controller should be tuned to the complex time-dependent characteristics of the total process system without the introduction of superfluous complex response from the actuator. Therefore, any nonlinear, time variable components to the mathematical model of a servomechanism must be regarded as a failure to precisely position the valve, and, therefore, a failure to control the system with the precision the system operator requires.

In an idealized simple servomechanism

J = Polar moment of inertia
B_o = Viscous damping parameter
e_a = Amplifier output voltage
c = Controlled voltage
r = Required voltage

Typically, such a servomechanism is represented as (4):

$$Jd^2c/dt^2 + B_o dc/dt + Kc = Kr$$

(diagrammed in Figure 1.)

In a practical actuator, since the viscous damping parameter is relatively small, and the mechanical effect of random pressure change must be considered, the equation becomes:

$$\Phi(Jd^2c/dt^2 + Kc) = Kr$$

Where Φ = normalized probability distribution.

Jd^2c/dt^2 is responsible for overshoot, and, ultimately, for motor burnout due to noncritical damping. Φ, of course, must be substantially eliminated from the equation. Φ introduces external "noise" to the system due to random line fluctuations that cause the loss of tight control on units using the "balance of forces" method of positioning, such as pneumatic actuators.

Jd^2c/dt^2 cannot be eliminated completely, but it can be reduced to a negligible ringing if a motor is operated at very low speeds. Using a constant-speed motor, a typical block diagram is shown in Figure 2.

This system has no true frequency response. dc/dt is merely the voltage change per second that is a constant due to the use of a constant speed motor. The u function is defined as follows:

$$u[e(v)] = \begin{cases} 1 \text{ if } E_a(s) \geq e_o \\ 0 \text{ if } -e_o < E_a(s) < e_o \\ -1 \text{ if } E_a(s) \leq -e_o \end{cases}$$

What this accomplishes is the virtual elimination of usual servomechanism response. With the definition of a discrete deadband, e_o, and with slow, constant speeds, very precise control can be realized. What is required is the following:

- An electronic control to achieve the u function
- A motor/power arrangement to achieve dc/dt = constant
- A mechanical system to eliminate Φ, and to minimize Jd^2c/dt^2

3. Experimental Procedure

The following is a brief discussion of the practical electronic actuator.

3.1 The Electronic Circuity

The electronic portion of the actuator utilizes several methods of control that differ from the electric actuator. One of these differences is the elimination of limit switches.

3.1.1 Limit Switches

The limit switch approach is the "brute force" approach, where the full current draw of a large, high RPM motor is physically broken. This approach, while effective, has several limitations:

- Limit switches are relatively difficult to adjust positionally.
- Limit switches are mechanical devices that can be damaged if they are overrun.
- Limit switches transcend electronic control.
- Limit switches do not allow a trickle current for an electronic "brake" in fail-safe applications.

The electronic approach is twofold. First, the motor is energized through the use of solid-state electronic switches. These electronic switches can be controlled directly from low-power microcircuits.

Second, the motor position can be known through a position-indicating transducer. Whenever the motor turns a valve up to its pre-adjusted final position, the electronic circuit automatically turns the motor off. Some of the advantages are as follows:

- There is no spark, or sudden break of current.
- The limits can be remotely adjusted.
- The electronic circuit controls the limits.

3.1.2 Design Simplicity

Based on the mechanical and electrical criteria derived in Section 2, the circuit does not demand very much complexity. This fact lends itself to the following advantages:

- Very small package
- Low cost
- High reliability
- Easy repair

3.1.3 ASCII Codes and MAP

The language of computer networking today is based on ASCII codes. With an RS232 serial input, the actuator performs virtually identically as either an analog or a pulsed-digital input. It can retain the above mentioned advantages while communicating on a higher technical level.

3.2 The Motor

Stepper motors are digital devices. They are designed and built with controlled magnitude steps. When one digital pulse is received, the actuator will clock one known angle. It is for this purpose that steppers were designed, and this is how they are used in most applications, such as in drives for large telescopes at observatories.

3.3 Mechanical Design

In conjunction with the motor, the overall mechanical design must all but eliminate any movement due to random line fluctuations -- or due to variations in controlled fluid flow. This is easy to accomplish with a stepper-type motor used in conjunction with a simple linear actuator or ball screw arrangement.(5) The intrinsic detent torque of the motor keeps the shaft from turning, and the friction of the linear actuator screw resists random variations in thrust from the valve. The ball screw assembly is used in conjunction with a spring and an electronic brake circuit.(6)

4. Results

4.1 Life Test

Actuators of the type described in this manuscript have been life tested under a variety of conditions. With an ambient temperature of 72°C (160°F) and an internal electronic temperature of 104°C (220°F) at 100% continuous operation duty cycle, the actuator operated for more than one million cycles before a minor failure occurred. Other actuators have been life tested against pressure at room temperature at more than four million operating cycles. Some actuators have experienced no failures even after eight years of continuous operation in the field.

4.2 Performance Tests

The actuators have been vibration-tested by the United States Coast Guard.

Factory Mutual Insurance Company has approved this type of actuator for intrinsically safe operation as it is both waterproof and explosion proof.

The positioning performance with 4-20 mA control is as shown in Figure 3. For digital results, a single CMOS pulse clocks the linear actuator 0.0127 mm (0.0005 in.) and on rotary valves, 0.09 degrees.

4.3 Dimensions

The physical dimensions of a complete electronic actuator can be seen in Figure 4.

4.4 Applications

With the use of this actuator, many field applications enabled older technologies to make a quantum leap of advancement in quality and efficiency. A few examples follow:

- For a 0.025 inch stroke on a steam turbine governor, electronic actuation has replaced pneumatic, thereby enabling much greater control and prevention of oscillations.
- On an offshore oil platform, such actuators aid in utilizing solar power and in enabling telemetry communication.
- On spray-water valves, precision electronics control the cooling rate of steam to produce a higher quality steel.
- In gasoline blending systems, electronic valves are accurate to one-tenth of an octane point.
- Electronic control on a blowing agent for polyurethane foam improves foam quality and controls consumption of neon gas.
- Electronic actuation is controlling the headbox in a paper manufacturing process and improving product quality while also reducing cost.
- The U.S. Coast Guard uses electronic digital actuators to steer small ships.
- In a leather manufacturing process, electronic valves produce dynamic titration of pH control.
- A better quality Mylar® is produced with an electronic actuator controlling the process vacuum.
- Very smooth pump speed control is achieved by replacing pneumatics with electronic control.
- In various manufacturing plants, electronic actuation is chosen simply because clean air is either not available or not desired.
- The actuator can be operated to -40°C, which is ideal for many cryogenic and outdoor winter applications where pneumatic units can freeze.
- Electronic actuators are used in several sky-rise heating systems, for temperature control.
- Electronic valves exhibit repeatable bubble-tight shutoff with precise digital control.
- Electronic valves control the fermentation process in several breweries.
- An electronic actuator is used by a major university to aid in modeling nuclear power plants.
- Electronic valves are now used in sprinkler systems for fire protection.

5. Discussion

5.1 Weakness and Limitation of Solid-State Actuation

5.1.1 Thrust per Unit Volume

One of the limitations of solid-state actuation is the fact that the total mechanical power must be generated by the motor. In a pneumatic device, power is proportional to the input pressure. Using standard pressures, a pneumatic actuator can deliver about 50% more force than can a comparable actuator driven by an electric motor.

In the last decade, the motor industry has advanced remarkably. In 1978, a pneumatic actuator could have produced approximately four times the thrust of an electric motor unit. Recent advances, however -- such as new samarium-cobalt motors -- have raised the power of electronics to equal the pneumatics. Recent advances in stepper motors include the improvement of motor winding technology, the development of magnet material such as AlNiCo 8 & SmCo, and the interstitial spacing of magnetic material between stepper teeth. In the near future, with more readily available neodymium rare-earth motors, which are much more powerful than SmCo motors, electric power may exceed pneumatic power.

5.1.2 Intrinsic Safety

Pneumatic devices have no heat-generating parts and no electric components that produce sparks. Therefore, in some extreme situations, pneumatics will perhaps always be used for safety reasons.(7) Some components in electronic actuators can get fairly hot, but none produce sparks. Since stepper motors are brushless, they are seldom a safety concern.

Therefore, electronic actuators have been considered intrinsically safe by Factory Mutual Insurance Company. Also, when enclosed in a NEMA 7, Class 1, Group D enclosure, even in the event of an internal explosion, ignited gas would not propagate outside of the actuator. Therefore, while solid state devices probably will never be used for all environments, they can certainly be used for normally hazardous environments where explosion-proof rating is required.(8)

5.1.3 Speed

A pneumatic actuator responds instantly to a change in pressure. However, the elapsed time between the electronic initiation of a desired position change and the valve actually coming to rest at that position can be as much as 10 or 12 seconds. Therefore, an electronic motor-driven actuator can, in many cases, make the desired movement in less time than it takes for an I/P transducer to modify the pressure to the desired value. Therefore, in an actuator-to-actuator comparison, pneumatics respond more quickly.

If pneumatic and electronic systems are compared, the electronic system adjusts the valve more quickly. Including the time a pneumatic system spends hunting, or homing in on a desired position pneumatically, tilts the comparison even more in favor of the electronic system.(9) Although I/P transducer technology has continued to improve during the past five years, it still remains the chief area of restriction for pneumatic actuator systems.

5.1.4 Fail-Safe Upon Loss of Power

Upon loss of power, a spring is used to close the valve in both pneumatic and solid state actuators. Since the pneumatic actuator is virtually hollow, and is based on the balance-of-forces principle, an internal spring unit is relatively simple to design and build. A pneumatic actuator, however, fails only on loss of air, not on loss of plant power, thus a solenoid must be de-energized to vent the diaphragm. In other words, no truly "fail-safe on loss of power" pneumatic actuator exists.

In contrast to the simple pneumatic spring actuator, the electronic spring actuator is a complex device. To eliminate friction, a ball screw arrangement must be utilized. This requires complex timing circuitry and demands that the motor be partially energized continuously.

Despite its drawbacks, this method is used in many applications. Now, with the introduction of computer systems, another method of fail-safing valves has recently been gaining acceptance. Due to the low-current draw nature of the electronic actuator motors, a small uninterruptible power supply (UPS) system has often been used to fail-safe critical or potentially hazardous applications heretofore specified only for spring devices. This pneumatic orientation has been changing as computer UPS systems become more common and gain more acceptance.

5.1.5 Cost

The pneumatic actuator is a simple device, while the electronic actuator is relatively complex. This difference is reflected in the cost of each. However, when buying the pneumatic actuator (assuming plant air is available), an array of filters, an I/P transducer, and various other elements must also be added to the shopping list. Also, if air is not available, a compressor and tank, as well as tubing must be installed. Devices such as feedback potentiometers are not standard -- they are "add-ons" as well.

In contrast, the electronic unit needs only to be wired directly to the computer or controller to be ready for use. Therefore, when comparing the installed cost, electronics are usually not prohibitively expensive.

The real value in electronic control is the greater precision and reduced maintenance. If, through greater precision, thousands of dollars can be saved and maintenance reduced, then the electronics can pay for themselves, despite their higher initial cost. Finally, the prices of electronics are very quantity sensitive; as production rates increase, prices typically decrease dramatically -- often in a fairly short time. In all likelihood, as time passes the cost of electronic actuation will be significantly reduced.

5.1.6 Implementation

Perhaps the biggest problem to the introduction of electronic valve actuation is the resistance to its use found in process plants. This reluctance is understandable, since there have been many past attempts to produce electronic actuators that can compete with pneumatics. The vast majority of these so-called "electronic" valves have been and still are electric actuators with electronic embellishments. These actuators usually cannot position accurately, and do not have an adequate duty cycle for constant use. Therefore, they frequently fail soon after they are installed.

Acceptance of true electronic actuators has been predictably slow, but surprisingly steady. The most important fact concerning this acceptance is this: The new actuators really are different. They must be accepted because they are the only product that can meet the new standards of performance.

5.2 The Future of Control

5.2.1 Networking and Manufacturing Automation Protocol (MAP)

Since the introduction of computer control and plant automation, it has been the direction of technology to create "smart" machines that understand ASCII codes and perform various simple tasks. Obvious examples can be found in robotic plant automation and "smart" sensors. At the present time, computer hierarchies that provide centralized control at the mainframe level, compartmentalized process control at the substation level, and single loop P.I.D. control at the controller level are used in many industrial processes. A "smart" actuator that understands MAP and can perform basic P.I.D. functions can eliminate an entire level of control. Prototypes of this kind of microprocessor-based actuator have been in existence for several years.(10)

5.2.2 Fiber Optics

The trend in communication technology is away from using electricity for long-distance communications. Microwave, radio, and fiber optic technology are finding their place in the arena of control. Radio controlled devices have been used in manufacturing facilities, microwave telemetry for off shore platforms is becoming more common, and fiber optics for actuator control may soon be a specific requirement for newly automated plants.(11) Recently introduced "smart" sensors for process plants already have fiber optic capability.

6. Conclusion

6.1 Recapitulation

A modified stepper motor, coupled with a sound and truly electronic design, plus adequate mechanical support -- all in an acceptable enclosure -- can add up to a new technology that addresses the process industry's need to make the fullest use of sophisticated computer control.

6.2 Ideal Applications

Some actuator requirements are ideal for electronic actuation:

- In precision valve applications, electronic actuation permits the valve to function to its highly precise capability.

- Digitally stepping electronic actuators permits two valves to respond with exact proportional movements, and therefore solid-state actuators are ideal for blending applications.

- Any proportioning system is an ideal application for electronic control.

- For remote applications where air is not available, or where tubing would pass through hot or cold areas that would cause process air to expand or contract, electronics are ideal.

- For any digital control signal system, electronics are the only viable alternative.

7. References

1. Kuo, B.C.: Automatic Control Systems. New Jersey, USA, Prentice-Hall, Inc., 1982, pp. 8-16.

2. Ball, K.E.: "Final Elements: Final Frontier," Intech, Nov. 1986, pp. 33-34.

3. Curtin, D.L.: "New I/Ps, P/Is Secure Pneumatics' Foothold Against Digital Onslaught." I & CS, April, 1987, pp. 63-67.

4. McCausland, I.: Introduction to Optimal Control. New York, USA, Robert E. Krieger, 1979.

5. Usry, J.D.: "Stepping Motors for Valve Actuation," Instrumentation Technology; Article is based on a paper presented at the 1976 ISA Computer Interface Instrumentation Symposium, Newark, DE.

6. Patent No. 4,084,120

7. Moffatt, R.A.; Hudson, A. J.: "Air Logic: A Good Choice for Hostile/Hazardous Environments." I & CS, April, 1987, pp. 71-74.

8. Fisher, T.G.: "Designing Fail-Safe Systems with Solid-State Instruments." Intech, October, 1986, pp. 51-54.

9. Wilson, F.D.: "DuPont Improves Process Management with Use of Smart Pressure Sensor." Chemical Processing, June, 1987, pp. 109.

10. Herbst, B: "Maintaining the Local Area Network". I & CS, November, 1985, pp. 27-30.

11. Zetter, M.: "On-Line Optical Spectroscopy Teams Up with Fiber Optics". I & CS, April, 1987, pp. 35-37.

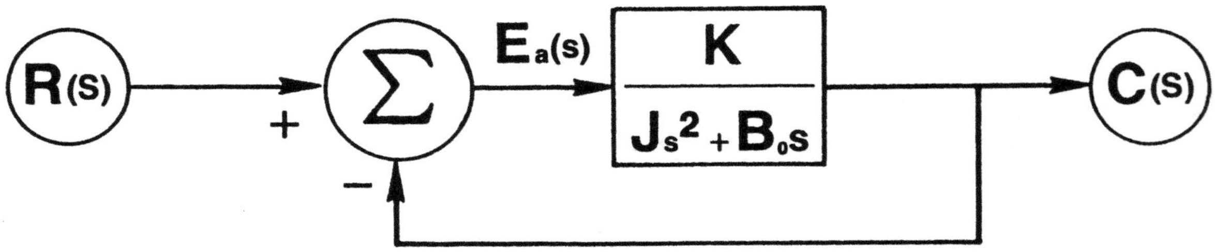

FIGURE 1 — BLOCK DIAGRAM OF SIMPLE SERVOMECHANISM

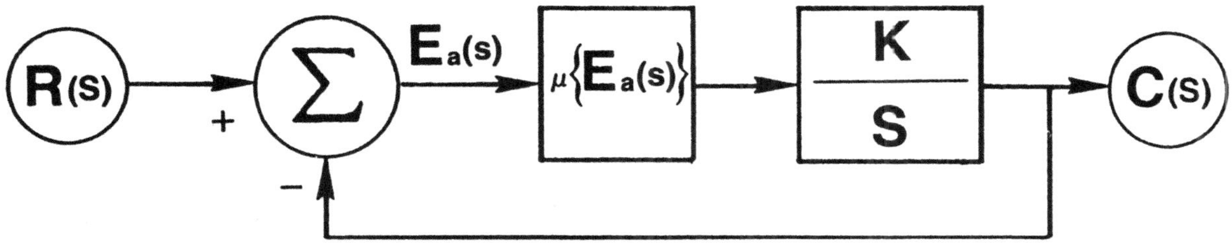

FIGURE 2 — BLOCK DIAGRAM OF ELECTRONIC ACTUATOR

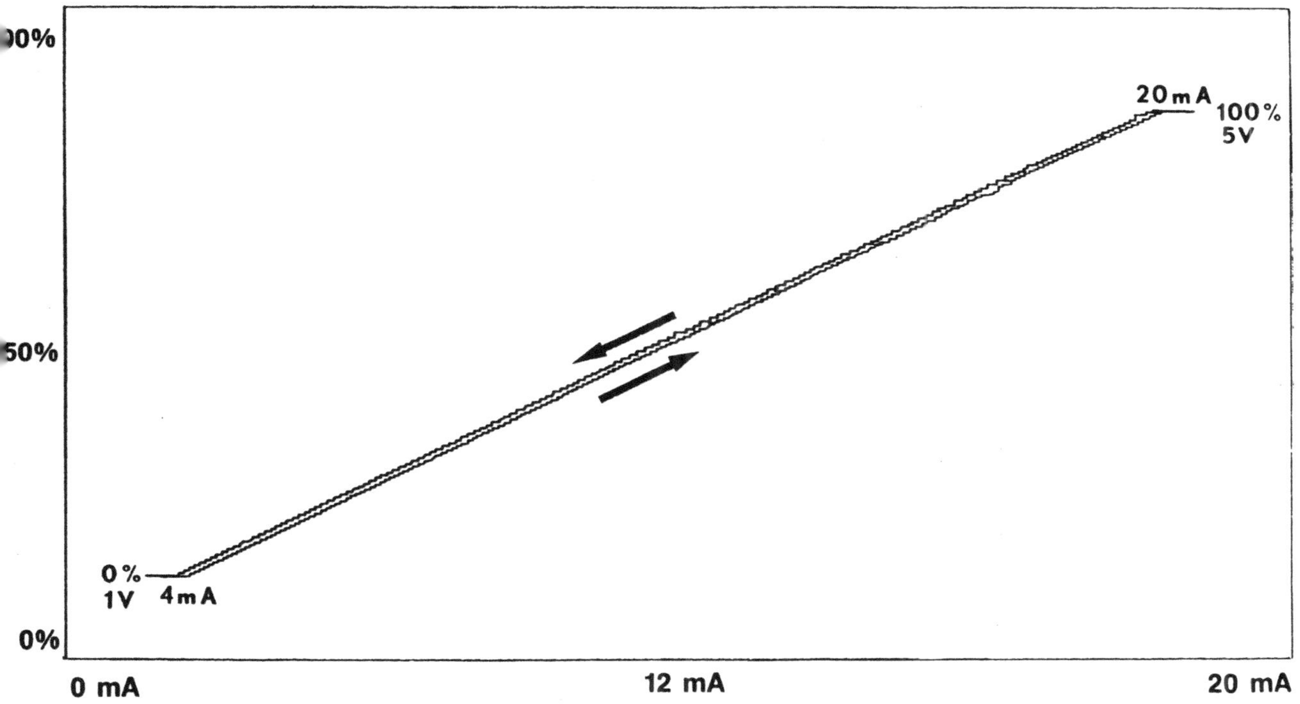

FIGURE 3 — RESPONSE TO SLOWLY VARYING SIGNAL

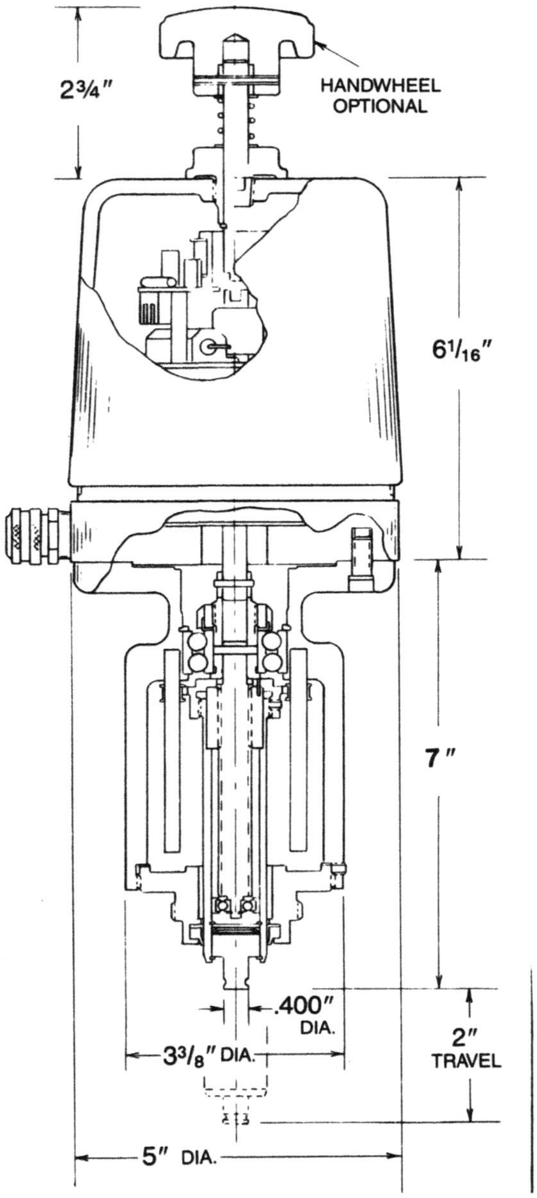

FIGURE 4 — P100 ACTUATOR LINEAR UNIT

2nd International Conference on

Developments in Valves and Actuators for Fluid Control

Manchester, England: 28-30 March 1988

PAPER C2

NEW TRENDS IN PNEUMATIC ACTUATOR DESIGNS

By Austin Pruden.

The Author has been in valve actuation for the last 25 years, is now Technical Marketing Manager for El-o-matic and lives in Holland.

SUMMARY

A fresh look at the problems of 1/4 turn valve actuation and some exciting new developments on a well established theme.

INTRODUCTION

In a field of modern plant control the pneumatic actuator has made a significant contribution in the automation of the new generation of soft seated 1/4 turn industrial valves. Providing greater reliability, lower installed costs with more compact and efficient assemblies.

With numerous new actuators coming on to the market each year, each seemingly a carbon copy of the historic market leaders, is it possible that significant advances in performance can be achieved in this already well established market sector?

To provide an answer it is maybe a good idea to take closer look at the task which confronts the average valve actuator in a modern plant situation.

WHAT AN ACTUATOR NEEDS TO DO

The actuator has to provide a torque output which is greater than the torque requirement of the valve on which it is mounted. A simple situation is a double acting actuator operating a 1/4 turn plug valve where both torques are fairly uniform throughout the 90 degrees of movement.

Held in Manchester, England. Organised and sponsored by BHRA, The Fluid Engineering Centre and co-sponsored by the British Valve Manufacturers Association and the Institution of Mechanical Engineers Process Industries Division

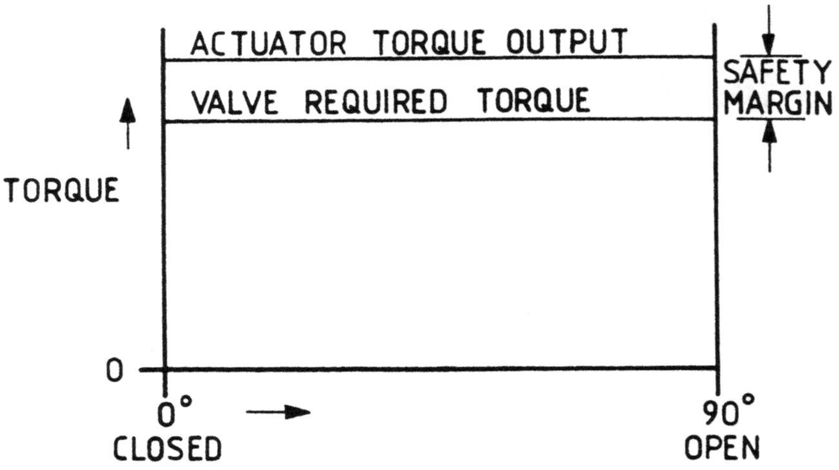

Fig 1: Simple 1/4 turn valve torque.

In practice this ideal situation rarely takes place as most soft seated valves when left in the closed position for any length of time will need a higher "breakout" torque for the first degree or so of its initial movement from the closed position than when moving from the open position.

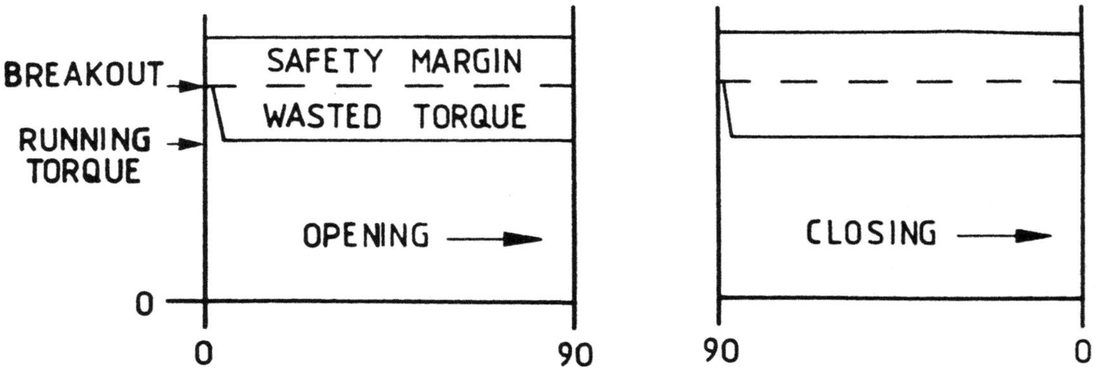

Fig 2: Break out and running torques.

By comparing the valve torque line with the actuator torque line it can be seen that there is a considerable area of wasted torque and this has been the target of many ingenious designs over the years. The most noticeable of these is the scotch yoke system which presents a higher torque at the start of stroke than in the mid section where it has to deal with the running torque.

Unfortunately actuators using this design tend to be more bulky and complicated, so much so that there are few which can equal the breakout torque of a modern rack and pinion actuator within the same envelope demensions.

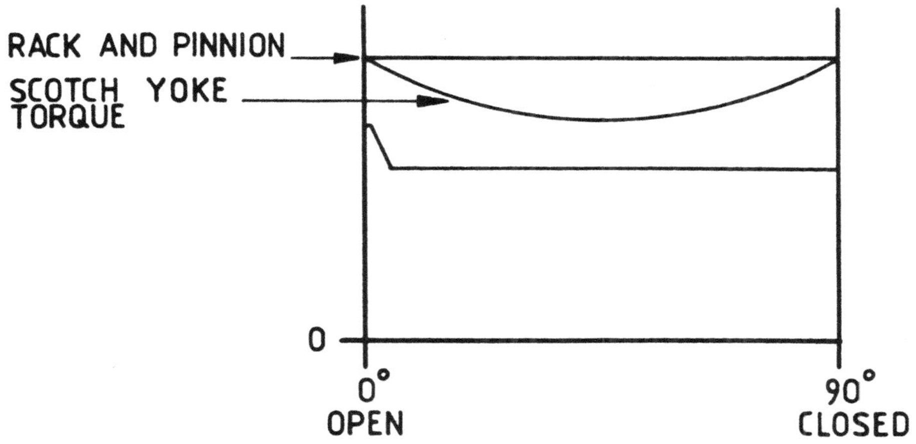

Fig 3: Torque from scotch yoke and rack and pinion actuators.

The task of providing a more efficient design is complicated still further by two other considerations:

1) The majority of actuator requirements are for "spring return" units.

2) The majority of 1/4 turn valves where actuation is required are Ball or Butterfly valve types where the torque pattern deviates still further from the linear and differs considerably from opening to closing.

For instance:

TORQUE PATTERNS FOR THE PRINCIPAL VALVE TYPES

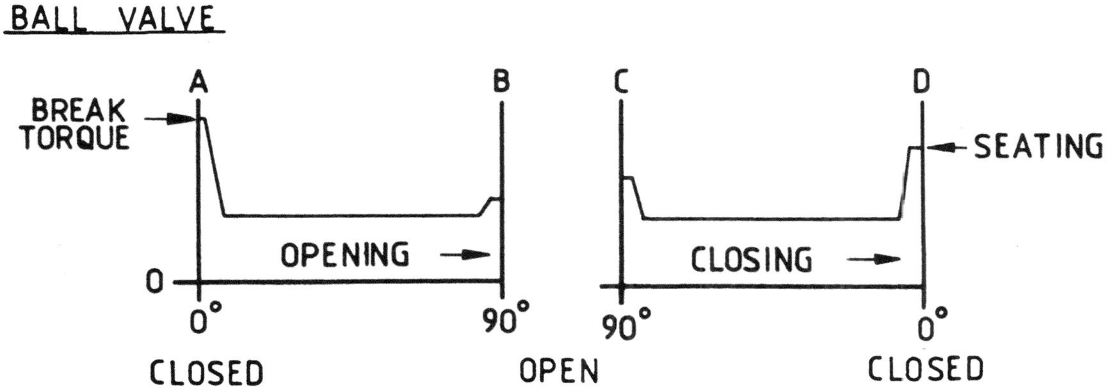

Fig 4.1: Ball valve torque.

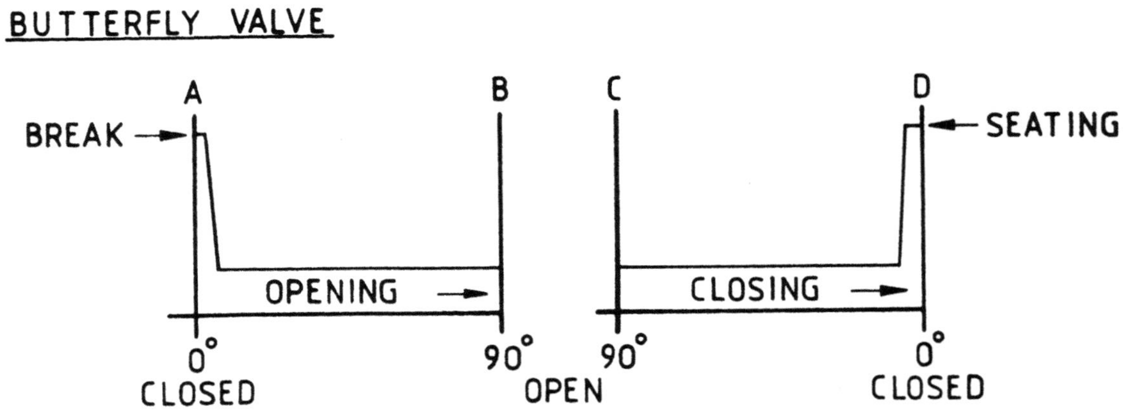

Fig 4.2: Butterfly valve torque.

Now lets see how the spring return actuator torque compares with these very varying torque patterns.

SPRING RETURN TORQUE MATCHING

In a double piston rack and pinion actuator, the opening stroke occurs when compressed air enters the centre chamber forcing the pistons outward against the springs. This air stroke first has to have enough piston forc to overcome the initial spring force, the remaining piston force is then converted by the rack and pinion into torque open the valve. This remaining torque decreases as the springs become more compressed throughout the stroke.

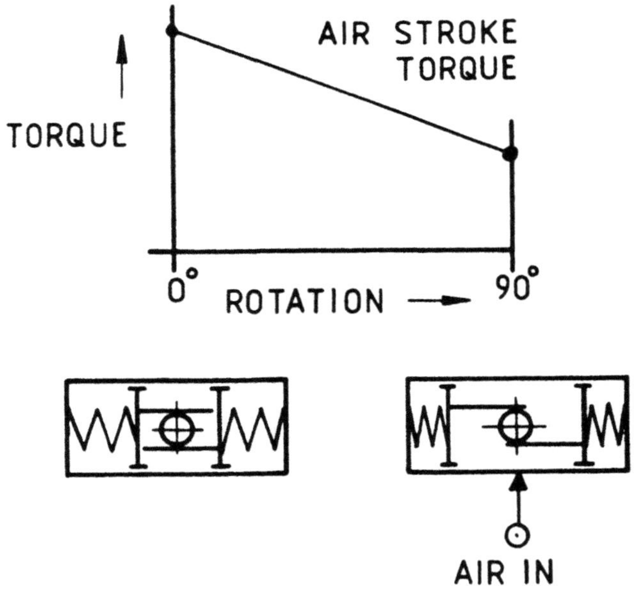

Fig 5: The air stroke.

On releasing the air the reverse takes place with high torque at the start of stroke when the springs are fully compressed and a lower torque at the end.

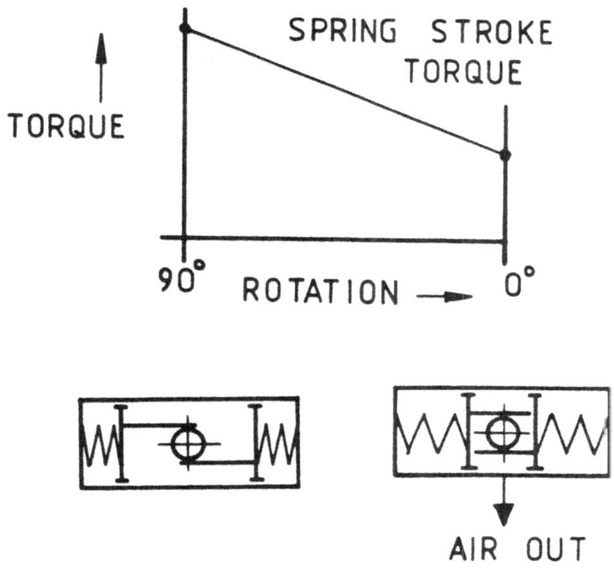

Fig 6: The spring stroke.

The relationship between the torque output of the actuator and the torque requirement of our two main valve types can now clearly be seen:

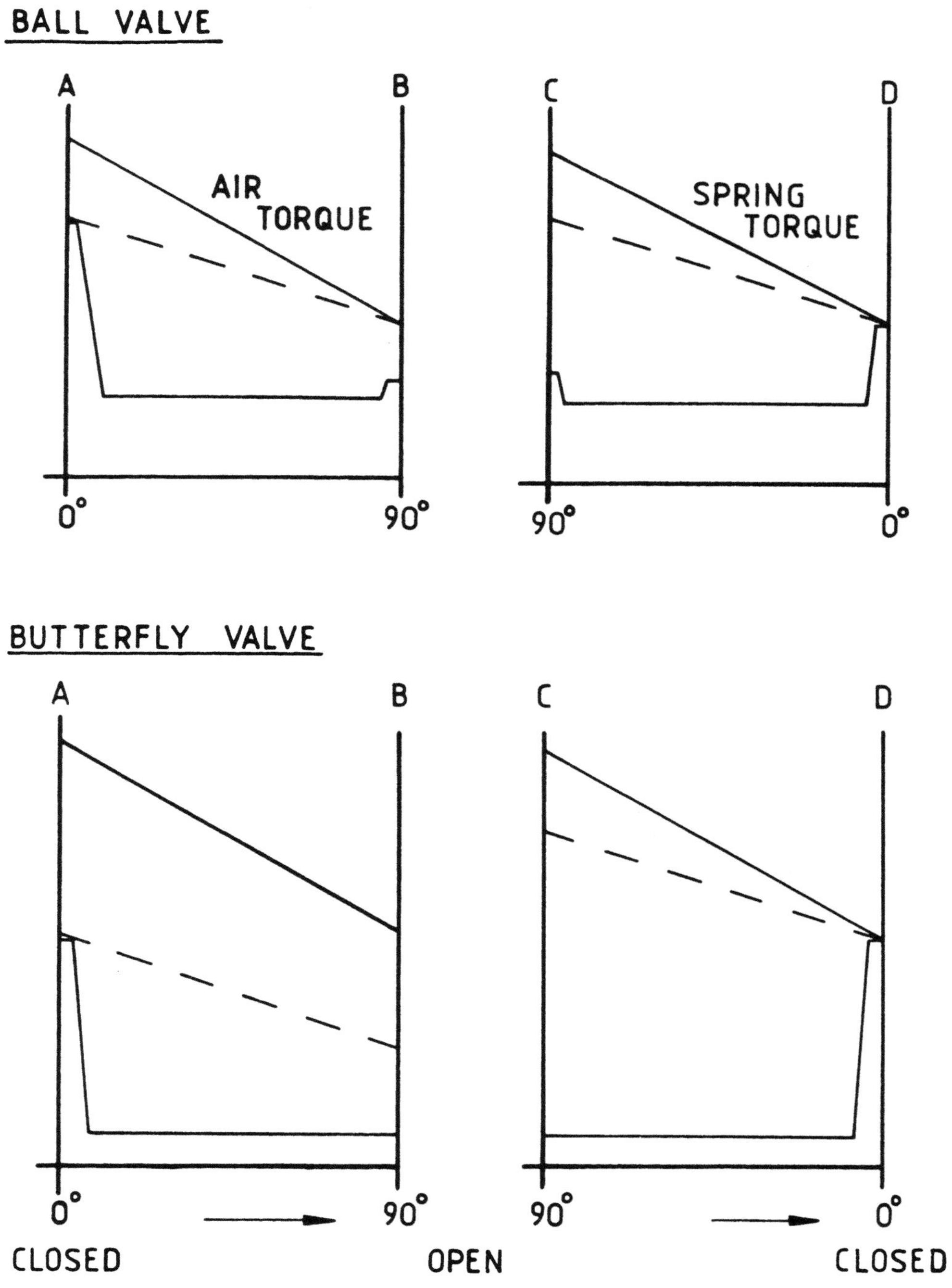

Fig 7: The spring slope.

It can also be seen from the above that there are significant advantages to be gained by reducing the slope of the torque line and, in the case of the butterfly valve, increasing the spring force and therefore the available torque at point "D". This is achieved at the expence of a lower torque at "A" and "B".

THE CONCEPT OF THE IDEAL SPRING

Most of the major manufacturers of pneumatic actuators are acutely aware of these concepts and are continually developing ways to make improvements in this area.

A new actuator now on the market goes a long way in achiving the ultimate in actuator design by concentrating on these requirements. Computer aided design featured significantly the in initial study and throughout the design and developement of the actuator springs.

Now anyone who has had anything to do with spring design will have been impressed by the long strings of formulas used, the quantity of variables involved and the fact that quite small changes in any of these variable makes an enormous difference in the final results.

This normally means opting for "something that comes near to the required results", or just going along with the spring manufacturing recommedtations.

So, whats wrong with that? I hear you say, well, the spring industry is involved with big volume, that means they try to get you to accept, as far as possible standard springs with normal characteristics. But the actuator spring is a long way from normal.

Take for example an internal combustion engine valve spring; cycling in the region of 1.000.000 cycles per day, compared with the actuator spring which may not need that in a complete life time and which may remain in the fully compressed state for the majority, of that life time.

The actuator spring design program uses nothing special in the way of formulas, indeed the standard text book formulas canbe used throughout, but the parameters used and the way the routines interact are very relevent to the special requirement of actuator.

Right from the start the actuator spring requirement works against that which best fits into a basic spring design. We need less slope in the compression line, a flatter rate spring. This is normally achived by increasing the overall length to working length ratio but if this is increased beyond a 5:1 ratio with its diameter the spring is no longer self supporting and there is a danger of buckling.

This can be clearly seen in some actuator designs where guidance has had to be introduced by allowing the spring to rub along its guiding recess, an undesirable feature as it introduces unpredictability both in performance and cycle life.

Unpredictibility, that's just what we don't need in todays actuators.

Next the problem of spring force: we need more spring force to cope with the high seating torque of the butterfly valve.

So we need to squeeze more spring force into as small a space as possible, remember space is at absolute premium in modern plant situation and space in an actuator housing costs money.
A cost which can be reduced by efficient spring design.

The modern micro computer can carry out thousands of calculations in a second and in the spring design program a continuing process is carried out, varying input parameters, comparing with previous results, all the time being manually guided towards the achievment of the ideal spring pack for that particular actuator size.

THE IDEAL ACTUATOR

This has resulted in an actuator range which is considerably superior to the majority of units currently available on the market today.

This new actuator series contains the following important features:

1. Compact
   Smaller than most actuators of the same torque output.

2. Lightweight
   Aluminium and high strength alloys used throughout the construction.

3. Stroke adjustment
   Stroke limiting screws are provided where they are most needed: at the end of the closing stroke.

4. High Torque
   Has an end of spring torque of 0.42 x DA torque, most other actuators rarely exceed 0.32.

5. Flexible design

This last point is vitally important to get the best out of any actuator design.

Not only does the spring requirement vary with valve type but also with the available air pressure.

This new actuator features 2 spring packs per actuator, one for each piston. Each spring pack contains 3 concentrically mounted springs, with the total load is shared between them in the following manner.

Outer spring	1/2
Middle spring	1/3
Inner spring	1/6

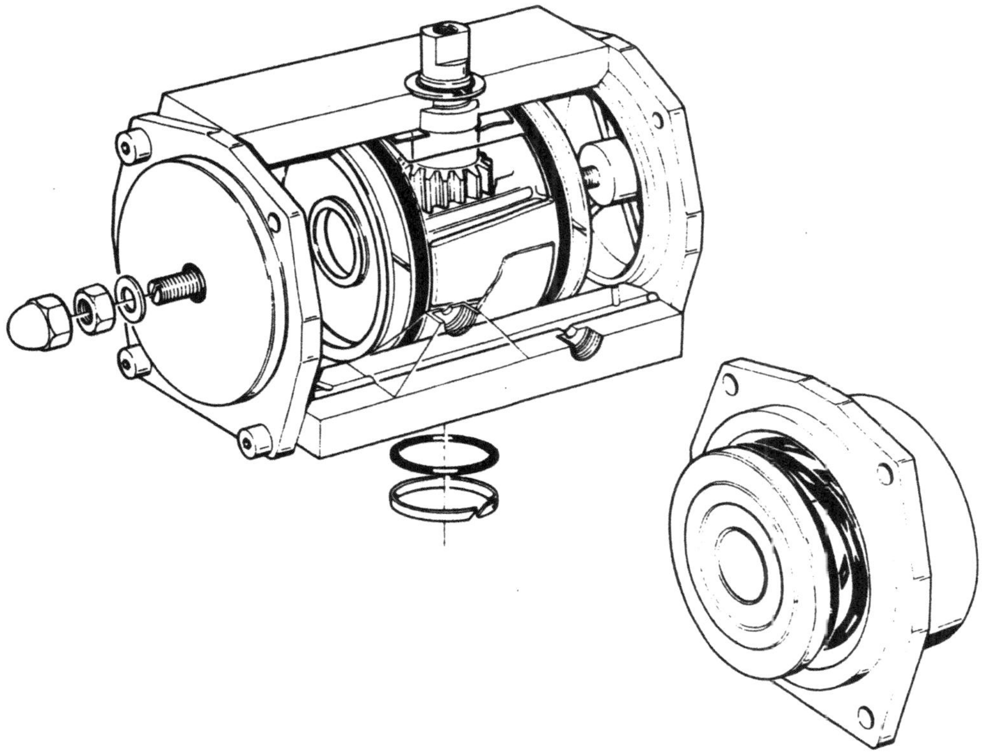

Fig. 8. The Ideal Pneumatic Actuator.

As each spring is totally, self suporting the spring pack may be assembled in a total of 6 possible variations to cater for a wide range of operating pressures. The resulting spring sets are numbered from 1 to 6 and are assembled as below.

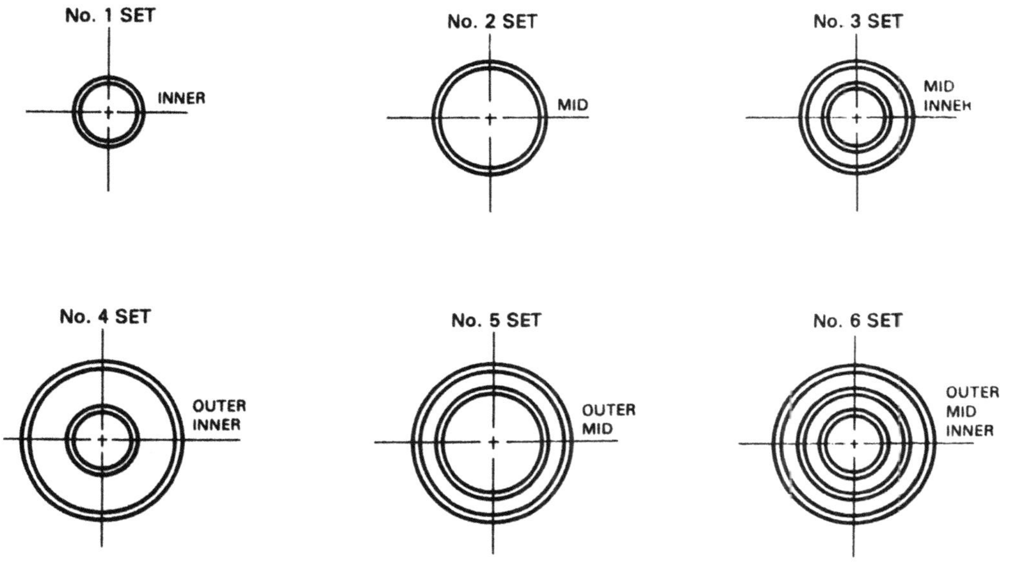

Fig. 9. Installation of springs.

Fig. 10. Operating pressures.

SET No.	Springs Used	Ball Valve	Butterfly Valve	
		F.O. or F.C.	F.C.	F.O
		operating pressure (bar)		
6	Outer+mid+inner	5	6	8
5	Outer+mid	4	5	7
4	Outer+inner	3	4	6
3	Mid+inner	2	3	5
2	Mid	1	2	4
1	Inner	1	1	3

The importance of the flexibility can be seen when you consider that the torque requirements vary not only with valve type, but also depending on mode: "Fail Open" (F.O.) or "Fail Closed" (F.C.).

This new actuator will provide significant benefits to those in the industry who need to automate those 1/4 turn valves in the 1/2" to 6" range as well as a wide variety of other rotary application where a powerful compact lightweight unit is needed.

2nd International Conference on

Developments in Valves and Actuators for Fluid Control

Manchester, England: 28-30 March 1988

PAPER D1

PLUG VIBRATIONAL TENDENCIES OF TOP GUIDED THROTTLING CONTROL VALVES

Henry Illing
B.M.E., P.E., Staff Engineer, DeZURIK, Sartell, Minnesota

Summary

This paper describes various vibration modes that can take place in control valves. Reasons for these are discussed with proposed solutions to avoid them in design as well as treatment in the field. Laboratory testing is described and results are shown that verify some of the proposed theory.

1.0 Introduction

Vibration problems in control valves continue to exist in many different forms and with various consequences. The Top Guided valve with its cantilevered plug sometimes shows a strong tendency for vibration on high pressure throttling applications.

This paper discusses the configurations and conditions that can produce vibration. It also discusses the various modes of vibration that can exist and the damage that may occur. It shows test results from a highly instrumented angle valve as well as details of the accelerometer and strain gauge instrumentation used.

Instability and vibration comparisons are shown on both flow-to-open and flow-to-close plugs.

2.0 Basic Types of Vibration

Vibration of control valve plugs can be classed in two categories; Lateral and Axial.

2.1 Lateral Vibration

Lateral plug vibration, as the name implies, consists of cyclic motion in a direction at right angles to the plug-stem centerline. The locus of this, as shown by test, is not a simple straight reciprocating path, but is elliptical or orbital.

The cause of this instability is shown in Figure #1. The steady state diagram on the left shows a hypothetical condition where the plug is perfectly aligned and the flowing stream has no turbulence or disturbances. The right hand diagram shows an

Held in Manchester, England. Organised and sponsored by BHRA, The Fluid Engineering Centre and co-sponsored by the British Valve Manufacturers Association and the Institution of Mechanical Engineers Process Industries Division

exaggerated eccentricity of the plug that causes asymmetric lateral loading. Add some stream turbulence and the stage is set for resonant or forced vibration.

Here is what takes place: At any given point in a flowing stream, total pressure consists of static plus dynamic pressure. These two are interchangeable depending on velocity with the dynamic pressure equal to $\rho V^2/2$ where ρ is the mass density and V is velocity. Furthermore, it is essential to realize that only the static portion of the total pressure creates force on the plug.

At first glance the diagram on the right appears to be wrong since we normally associate greater restriction with increased velocity. But here the valve is relatively close to the seat with nearly equal pressure around the periphery upstream of the restriction. Therefore, it is the coefficient of discharge that affects velocity and, taking small incremental areas, there is obviously a lower discharge coefficient on the right due to the higher ratio of wetted area to flow stream cross section. Also, immediately downstream of the maximum restriction, some diffusion (pressure recovery) is taking place. With a smaller mass rate of flow on the right this will take place further upstream producing a greater static pressure over areas on the right. The combined affect of these two (discharge coefficient and pressure recovery) creates higher static pressure on the more restricted side with a resultant force that pushes to the left. If the guide clearance were zero and the stiffness of the guide system were infinite, this would be of little consequence. In the real world, guide systems must have some clearance and guide stiffness is always some finite quantity. Improper consideration of these two factors in the design stage is the single most common cause of lateral vibration problems. Looking at the eccentric condition of Figure #1 there would still be no problem if the flowing stream were without disturbances of any kind. However, except for extremely low pressure drops with flow rates down in the laminar flow range, there is always some turbulence present in the flow path. A very strong exciting force is the severe turbulence generated by either vaporization/cavitation in liquid service or oblique/normal shock in compressible flow. Any of these upsetting forces contain an almost infinite number of exciting frequencies that the plug can respond to with the resultant reversal of the force illustrated in Figure #1. Somewhere in the broad spectrum of frequencies, one can coincide with the valve's natural frequency and, if its energy level is high enough, the first order fundamental mode of vibration (resonance) will begin.

There can also be random cyclic forces from cavitation or shock that are strong enough to cause severely damaging forced lateral vibration of the plug that is independent of resonance.

Figure #2 shows a comparison of the surface areas that are exposed to these forces for flow-to-open and flow-to-close. These diagrams show the plug near the seat where the worst problems of vibration usually occur. Note the large difference in plug area that is exposed to cyclic forces when these two are compared. It is believed that it is this area difference that causes the greater lateral excitation and vibration of flow-to-close valve plugs during throttling operation.

2.1.2 Consequences of Lateral Vibration

2.1.2.1 Guide System Damage

Either resonant or forced lateral vibration causes high cyclic impact loading on the guide system. The locus of vibration is not straight back and forth as implied by Figure #3 but, as previously stated, is known to be elliptical so that impact forces have both normal and tangential components. The tangential component and resultant rubbing causes wear, galling and premature failure. These tangential forces create even more problems from the moments that are created. These cyclic moments can unwind threaded guide bushings as well as threaded stem and accessory connections.

2.1.2.2 Outside Damage

Accessories such as positioners can pick up sympathetic vibration with resulting short life and/or degraded performance.

2.1.2.3 Plug-Seat Ring Contact

The most damaging results of lateral vibration occur when amplitudes are large enough to allow contact of the contour of the plug with the inside diameter of the seat ring. Figure #4 shows a diagram of this. With the larger moment arm of the plug outside diameter the moments are much larger than those in the guide area mentioned above and unwinding tendencies are proportionally greater.

Scuffing of the plug contour and seat ring inside diameter causes deterioration as depicted in Figure #5. Equal percent plugs are more susceptible to this than linear contoured plugs because of the long, close annular clearance. In a short time the seat seal is lost because of displaced material from the peening and galling of this action and there can also be a degraded flow characteristic near the seat caused by dimensional change.

2.1.2.4 Stem Fatigue

In addition to torsional moments, seat ring to plug contact causes reverse bending moments in the stem that cause either low cycle or high cycle fatigue usually at the first thread outside of the plug post. Figure #6 shows a diagram of the reversing or rotating impact couple that produces this at point #2 as well as the cantilever reverse bending failure that can occur at point #1.

It's interesting to note that this type vibration has caused plug post fatigue above the bottom flow-to-close plug in double seated valves as shown in Figure #6a.

2.1.3 Early Detection in the Field

Lateral vibration is usually audible. It is characterized by a rattling, humming or buzzing sound and can be confused with cavitation noise in liquid service. It can be felt by placing the fingers on the valve stem above the packing when conditions permit. If resonance exists, it can be easily detected by placing a vibration pickup on the stem above the packing and doing a frequency scan. The plot of acceleration vs. Hz will show a sharp spike where the resonance occurs. A resonant condition almost always means greatly reduced life of trim parts and accessories.

2.1.4 Prevention

Both prevention in design and cure in the field are difficult for this mode of vibration.

2.1.4.1 Design Precautions

In the design phase, the first order natural frequency can be roughly estimated by the formula in Figure #3. This simplified formula does not consider guide clearance, elastic rebounding, damping or stem weight but still can give a rough approximation for preliminary design. The calculated natural frequency can be used to determine g loading and amplitude, assuming sinusoidal motion. Generally g loading under 10 is acceptable providing that the associated amplitude does not allow the plug contour to contact the seat ring at positions near the seat.

Modeling a new design after a successful existing design, keeping clearances and natural frequencies about the same is good practice when enough similarity of geometry and service conditions exists.

However, the only positive way of proving a new design is testing under actual or simulated conditions.

2.1.4.2 Field Solutions

In the field there is little that can be done about this vibration mode other than trying reverse flow (turning the valve around) or installing a second valve either in series to lower the pressure drop or in parallel (split ranging) to change the operating stroke position where the vibration occurs.

2.2 Axial Vibration

Axial vibration is cyclic motion alternately toward and away from the valve seat. It can be separated into two forms although it is possible for all modes to exist simultaneously. These two types can be thought of as high frequency and low frequency.

2.2.1 High Frequency

This type is most common in flow-to-close unbalanced plugs and usually involves the plug striking the seat. In principle, the driving forces are similar to those producing lateral vibration since it too involves exchange of static and dynamic pressure. It can occur in either liquid or gas service.

The sequence goes like this: As the flow-to-close plug approaches the seat there is a rise in ΔP accompanied by increased velocity further lowering the static pressure, raising the effective ΔP across the plug. This divergent (self generating) process continues until the plug strikes the seat. At this point all flow is cut off and Bernoulli effects drop to zero whereupon the plug rebounds elastically off the seat helped by the elastic actuator and the cycle begins over. The frequency far exceeds the control response of the actuator so that positioners and other accessories are of little help. Frequencies can range from 20 to 7000 Hz.

2.2.1.1 Consequences

Although this mode is usually less damaging than lateral it can still cause rather rapid deterioration of parts.

The seat surfaces become wide, misshapen and rough over a relatively short period of time leading to premature and excessive seat leakage.

The packing and guide system life is greatly reduced and frequent retightening of the packing is required.

Like lateral vibration, this can damage accessories through sympathetic vibration, resulting in short life and degraded performance.

2.2.1.2 Detection

Usually this vibration mode is audible, and like lateral it may be heard as a rattle, hum, or buzz. It is hard to audibly distinguish axial from lateral but a vibration pickup, attached to the stem, will show much greater acceleration in the axial direction than in the lateral.

2.2.1.3 Prevention in Design and Cure in the Field

In the design stage the prediction and prevention by analytical methods is laborious and without credibility due to myriad unknowns, i.e. elastic rebound characteristics, effective suction area, installed system characteristic and others. Therefore, the designer needs to use the maximum practicable spring rate in the actuator and must depend heavily on geometrical similarity of existing successful designs. One small design configuration change that is usually effective on the flow-to-close plug is a short, 2 mm to 4 mm (.050 to .100 inches), straight section on the valve contour adjacent to the seat. This may adversely affect rangeability and C_v but can be a good tradeoff if axial vibration is suspected. This desensitizes the plug to Bernoulli effects near the seat and eliminates ΔP changes that occur during the last few millimeters of plug closure.

When axial vibration shows up in the field the best solution, when possible, is to turn the valve around and run flow-to-open even if a larger actuator is required. If this is impossible, a hydraulic snubber may control the vibration. These are rather large viscous dampers consisting of a piston 3 to 4 inches in diameter that is housed by a cylinder and reservoir and has a needle valve to adjust flow of oil from one side of the piston to the other, as shown in Figure #7. Stem movement forces oil through the adjustable orifice to give the required double acting damping.

3.2.2 Low Frequency Axial Vibration - .5 to 5 Hz

This mode of vibration can be the most damaging and dangerous of all types. The reason for this is the very large sudden changes of velocity in the upstream and downstream pipe. The forces generated come from the simple F = Ma formula where F is the force generated in KG, M is the mass of the fluid in the pipe length upstream and downstream, and a is the acceleration or deceleration of the fluid produced by plug excursions. Installations with long liquid legs upstream and downstream and flexible or weak piping supports can be extensively damaged by the large forces and movements produced. Needless to say, the liability risk to property and operating personnel is large.

This type vibration can occur at any place in the valve stroke but is most common in the first 50%. If the stem amplitude becomes large enough for the plug to impact on the seat, water hammer (in liquid service) can occur with pressure spikes upstream that are high enough to overpressurize pipe and fittings.

2.2.2.1 Causes

In a spring diaphragm actuated valve the causes of this are adverse gradient and divergent fluid forces on the plug. It can happen in either balanced valves such as the double ported or balanced cage styles or it can occur with the flow-to-close plug. It usually manifests itself as undesirable stem oscillations with a frequency that is too fast to be arrested by a valve positioner. The conditions that lead to this instability are shown in Figure #8. The adverse gradient section of the stroke indicates a condition where the rate of change of the stem force and the actuator spring rate are equal but with opposite signs so that they cancel each other. This allows the plug to hunt or oscillate through this range of stroke "in search" of equilibrium. As the plug approaches the seat in a flow-to-close valve, oscillations can become strong and divergent and impact with the seat can occur. This divergence can be greatly aggravated by a pipe system characteristic that forces a large ΔP increase on the valve as it is closed.

2.2.2.2 Prevention in Design

To prevent low frequency oscillation of either balanced or flow-to-close plugs, the valve designer needs data on stem load vs. stem position at the expected service pressure drop. This information can be plotted and the maximum slope selected from the curve. This slope represents stem force rate of change and it is necessary to select a spring with a substantially greater rate of change. If this is impossible because of practical spring sizes then a hydraulic snubber, as previously described, should be used. The ultimate, and much more expensive, solution would be a non-elastic mechanical or hydraulic actuator.

Use of a properly sized double acting air cylinder actuator has some advantage over spring diaphragm types. The coulomb damping of these actuators gives some advantage (albeit very small) and the high non-linear spring rate of the double acting higher pressure air helps avoid the adverse gradient and divergent force situations. To be most effective, these should be designed so that the piston is very close to the bottom closure (cylinder head) when the plug is on the seat. By doing this any large undesirable rapid stem movements create rapid compression or decompression of the air in this cavity adding substantially to the high spring rate needed to avoid adverse gradient and maintain stability.

2.2.2.3 Field Solutions

Any of the above actuator configurations, including snubbers, can usually be changed in the field without removing the valve from the flow line. A much heavier spring sometimes works. One or more of these should be done immediately upon the first sign of spurious stem-plug movements.

3.0 Why Use the Flow-to-Close Configuration

The obvious answer to the question for balanced plugs is the large reduction in actuator size afforded by these. The adverse gradient presents only a small problem here because of the intentionally designed small unbalanced areas. The unbalanced flow-to-close plug, however, has full port unbalance built in. Comparatively speaking, this produces enormous variable forces and a large propensity to exhibit the previously cited problems. The risk of these must be balanced against two distinct advantages.

3.1 Actuator Size

The actuator size requirement is reduced by the aiding rather than opposing plug force. Depending on pressure drop and plug size this alone can be enough to give the often required 0.88N per linear cm (50 lbs. per linear inch) force to seal the valve seat for Class IV shutoff. The actuator then only needs to furnish the quantity of force needed to open the valve against pressure drop forces.

3.2 Cavitation Damage

On liquid service it is often an overwhelming advantage to allow cavitation and resultant damage to take place downstream of the seating areas. This can greatly extend the life of the seat seal. In applications where seat leakage is important this can be a large asset. In angle valves it can place the cavitation envelope completely downstream of the valve and avoid body damage as well.

3.3 Slurry Service

Less seat damage and sometimes longer body life is achieved in erosive slurry service by running flow-to-close.

4.0 Recent Tests

In 1985, laboratory tests were undertaken as a joint venture between DeZURIK and the E.I. DuPont Company to determine the feasibility of running a 200 mm size throttling angle valve flow-to-close with a double acting 500 mm size pneumatic actuator. Up until this time much of the foregoing was conjecture based on damage patterns, high pressure destructive endurance testing, external vibration tests and field experience.

The large physical size of this valve made it possible to install relatively sophisticated instrumentation as shown in Figure #9. Hollow stem and plug construction allowed dry installation of stem bending strain gauge bridges S11, S12, and S13 and also valve plug accelerometers XPLA, YPLA, and ZPLA for measuring acceleration in the X, Y and Z axis. Bently-Nevada proximity probes were installed through bulkhead fittings in the body wall at 90° to each other so that stem motion could be measured in the X and Y directions. A 20 turn spring loaded potentiometer was installed for stem position readout. Additionally, accelerometers XFLA, YFLA, and ZFLA were installed on the outside diameter of the mating flanges to measure X, Y, and Z accelerations of the valve and piping.

4.1 Bench Testing

Before installing the valve in the test line, a shaker was attached to the valve plug and an exploratory run was made to search out the plug's lateral fundamental resonance. This varied from 75 Hz near the seated position (maximum overhang) to 156 Hz near the wide open position. This was done so that resonance could be separated from harmonics, and forced and random vibrations.

4.2 Flow Testing

4.2.1 Test Procedure

The test media was ambient temperature water with a nominal inlet pressure of 1920 KPa (280 psig) and with the outlet pressures ranging from 138 KPa (20 psig) to 1653

KPa (240 psig). Stem position was varied from 100% stroke, 19 cm (7.5 inches) to 2% open, 3.8 mm (.15 inches) in steps of 10% except for the 5% and 2% positions near the seat. Flow was measured with a low head loss venturi style flow meter.

The flow-to-open condition was run first since it was expected to be least damaging to the instrumentation.

4.2.2 Flow-to-Open Results

The valve was run through a large matrix of pressure drop and stem positions with two different bushing clearances. New bushing clearance was 0.127 mm (.005 inches) and the simulated worm bushing clearance was .38 mm (.015 inches). The maximum peak g level was 1.33 with the tight bushing and 1.89 with the loose one. Bending stress was 1902 KPa (276 psi) with the tight guide and 3169 KPa (460 psi) with the loose one. The Bently probes showed negligible motion and the flange accelerometers only showed background noise.

4.2.3 Flow-to-Close Results

The valve was run with the tight bushing with the following results:

Lateral vibration was significantly greater for all conditions tested. The test had to be discontinued near the 20% open position because of violent low frequency axial vibration.

Prior to shutdown the maximum peak g level recorded was 13.55 at 35 Hz and 44854 KPa (6510 psi) stress.

Motion measured on the Bently probes was 1.02 mm (0.04 inches) near the guide bushing compared to near zero for flow-to-open: Reading this motion on a Lisajous mode showed an elliptical orbit with a rotating axis.

Axial motion which had been negligible in the flow-to-open test jumped to over 10.2 mm (.4 inches), as shown in Figure #10. This occurred at the 13% stem position and began as two rather mild excursions ten seconds apart and progressed to the cyclic mode of about 3 Hz and 10.2 mm (.4 inches) amplitude about 2 seconds later. The sharp upward slope on the right shows the immediate manual opening and shut down.

It is worth noting that this transient literally shook the building and might have destroyed a conventionally supported piping system that had not been prepared for this event by heavy chain lagging of the pipes.

5.0 Conclusions

The above tests as well as reported incidents in the field and other lab tests indicate that running unbalanced throttling control valves in the flow-to-close direction involves considerable risk unless associated problems are anticipated and proper precautions taken.

The temptation to run this way can be strong because of the economic advantage of smaller actuators as well as the seat life extension under cavitating conditions. However, those responsible for applications and specifications should weigh these advantages carefully against the pitfalls before becoming committed to this style of operation. When vibration problems are encountered in the field, immediate action is called for to avoid costly repair and downtime as well as safety risks to operating personnel.

Acknowledgments

The author wishes to thank Messrs. Walter Haentjens of Hazelton Pump for use of his facility, and DuPont's James W. Hutt and Dr. S.T. Myrick for their extremely competent technical work that made the high pressure flow data possible.

References

(1) Den Hartog, "Mechanical Vibrations"
(2) My Klestad, "Mechanics of Vibration"
(3) H.D. Baumann, "Control Valve Noise, Its Cause and Cure"
(4) E.E. Allen, "Control Valve Noise" (2nd Edition of ISA Handbook of Control Valves)

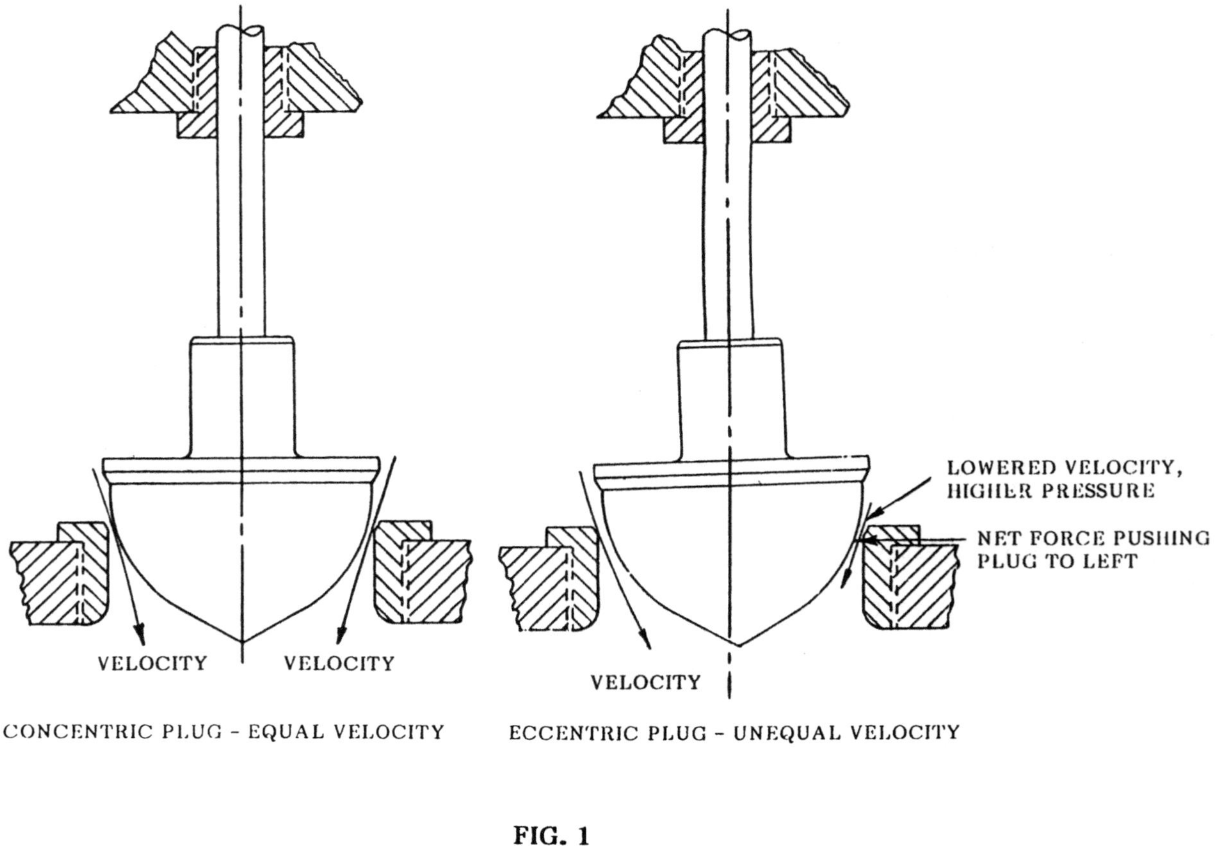

FIG. 1
EFFECT OF UNEQUAL VELOCITY ON OPPOSITE SIDES OF A THROTTLING CONTROL VALVE PLUG

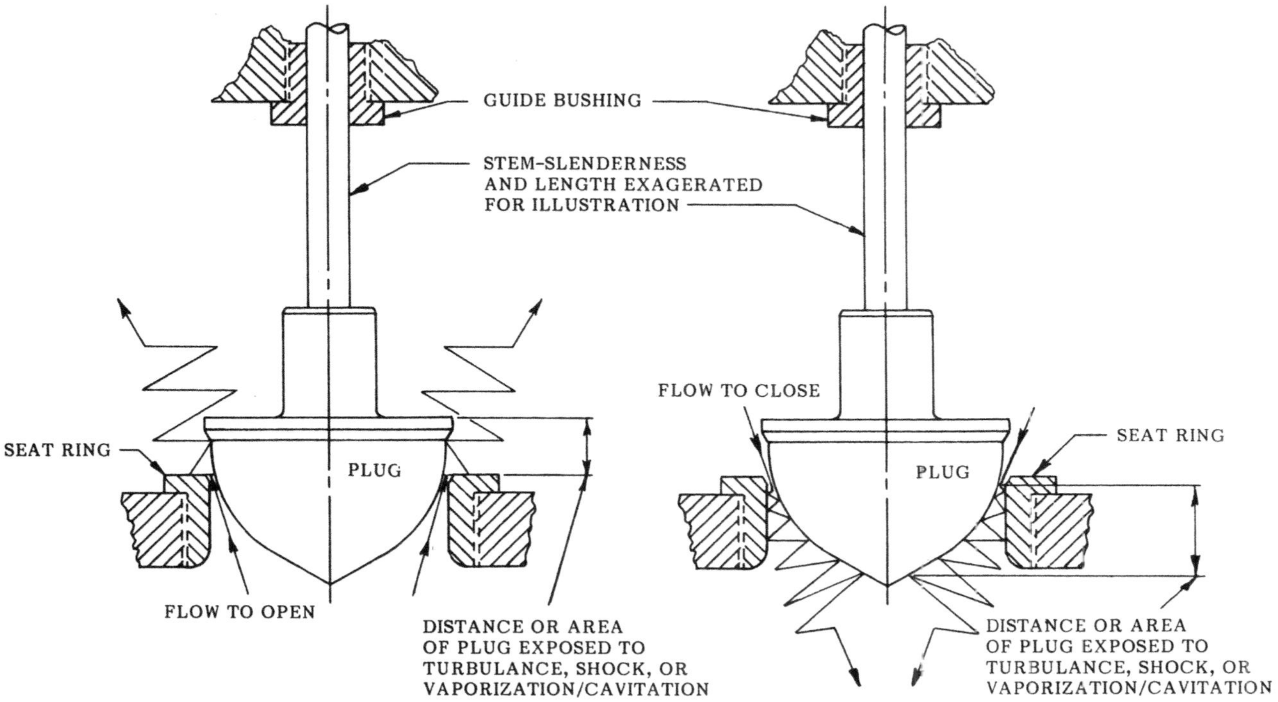

FIG. 2
PROPOSED COMPARISON OF EXCITING FORCES ON A THROTTLING CONTROL VALVE PLUG

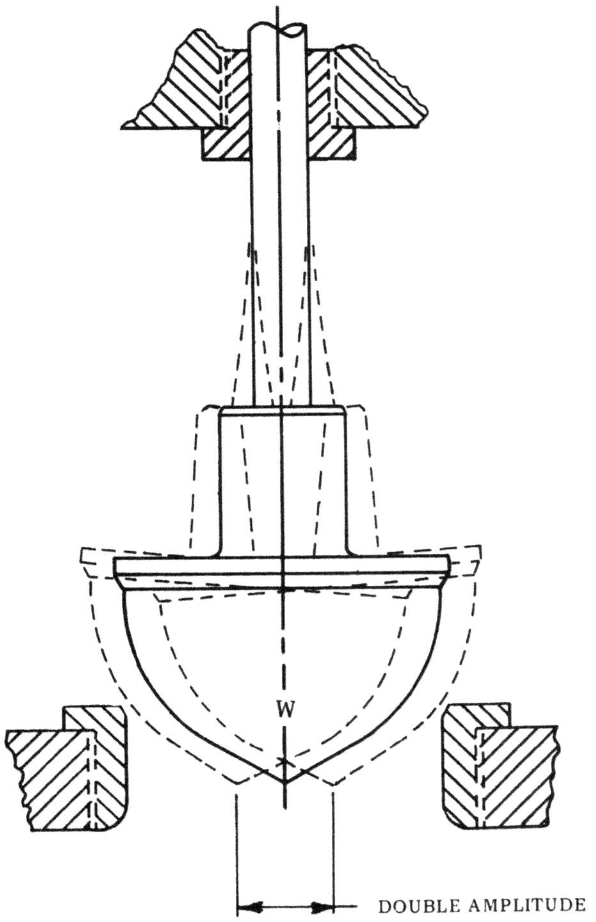

NEGLECTING GUIDE CLEARANCE, REBOUNDING AND PLUG POLAR MOMENT:

$$Wn = 0.159\sqrt{\frac{KG}{W}}$$

Wn = NATURAL FREQUENCY HZ
K = LATERAL SPRING CONSTANT FOR A CANTILEVER BEAM
W = WT. OF PLUG
G = GRAVITY CONSTANT

FIG. 3
SIMPLIFIED 1ST ORDER FUNDAMENTAL FREQUENCY OF A CONTROL VALVE PLUG

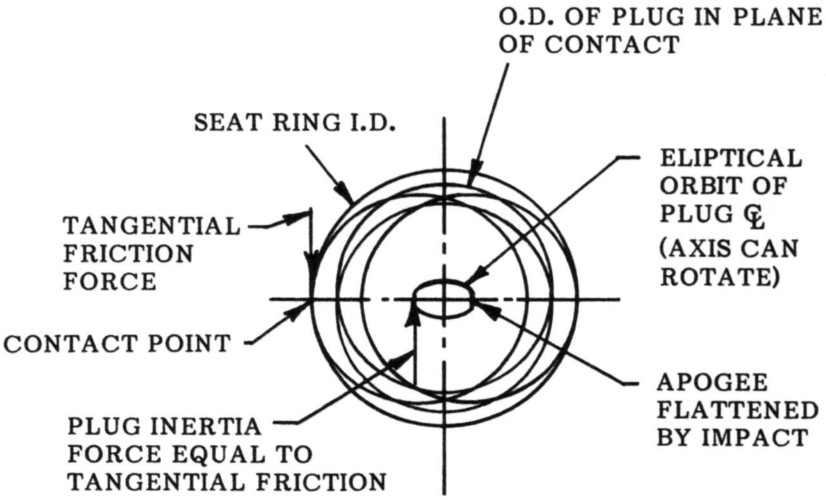

FIG. 4
DIAGRAM OF FRICTION COUPLE THAT PRODUCES STEM CONNECTION UNWINDING

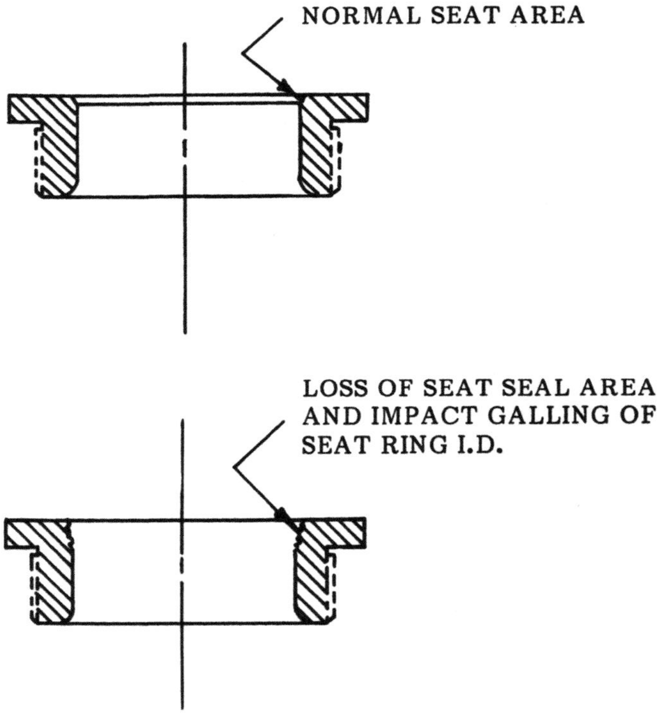

FIG. 5
SEAT RING DAMAGE FROM LATERAL VIBRATION

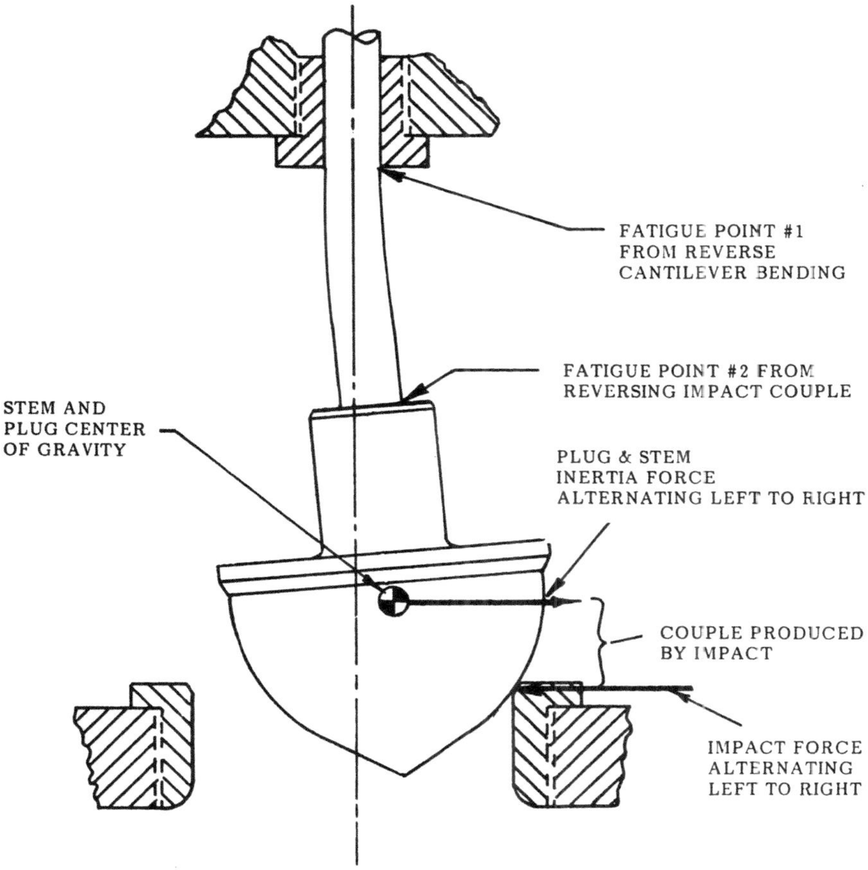

FIG. 6
SOURCE OF FORCES CREATING STEM BENDING FATIGUE DURING LATERAL VIBRATION OF PLUG

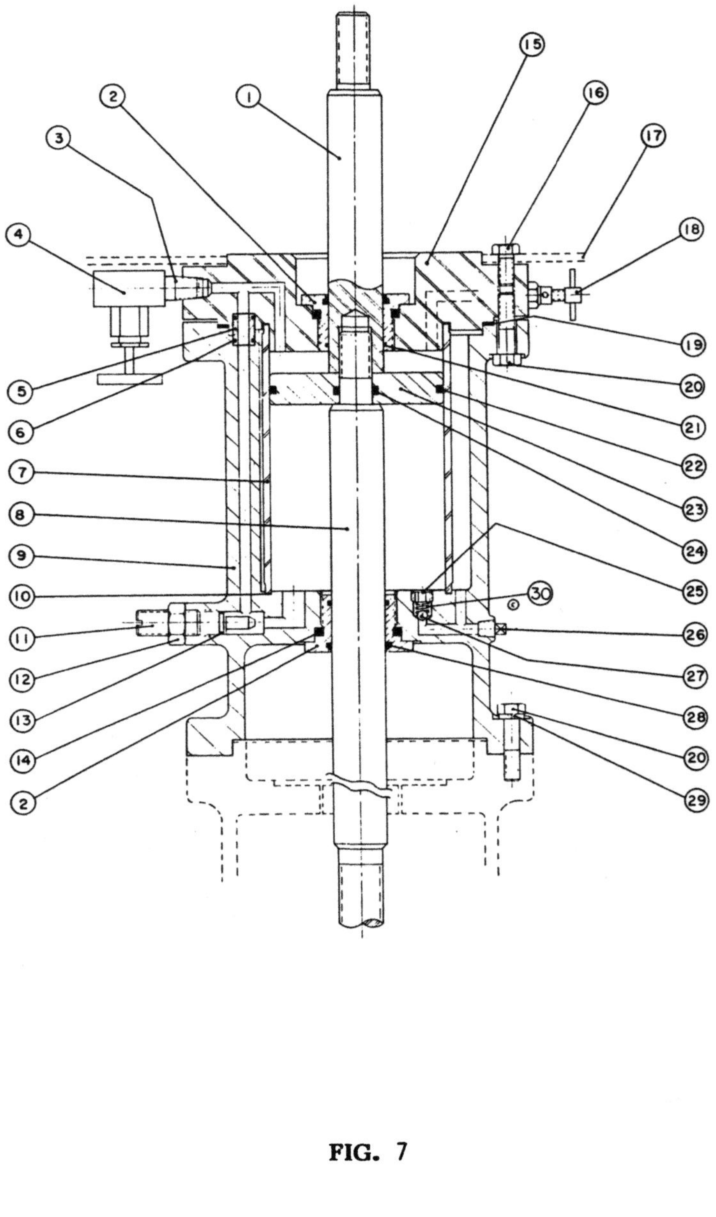

1
2
3
4
5
6
7
8
9
10
11
12
13
14
2
15
16
17
18
19
20
21
22
23
24
25
30
26
27
28
20
29

FIG. 7

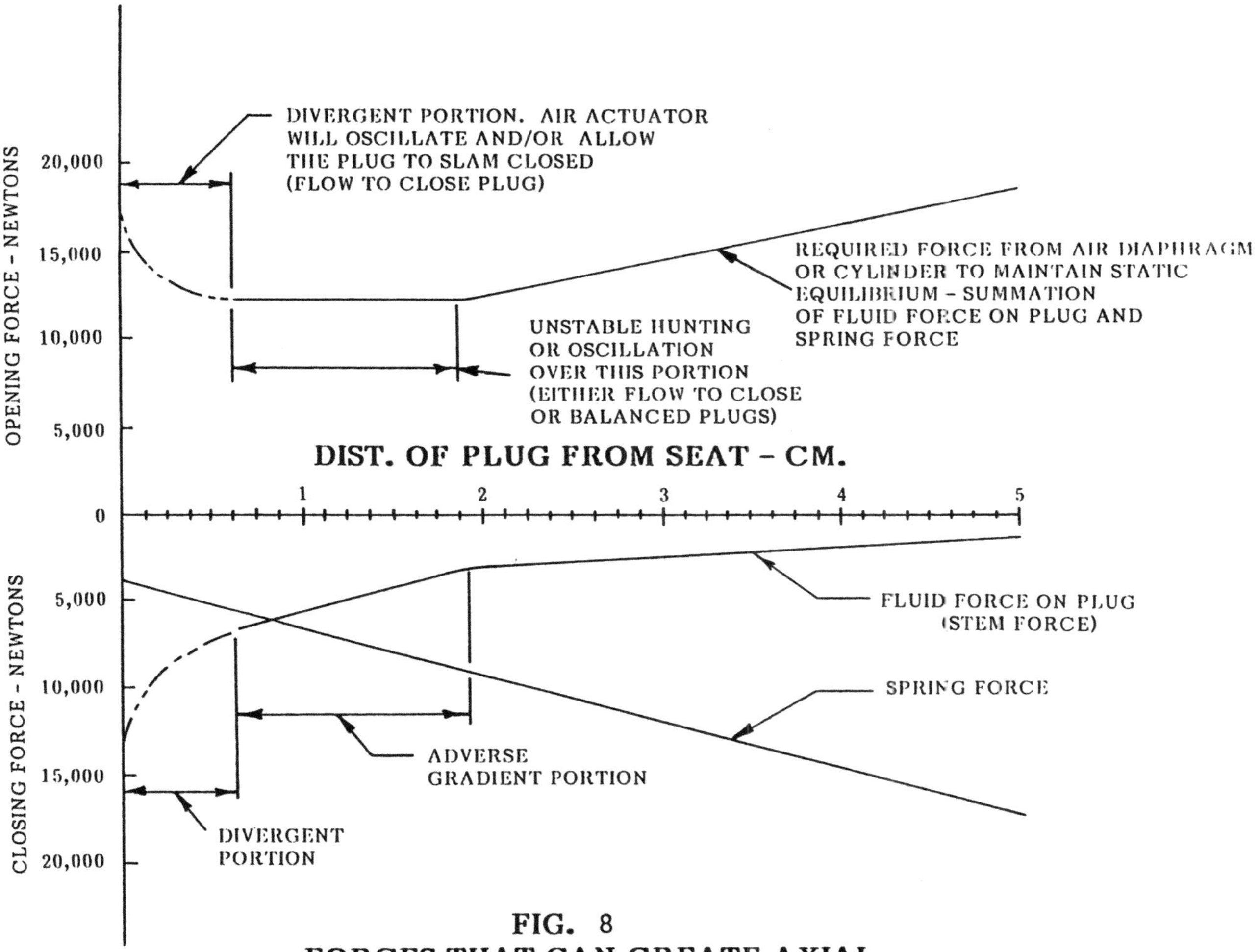

FIG. 8
FORCES THAT CAN CREATE AXIAL INSTABILITY AND/OR VIBRATION IN A FLOW TO CLOSE PNEUMATICALLY ACTUATED CONTROL VALVE

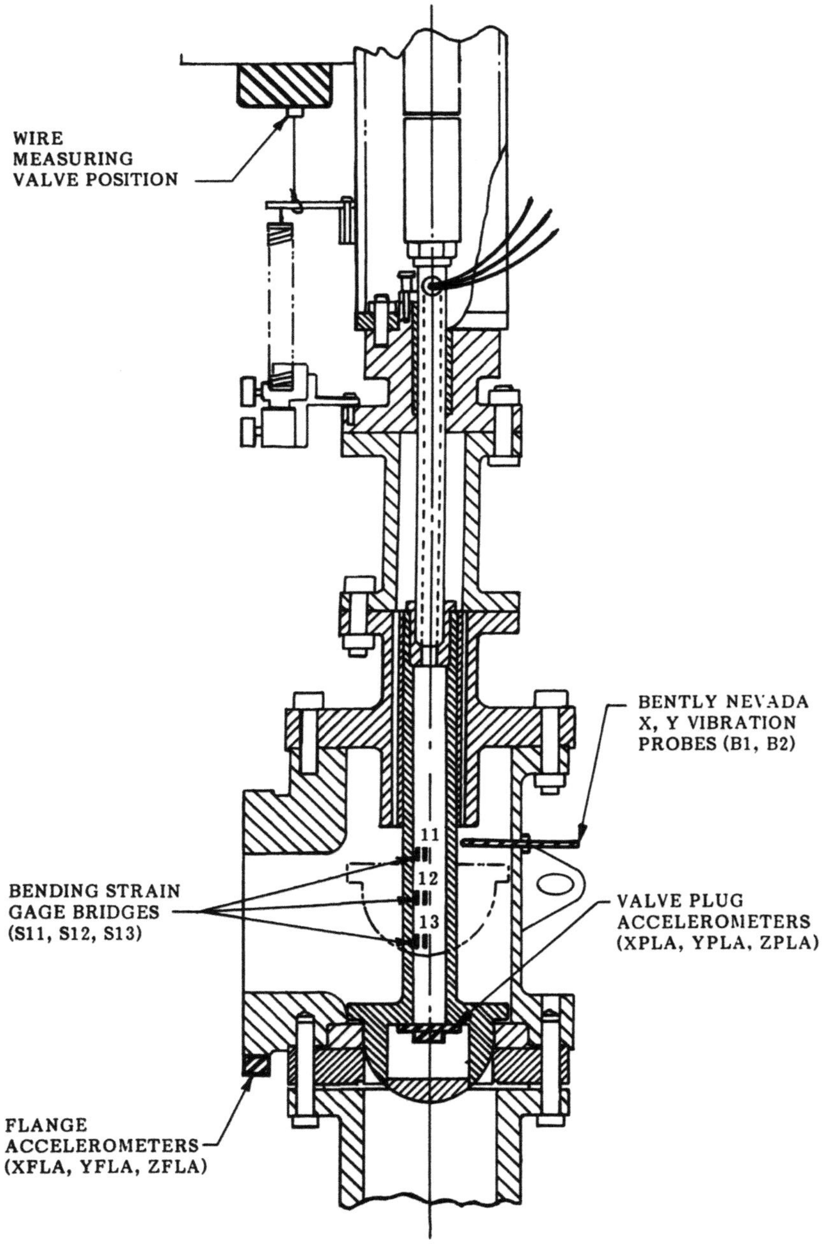

FIG. 9
INSTRUMENTATION INSTALLED
ON 8" ANGLE STYLE TEST VALVE

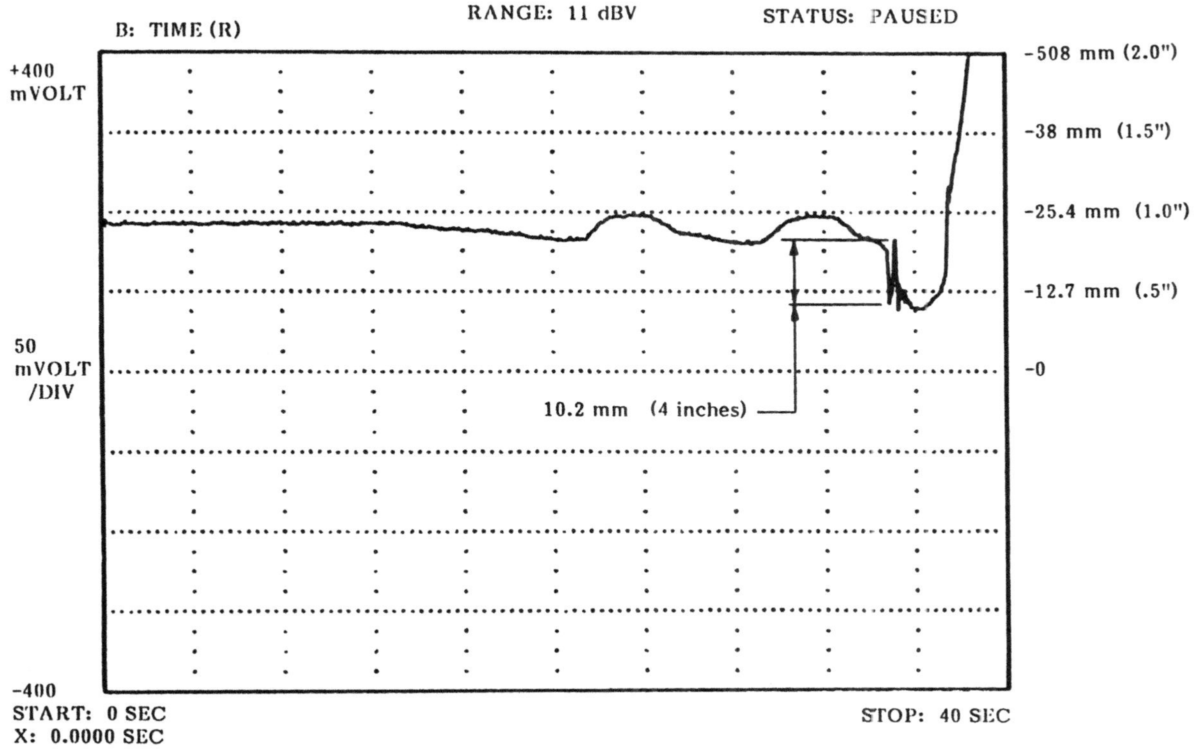

FIG. 10
COPY OF X-Y PLOT READING OF STEM POSITION VS TIME. FLOW TO CLOSE OPERATION

2nd International Conference on

Developments in Valves and Actuators for Fluid Control

Manchester, England: 28-30 March 1988

PAPER D2

A ONE DIMENSIONAL QUASI-STEADY MODEL FOR PREDICTION OF SAFETY VALVE LIFT CHARACTERISTICS

F.K.K. Lung
Marchwood Engineering Labs., CEGB, Marchwood, Southampton, SO4 4ZB.

SUMMARY

A one dimensional quasi-steady model is presented for predicting spring loaded safety valve characteristics. To extend the applicability of the model to two-phase and subcooled water operation, the thermodynamic properties are determined from steam tables directly. Based on the information published in the literature, valve characteristics for a 6 inch spring loaded safety relief are constructed. Numerical experiments are carried out to study the valve behaviour under different operating conditions. The effects of pressure ramps and ring settings on the mass flow rates, percentage blowdown and reseat pressure are discussed in some detail. Design parameters such as the maximum amount of lift allowed, seat area and back pressure on the valve disc are shown to be important for satisfactory valve performance.

Held in Manchester, England. Organised and sponsored by BHRA, The Fluid Engineering Centre and co-sponsored by the British Valve Manufacturers Association and the Institution of Mechanical Engineers Process Industries Division

1. NOMENCLATURE

A	area
F	force
k	spring constant
m	mass of moving parts
$\dot{M}$	mass flow rate
P	pressure
t	time t
y	valve lift
u	velocity
ρ	density
θ	angle

Subscripts

1	nozzle area
2	seat area
3	area of the disc which is not covered by the nozzle area and seat area
b	area at which back pressure is acting upon
e	flow exit plane Q (see Figure 1)
s	spring
t	throat
w	gravitation
Δ	increment

2. INTRODUCTION

Valves are important structural components in power plants. Poor valve performance can also be costly to power generation. In a pressurised water reactor (PWR) power station, pressure relief devices play a crucial role in both operational and safety aspects of the running of the plant. Information on safety valves such as the mass rate of discharge, accumulation pressure and the amount of blowdown is important as it defines the capabilities and limitations of a valve. As this information is embedded in the lift charateristics of a valve, the study of valve lift behaviour is extremely beneficial.

Over the years, various workers have examined valve characteristics experimentally. Narabayashi et al (1) have set up scaled models to investigate the safety valve performance under two-phase flow conditions. Sallet (2) used a Plexiglas scaled model to find out the internal flow field in a valve. He observed that the flow coefficient depends on the amount of valve opening. At the CEGB Marchwood Engineering Laboratories, a full-scale valve test facility has been built for testing both PWR and AGR (advanced gas-cooled reactor) safety and relief valves (3).

There also has been much interest in the analytical prediction of valve behaviour. Singh (4), Teske et al (5), Langerman (6) and Parker (7) have presented models for simulating spring loaded safety valve characteristics. They have assumed the flow passage to be that of a convergent/divergent nozzle and used isentropic gas relationships to find the solutions. Modification of the momentum equation was necessary for the prediction of two phase and subcooled water flows.

In the present paper, particular interest is given to spring loaded safety valves for use in PWR primary circuit. Through an analytical valve model, the implications of various valve design parameters, operational conditions and control settings on the valve performance are examined in detail. It is hoped that this study can deepen the understanding of the operation of these valves for both valve designers and users.

3. THE VALVE MODEL

Figure 1 shows the schematic diagram of the valve and the forces associated around the valve disc area when it has been lifted. Assuming there is no damping, the equation of motion of the valve disc can be written as:-

$$m\ddot{y} = F_1 + F_2 + F_3 - F_b - F_w - F_s \qquad (1)$$

where m is the mass of the moving parts and y is the valve lift. F_1, F_2 and F_3 are the forces acting under the valve disc due to the fluid in the nozzle area (A_1), the seat area (A_2) and the area exposed to the exit plane (A_3) respectively. F_b, F_w and F_s are the back pressure acting on the disc, the gravitational force on the moving parts and the spring force respectively. If the time step, Δt, is small, the forces acting on the valve disc between adjacent times are roughly the same. Equation (1) can be solved analytically and gives the following solution at time t:-

$$y(t) = G(t)\cos\sqrt{\frac{k}{m}}\,\Delta t + H(t)\sin\sqrt{\frac{k}{m}}\,\Delta t + \frac{F(t-\Delta t)}{k} \qquad (2)$$

$$\dot{y}(t) = -G(t)\sqrt{\frac{k}{m}}\sin\sqrt{\frac{k}{m}}\,\Delta t + H(t)\sqrt{\frac{k}{m}}\cos\sqrt{\frac{k}{m}}\,\Delta t \qquad (3)$$

where

$$F(t-\Delta t) = F_1(t-\Delta t) + F_2(t-\Delta t) + F_3(t-\Delta t) - F_b(t-\Delta t) - F_w(t-\Delta t) - C \qquad (4)$$

$$G(t) = y(t-\Delta t) - \frac{F(t-\Delta t)}{k} \qquad (5)$$

$$H(t) = \dot{y}(t-\Delta t)/\sqrt{\frac{k}{m}} \qquad (6)$$

k is the spring constant. C is the initial compression of the spring which controls the set pressure of the valve. Assuming the initial back pressure on the disc is atmospheric, C can be calculated from:-

$$C = P_{set}A_1 + \tfrac{1}{2}(P_{set} + P_{atm})A_2 - P_{atm}(A_1 + A_2) \qquad (7)$$

where P_{set} and P_{atm} are the set pressure and the atmospheric pressure respectively. The second term in the right hand side of equation (7) represents the force required to overcome the lifting force due to the build up of pressure gradient across the valve seat.

The force under the seat, F_f, can be calculated by:-

$$F_f = F_1 + F_2 + F_3$$
$$= P_1A_1 + \dot{M}_1u_1 + \bar{P}_2A_2 + P_3A_3 + \dot{M}_eu_e\sin\theta_e \qquad (8)$$

where $\dot{M}_1$ and $\dot{M}_e$ are the mass flow rates, u_1 and u_e are fluid velocities at the nozzle and the exit plane near the adjusting rings respectively. The mass flow rates can be determined from $\dot{M} = \rho A u$ when the thermodynamic properties along the flow passage are determined. $\bar{P}_2$ is the mean pressure on the seat area and θ_e is the flow exit angle measured from the abscissa. The flow exit angle depends on the adjusting ring settings and is determined in a similar manner as described in Teske et al (5).

In order to calculate F_f, it is necessary to find out the pressure and velocity variations under the valve disc. The fluid flow through the valve is assumed to be isentropic and the geometry of the flow passage is modelled by a convergent/divergent nozzle.

The thermodynamic properties along this flow passage are calculated through an iterative process by using the CEGB steam table routines (8). This method of calculating thermodynamic properties has two advantages, namely, (1) There is no need to assume an isentropic (or polytropic) index for the gas relationships. For real gases, a constant polytropic index has a limited range of validity. (2) There is no need to modify the equation of state for two phase or subcooled water flow calculations as long as the fluid in the flow passage remains in thermodynamic equilibrium. Ransom et al (9) found that the homogeneous assumption is valid for subcooled choking processes.

Since the flow geometry inside the valve is complex, pressure losses around the seat region are inevitable. Langerman (6) suggested that the total stagnation pressure loss, ΔP_o, at the exit plane of the discharge can be approximated by:-

$$\Delta P_o = \left(1 - \frac{A_t}{A_e}\right)^2 \frac{\rho_t u_t^2}{2} \qquad (9)$$

where A_t, ρ_t and u_t are area, density and velocity at the throat respectively. A_e is the flow area at the exit plane.

Depending on the valve design, the magnitude of the back pressure on the valve disc can be very different. The back pressure can be affected by the outlet pressure through the bleed hole (see Figure 1) and P_3 through the clearance between the guide ring and valve disc. In this particular model, the bleed hole is assumed to be blocked and the back pressure is a function of P_3. Since the clearance is small, it will take some time before the back pressure senses the effect of P_3. As there is no better information concerning back pressure available, it is arbitrarily assumed to be equal to:-

$$P_b(t) = [3 \times P_b(t - \Delta t) + 0.8 \times P_3(t)]/4 \qquad (10)$$

In equation (10), the back pressure at time t is weighted towards that at time t-Δt and only 80% of the pressure P_3 is sensed by $P_b(t)$ at any instant. Unless stated otherwise, equation (10) is used to model the back pressure in the valve:-

4. THE HYPOTHETICAL SPRING LOADED SAFETY/RELIEF VALVE AND THE NUMERICAL VALVE TEST

Due to reasons of proprietors' interest, it has not been possible to obtain exact dimensions of any safety/relief valves. However, Hangerman (6) and Teske et al (5) did provide certain information on the characteristics of a Crosby 6M6 valve. Based on their information, a hypothetical valve was set up and the characteristics are summarised in Table 1. The guide and nozzle rings are assumed to be adjustable and their respective positions, A and B are measured from the X-axis as shown in Fig.1. It must be noted that valve performance analyses based on the characteristics laid down in Table 1 may not necessarily reflect the behaviour of the Crosby valve.

In order to compare the effects of different valve parameters, a standard numerical test condition is set up. The hypothetical valve is assumed to be mounted on top of a pressure vessel with starting test conditions of 170 bar and 360°C. The set pressure of the valve is taken to be 172 bar. Unless stated otherwise, the numerical valve test is carried out in the following manner:-

(1) The pressure vessel is compressed isentropically at a pressure ramp rate of 2 bar s^{-1}.
(2) The valve lifts at 172 bar.
(3) The compression is stopped when the pressure vessel reaches a pressure of 176.9 bar.
(4) A negative pressure ramp follows after that at a rate of 2 bar s^{-1}.

5. RESULTS AND DISCUSSION

5.1 Lift characteristic of the hypothetical valve

Figure 2 shows the lift characteristics of the hypothetical valve. The guide and nozzle ring settings are 0 and -5.2 mm from the X-axis respectively. The solid line in Figure 2 represents the valve lift position and the dotted line is the vessel pressure. The valve starts to open when the pressure reaches 172 bar and takes about 40 ms to reach the maximum rated lift. The valve starts to close at 5.48 s when the pressure in the vessel drops to 170.9 bar. After that, the valve lift position decreases slowly with the falling pressure and when the pressure falls to 169.3 bar, the valve shuts abruptly. This valve closure phenomenon was also observed by Teske et al (5) and Langerman (6) on the Crosby valve.

This abrupt valve closure can be explained by considering the movement of the throat position in the fluid flow passage. When the valve is at maximum rated lift, the minimum flow area occurs in the flow exit plane (location Q in Fig.1). As the lift decreases, the flow area in the inner edge of the seat (location P) decreases. When the flow area at P becomes less than that at Q, the throat position shifts from Q to P. The flow between P and Q becomes supersonic and there is a large decrease in static pressure in this region whilst the back pressure remains at around 100 bar. This sudden force reduction causes the rapid valve closure.

5.2 Effect of adjustable ring settings

The effect of three different guide ring settings, namely, 0, -2.6 and -5.2 mm at two different nozzle ring settings, namely, 0 and -5.2 mm on the valve performance is investigated. Since lift characteristics for all these ring settings resemble that presented in Fig.2, they are not presented here. Although the computational results suggest that the ring settings do not have a significant effect on the pressure at which maximum rated lift is attained, they do affect the reseat pressure and the mass flow rate of the valve. Figure 3 shows the variation of percentage blowdown, pressure at which closing begins and the reseat pressure at these different ring settings. The percentage blowdown is defined as:-

$$\text{Percentage blowdown} = \frac{\text{set pressure - reseat pressure}}{\text{set pressure}} \tag{11}$$

Figure 3 clearly shows that as the guide ring position moves downward, the amount of blowdown increases. At a nozzle ring position of -5.2 mm, the percentage blowdown increases from 1.6% to 6.6% as the guide ring changes from 0 to -5.2 mm. The amount of blowdown also increases as the nozzle ring moves upward. At a guide ring setting of 0 mm, the amount of blowdown increases from 1.6% to 7.2% as the nozzle ring moves from -5.2 to 0 mm.

It is interesting to note that the difference between the pressure at which the valve starts to close and the reseat pressure increases as the guide ring setting decreases. Similar behaviour can also be obtained with an increase in nozzle ring setting. The effect of ring settings on the duration of reseating time can be explained by the geometry of the fluid flow passage through the valve and the dynamic force which the fluid exerts on the valve disc i.e. the reaction force increases as the flow exit angle θ_e increases.

The two adjustable rings serve two purposes. They alter the area and angle of flow at the exit plane Q (see Fig.1). Since the throat occurs at Q when the valve is at full lift, the ring settings also governs the discharge rate of the valve. Figure 4 shows the mass flow rate at various ring settings when the vessel pressure is 173 bar. It shows that the discharge rate decreases as the guide ring moves downwards. If the mass flow rate remains constant, an increase in the exit flow angle increases the dynamic force at the exit. This would help to extend the blowdown period. Perhaps the most important parameter in controlling the amount of blowdown is the exit area. As the flow at Q is choked during reseating, a decrease in the exit flow area would delay shifting of the throat position from Q to P and prolong the blowdown time.

5.3 Effect of pressure ramp on full lift and reseat pressure

With guide and nozzle rings set at 0 and -5.2 mm respectively, the effect of pressure ramp on the valve behaviour is studied. Table 2(a) shows that the pressure at which the valve achieved full lift is slightly lower when the pressure ramp is lower. As for the reseat pressure, a decrease in the downward ramp rate reduces the blowdown and the valve reseats at a slightly higher pressure. Table 2(b) shows the reseat pressures at downward ramp rates of 2 and 20 bar s^{-1}. This valve behaviour depends on how fast a valve can respond to a change in pressure and can be quantified by the force distribution and momentum effects around the valve disc and the spring constant.

5.4 Effect of an increase in maximum rated lift

Figure 5 shows the lift characteristic with guide and nozzle rings set at 0 and -5.2 mm respectively and the maximum rated lift increased to 17 mm. The valve takes 40 ms to reach the maximum rated lift but quickly drops down to a value of 14.3 mm at 240 ms as the back pressure increases. After that, the lift position increases throughout the upward pressurisation period to a maximum value of 15.5 mm then decreases with the downward pressure ramp. The valve reseats at a pressure of 169.32 bar. This reseat pressure is the same as that with maximum rated lift of 13.7 mm. Since the mass flow rate is controlled by the throat position which is at the exit plane Q, an increase in the maximum rated lift does not provide any advantage at all.

5.5 Effect of seat area, A_2

Figures 6 to 8 show the valve lift behaviour when the seat area is increased by 50% to 11.9 cm^2. With the nozzle ring sets at 0 mm, Figs. 6 to 8 show the variations in lift characteristics at guide ring settings of -5.2, -2.6 and 0 mm respectively. It is clear from these traces that valve lifting becomes more difficult as the guide ring moves upwards. Indeed, Fig. 8 does not show any steady lift at all. Oscillations prior to lifting of the valve can be explained by considering the force balance on the valve disc. As the valve begins to lift, the sonic line occurs in the inner edge of the seat and the static pressure in the vicinity of the nozzle and the seat drops to the critical pressure. Because of a much larger opposing force in the spring, although the valve lifts when the set pressure is reached, it cannot lift to a position where the mass flow rate is sufficient to increase the dynamic force to compensate the force reduction due to the drop in static pressure. The valve then shuts abruptly and the whole process repeats. Figures 6 to 8 also show an increase in the oscillation amplitude and a decrease in oscillation frequency as the vessel pressure increases. The frequency of oscillation drops from a few hundred Hertz to around 20 Hz just prior to full lift. The increase in amplitude is due to the increase in the driving force in pushing the valve open. Unless the stagnation pressure at the nozzle is high enough to lift the valve to a position where the mass flow is sufficient to create a substantial dynamic force to sustain the lift, the valve cannot open successfully. For the same reason, as the lift amplitude increases, the amount of dynamic force increases which sustains the valve's opening for a longer period of time and hence the drop in frequency.

These oscillations are clearly not desirable as they can cause damage to both the disc and seat of the valve. The seat must be carefully designed in order to minimize this risk. Setting the amount of compression on the valve correctly can also reduce the chance of inadvertent valve oscillations.

5.6 Effect of back pressure variation

Since British Standard BS6759 Part 1 Appendix B requires the maximum outlet pressure at the flange to be either 12% of set pressure or 17 bar, it might be useful to investigate the case where the back pressure in the valve disc is equal to the outlet pressure. Figure 9 shows the lift characteristic of the valve when the back pressure is set to be 10% of the vessel pressure. The guide and nozzle ring settings are 0 and -5.2 mm respectively. In this case, the valve reseats at 134.3 bar with a blowdown of 22%. This clearly demonstrates that the back pressure has an important effect on the valve behaviour. For this valve to operate satisfactorily under a lower back pressure, significant dimensional changes have to be made.

6. CONCLUSIONS

(1) A valve model has been developed for predicting the valve characteristics of a 6" spring loaded safety relief valve under various valve operating conditions.

(2) Operational aspects of the valve, namely, system pressure ramp rates, nozzle and guide ring settings are studied in some detail. Although the pressure ramp rates do not affect the set pressure of a valve, an increase in the positive ramp rate can reduce the time for the valve to reach maximum rated lift and an increase in the negative ramp rate can lower the valve reseat pressure. The reseat pressure and the percentage blowdown are affected by the nozzle and guide ring settings. The percentage blowdown can be increased by lowering the guide ring and raising the nozzle ring.

(3) Valve design aspects concerning the alteration of amount of maximum lift, seat area and back pressure are discussed. It is found that if the throat position of the valve occurs at the exit plane Q when the valve is at maximum rated lift, an increase in the maximum rated lift in the valve does not alter the set pressure and blowdown characteristics. An increase in the seat area could lead to pre-lift oscillations and a decrease in the back pressure could extend the blowdown period.

6. ACKNOWLEDGEMENT

This paper is published by permission of the Central Electricity Generating Board. The author would like to thank various valve manufacturers for their cooperation.

7. REFERENCES

1. Narabayashi,T., Nagasaka,H., Niwano,M. and Ohksuki,Y.: "Safety Relief Valve Performance for Two Phase Flow". J. Nucl. Sci. Tech., 23(3), 1986, pp.197-213.
2. Sallet,D.W.: "Thermal Hydraulics of Valves for Nuclear Applications". Nucl. Sci. Eng., 88, 1984, pp.220-244.
3. Airey,D.R., Richards,D.J.W., and Bryant,S.: "Valve Blowdown Test Facility for Testing High Flow Rate Safety/Relief and Block Valves". ASME Paper No. 87-PVP-7, 1987.
4. Singh,A.: "An Analytical Study of the Dynamics and Stability of a Spring Loaded Safety Valve". Nucl. Eng. Design, 72, 1982, pp.194-204.
5. Teske,M.E., Bilanin,A.J. and Hecht,A.M.: "Coupled Valve Dynamics Model for Steam, Two-phase, and Subcooled Discharge Flow Condition". EPRI NP-3493, 1984.
6. Langerman,M.A.: "An Analytical Model of a Spring Loaded Safety Valve". ASME paper no. 83-NE-19, 1983.
7. Parker,G.J.: " 'Pop' Safety Valve: A Compressible Flow Analysis". Int. J. Heat and Fluid Flow, 6, 1985, pp.279-283.
8. Cullen,M.G.: "CEGB Revised Steam Table Package-Pollak Formulation, Vol. 1, Reference Routines". CEGB Report no. CC/P662, 1982.
9. Ransom,V.H., Wagner,R.J., Trapp,J.A., Carlson,K.E., Kiser,D.M., Kuo,H-H., Chow,H., Nelson,R.A. and James,S.W.: "RELAP5/MOD1 Code Manual, Volume 1: System Models and Numerical Methods". Report No. NUREG/CR-1826, EGG-2070, EG and G Idaho, Inc., Idaho Falls, Idaho, March 1982.

TABLE 1

CHARACTERISTICS OF THE MODAL VALVE

Mass of moving parts	27 kg
Spring constant	2900 N mm^{-1}
Nozzle area	2730 mm^2
Maximum rated lift	13.7 mm
Total seat area in contact with seat when lift = 0	790 mm^2
Radius of the disc	47 mm
Total disc area subject to back pressure	2090 mm^2
Valve bore diameter at inlet	152.4 mm
Valve bore diameter at outlet	152.4 mm
Set pressure of the valve	172 bar
Mean thickness of the nozzle ring	10 mm

TABLE 2

EFFECT OF PRESSURE RAMP ON FULL LIFT AND RESEAT PRESSURE
(Nozzle ring = -5.2 mm, Guide ring = 0 mm)

(a) Effect of upward pressure ramp on full lift pressure

Pressure ramp (bar s^{-1})	Pressure at full lift (bar)
0.2	172.01
2.0	172.08

(b) Effect of downward pressure ramp on the reseat pressure

Pressure ramp (bar s^{-1})	Reseat Pressure (bar)
2.0	169.32
20.0	168.70

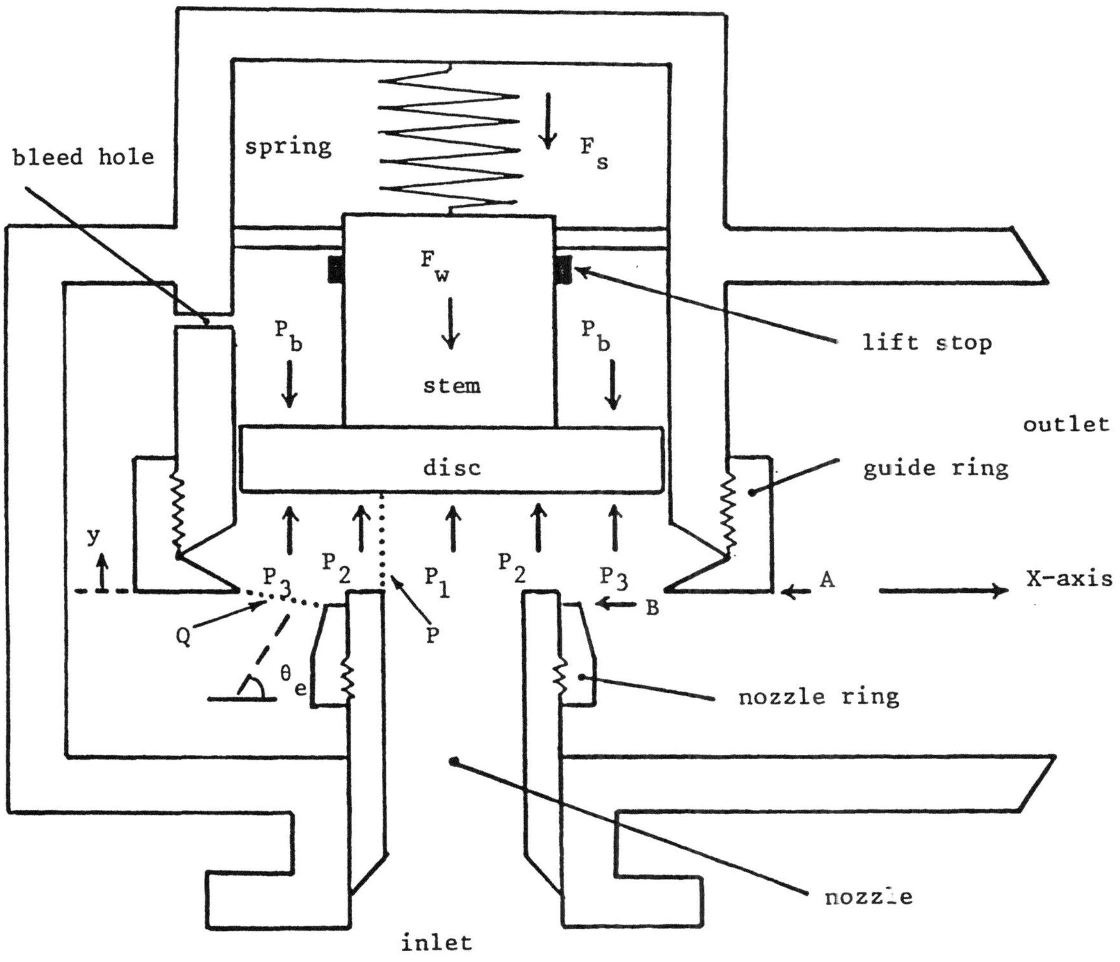

FIGURE 1: SCHEMATIC DIAGRAM FOR THE VALVE MODEL

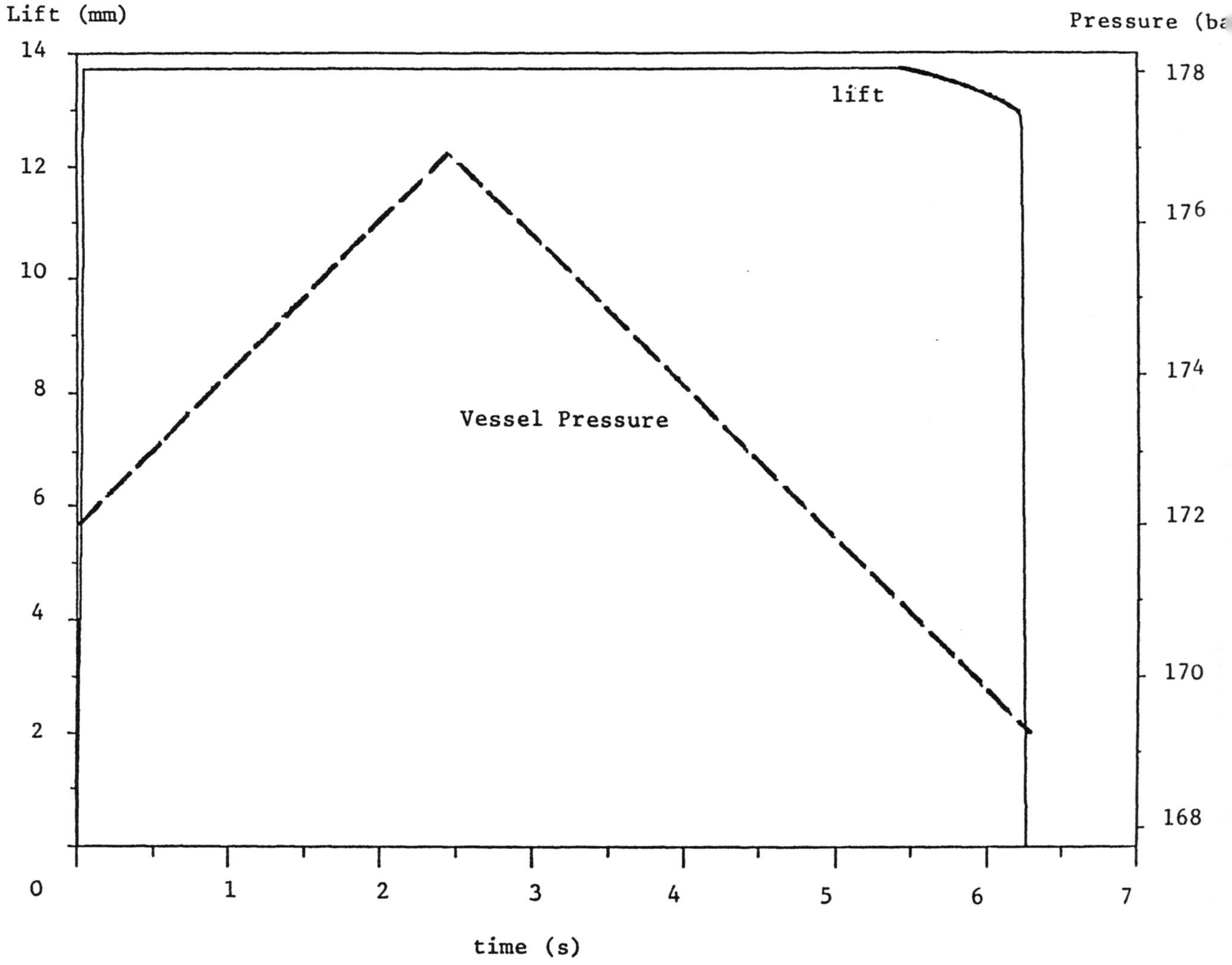

FIGURE 2: VARIATION OF LIFT CHARACTERISTIC WITH TIME (NOZZLE RING = -5.2 mm, GUIDE RING = 0 mm)

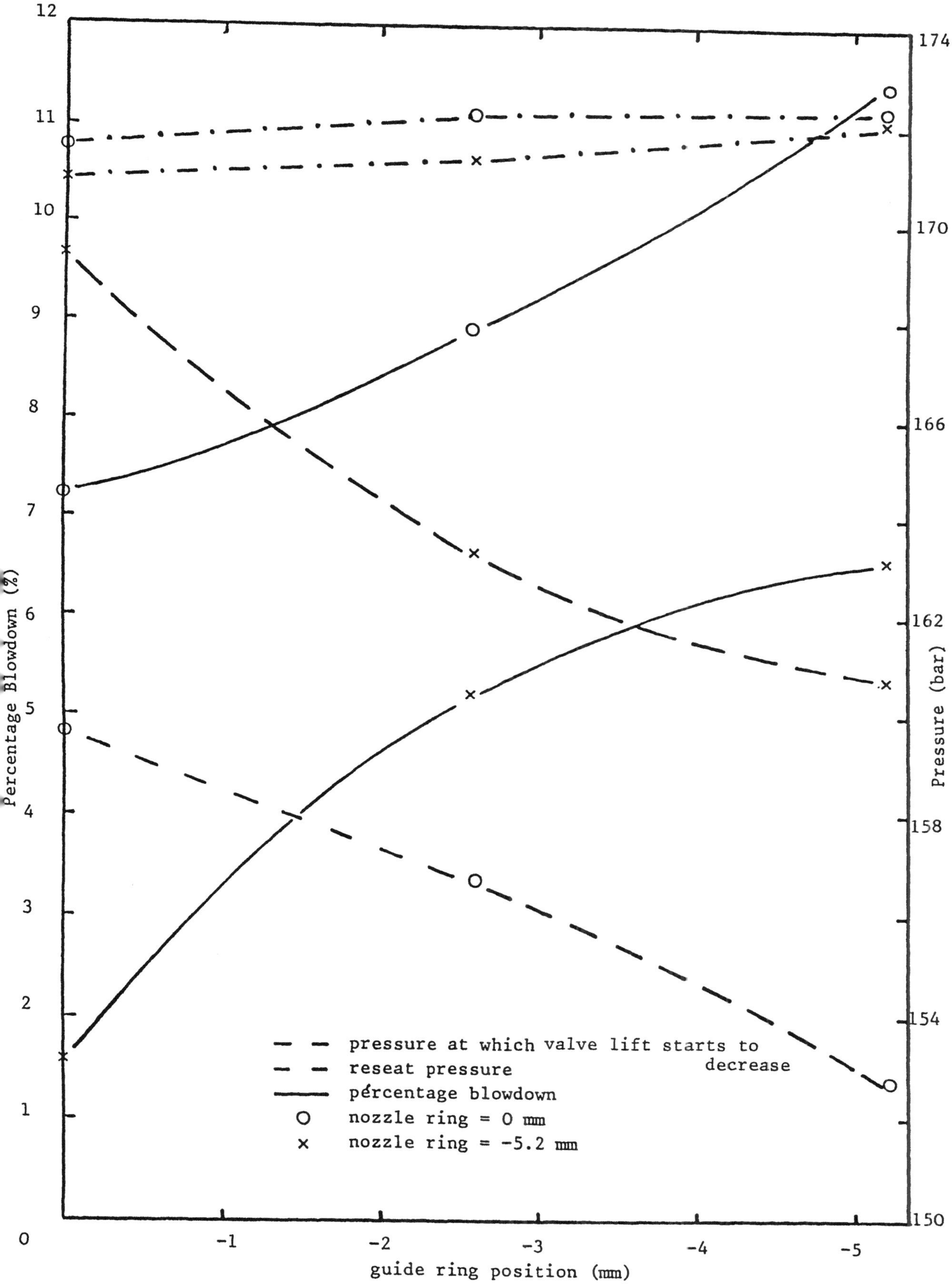

FIGURE 3: VARIATION OF PERCENTAGE BLOWDOWN AND RESEAT PRESSURES AT VARIOUS RING SETTINGS

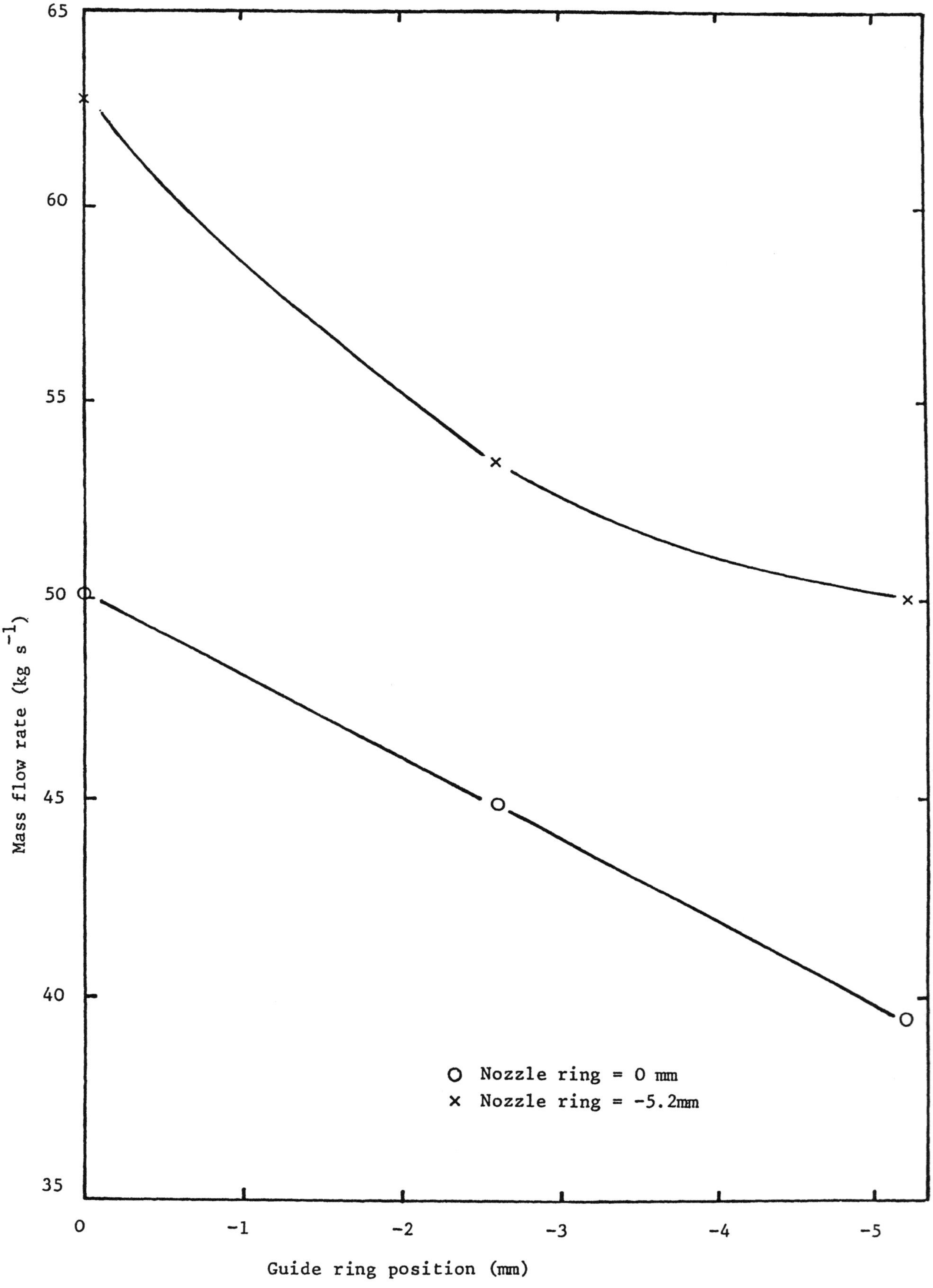

FIGURE 4: VARIATION OF MASS FLOW RATE WITH VARIOUS RING SETTINGS (Vessel pressure = 173 bar)

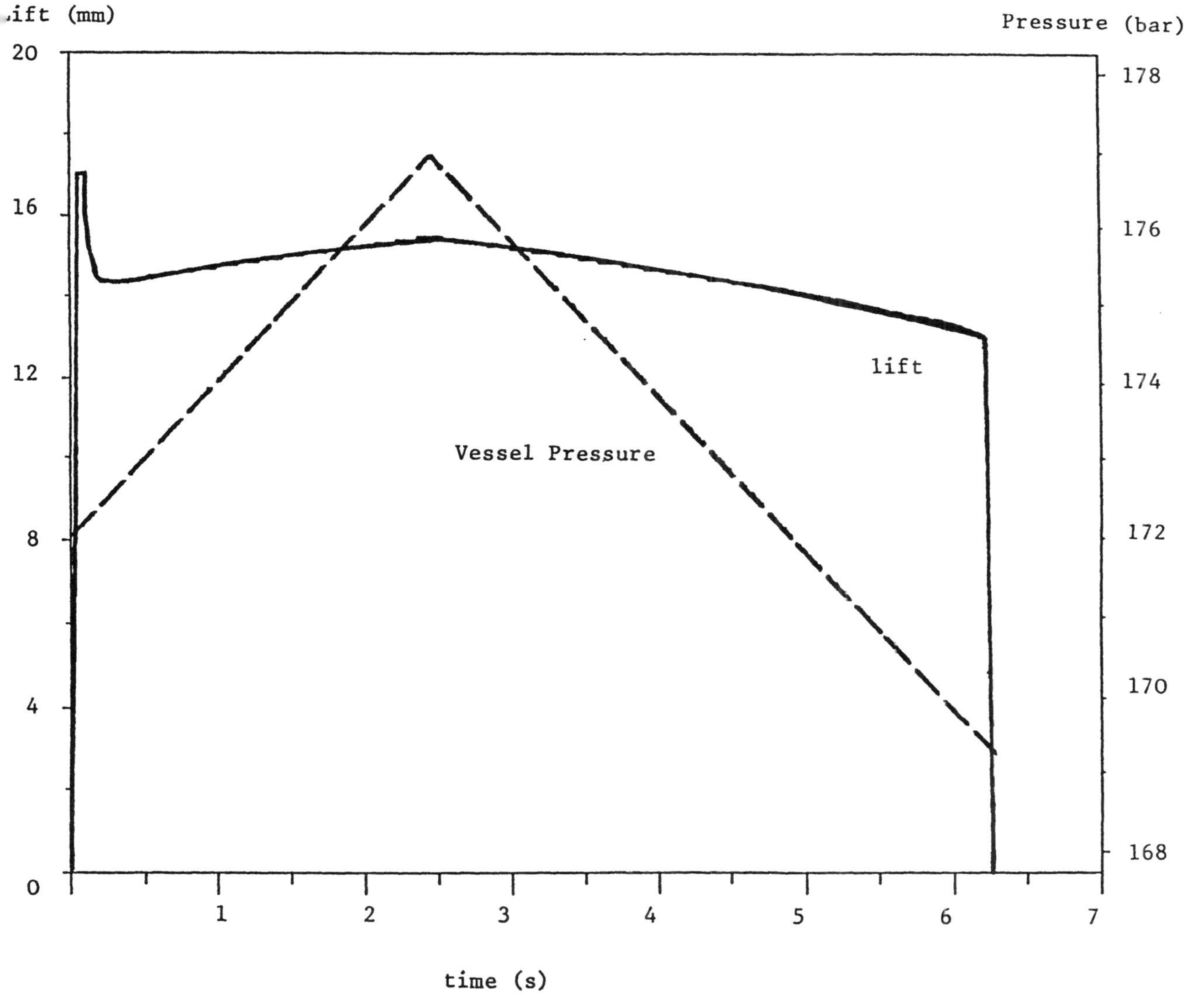

FIGURE 5: VARIATION OF LIFT CHARACTERISTIC WITH TIME WHEN THE MAXIMUM LIFT IS SET TO 17 mm. (NOZZLE RING = -5.2 mm, GUIDE RING = 0 mm)

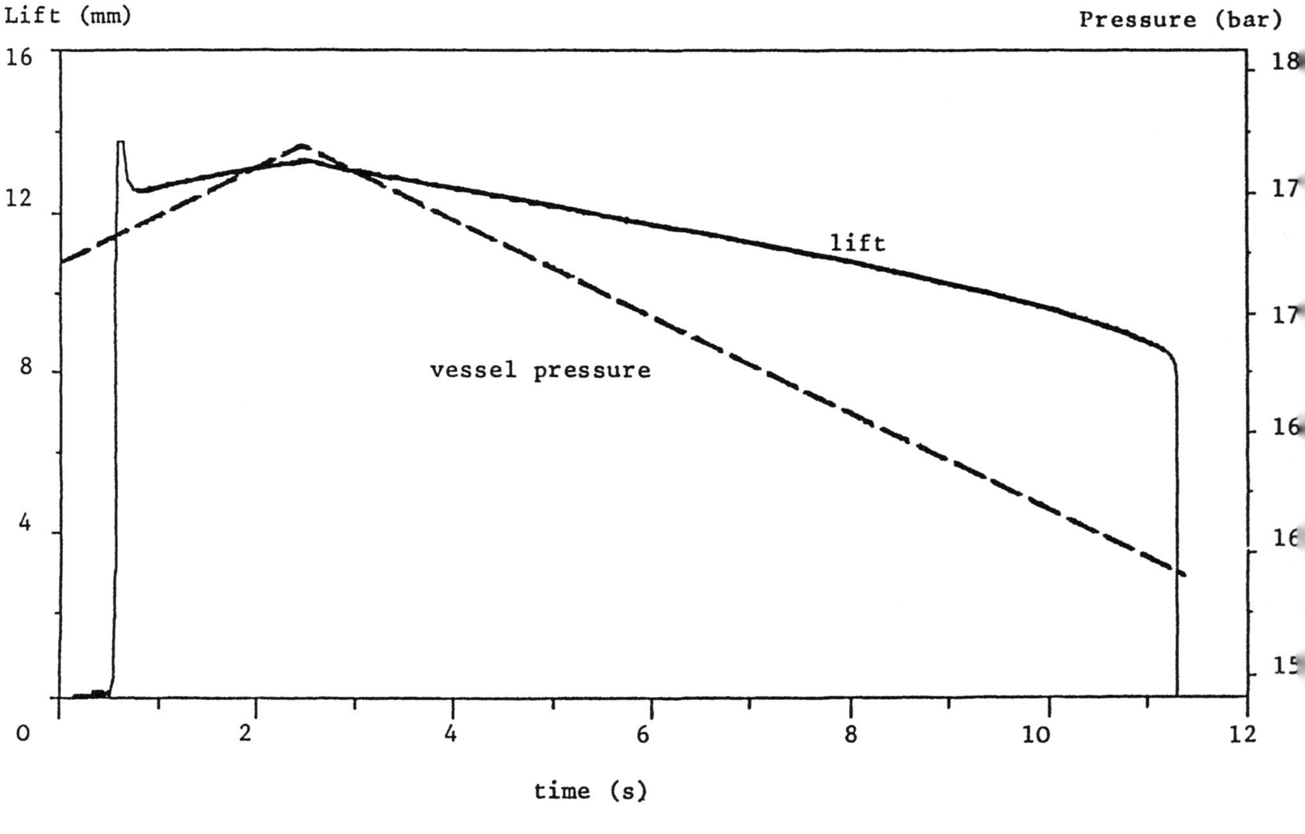

FIGURE 6: VARIATION OF LIFT CHARACTERISTIC WITH TIME WHEN THE SEAT AREA IS INCREASED TO 11.9 cm^2 (Nozzle ring = 0 mm, Guide ring = -5.2 mm)

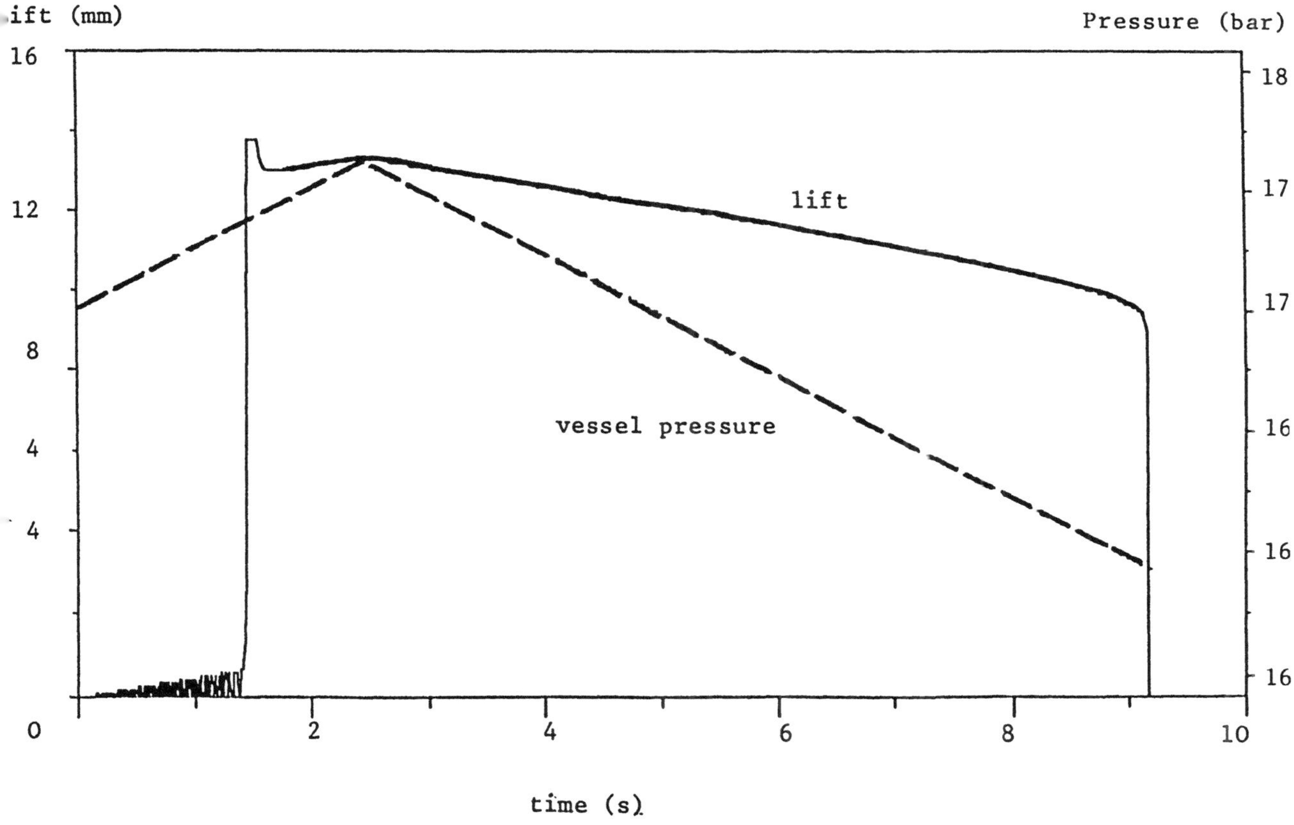

FIGURE 7: VARIATION OF LIFT CHARACTERISTIC WITH TIME WHEN THE SEAT AREA IS INCREASED TO 11.9 cm^2. (Nozzle ring = 0 mm, Guide ring = -2.6 mm)

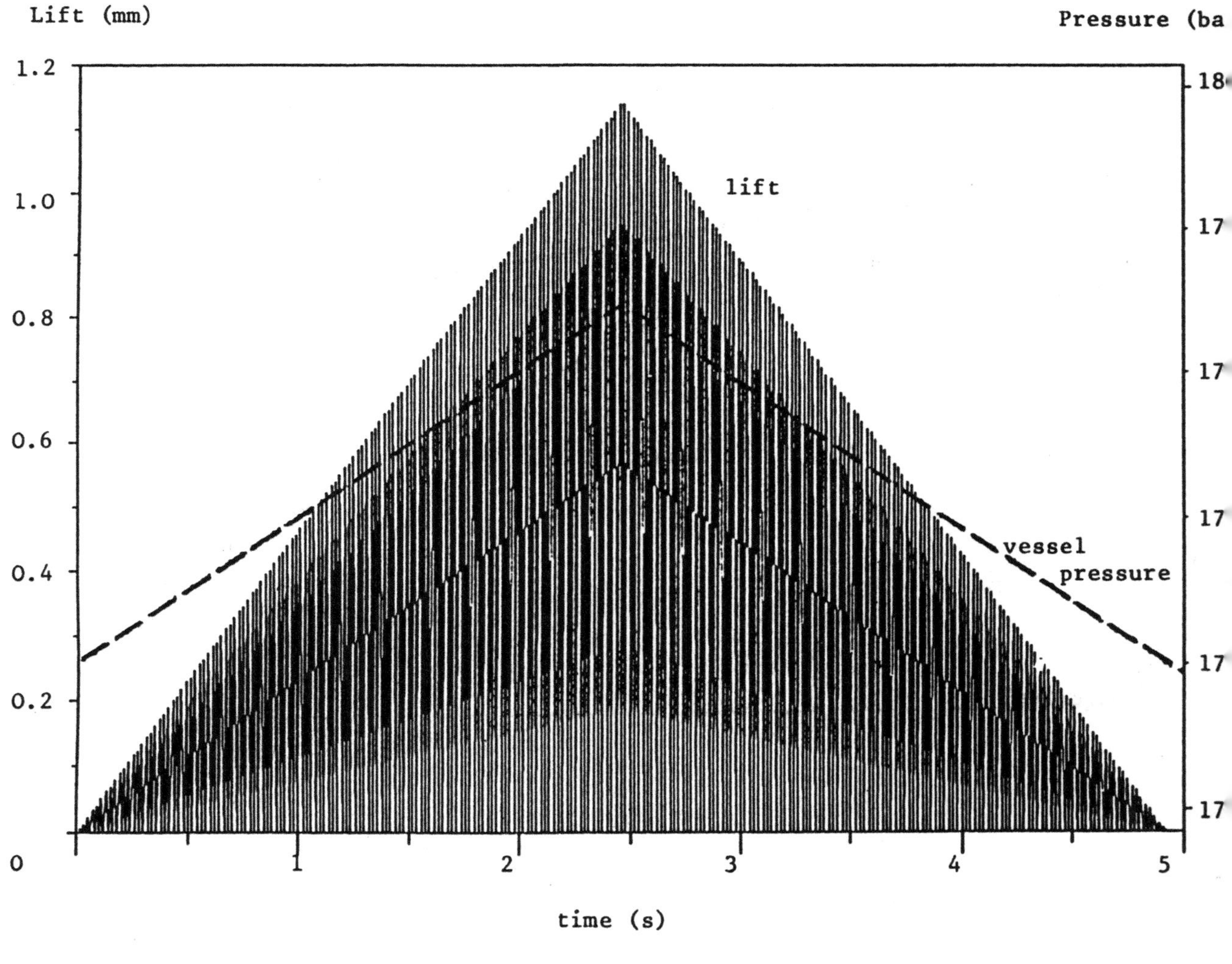

FIGURE 8: VARIATION OF LIFT CHARACTERISTICS WITH TIME WHEN THE SEAT AREA IS INCREASED TO 11.9 cm^2. (Nozzle ring = 0 mm, Guide ring = 0 mm)

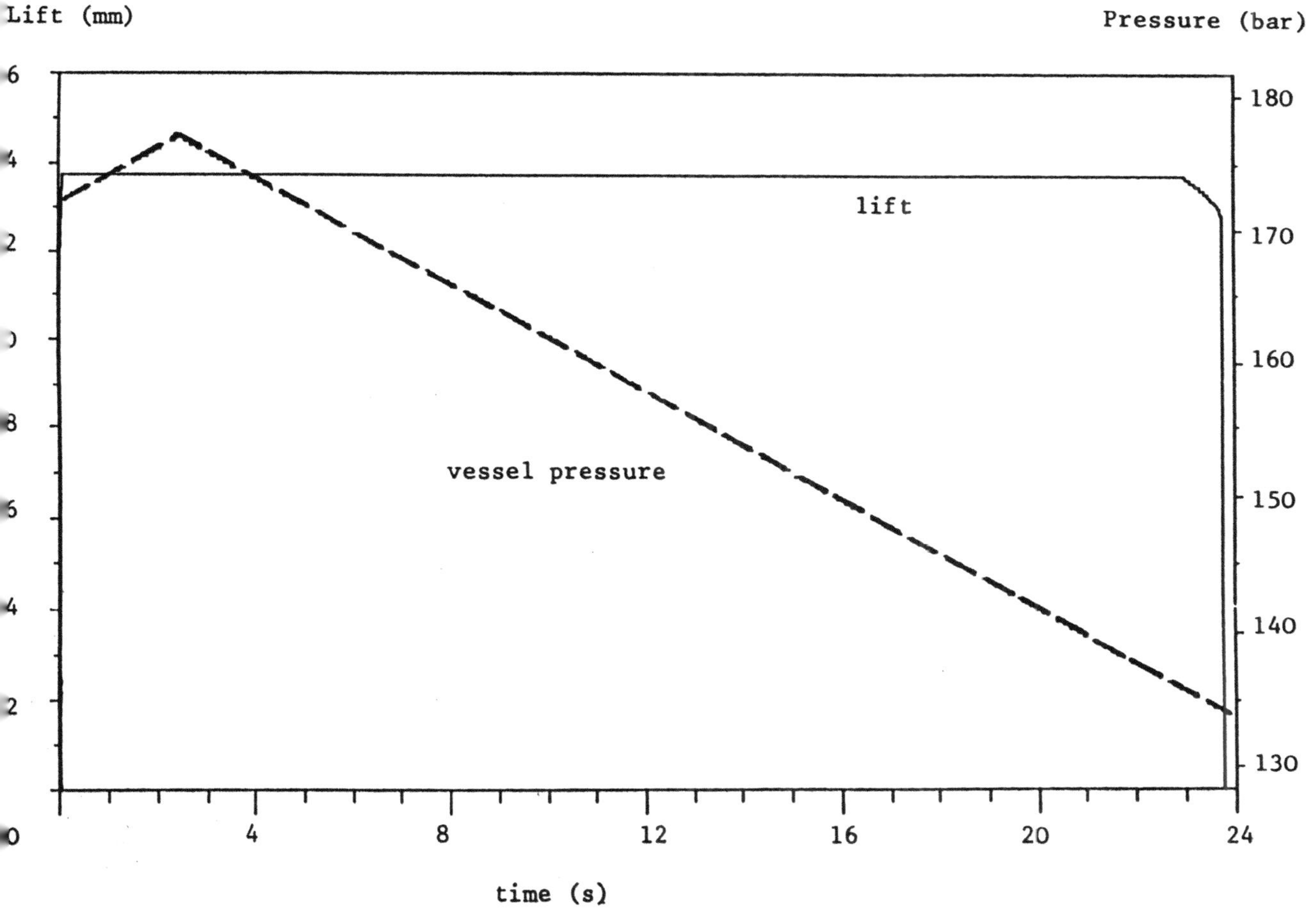

FIGURE 9: VARIATION OF LIFE CHARACTERISTIC WITH TIME WHEN THE BACK PRESSURE IS EQUAL TO 10% VESSEL PRESSURE. (Nozzle ring = -5.2 mm, Guide ring = 0 mm)

2nd International Conference on

Developments in Valves and Actuators for Fluid Control

Manchester, England: 28-30 March 1988

PAPER D3

USE OF DAMPING DEVICES TO CONTROL THE STABILITY OF SPRING LOADED SAFETY VALVES OPERATED IN LIQUID SERVICE.

J.M. Henry, P. Coppolani
Pump and valves department, Framatome, Paris la Défense, France

Summary

The operation of spring loaded safety valves in liquid service often involves high frequency chatter of the disc combined with considerable pressure pulsations. These oscillations may be very dangerous for valves and pipes.

This paper presents the different means which have been investigated to solve this instability problem. Finally hydraulic damping has been considered the best solution.

A new hydraulic damping device specially designed for safety valves is described. The experimental results show the high efficiency of this device.

1. Introduction

1.1 Mechanical safety valves instability problem

Mechanical safety valves vibration and high amplitude pressure oscillations able to induce damage to the valves, the piping and their supports frequently occur at the opening of valves which are not fitted directly to the capacity to be protected. This has become a real problem on pressurized water reactors where these valves cannot be installed as advised by codes for tightness, maintenance and safety reasons and because these valves must be periodically tested to verify their proper working.

This instability problem occurs in liquid service with mechanical safety valves (fig.1). It is mainly related to the coupling between acoustic wave propagation in the pipe and the mass/spring system of the valve.

1.2 Research and development program

FRAMATOME has undertaken an extensive research and development program to solve this instability with the help of the COMMISSARIAT A L'ENERGIE ATOMIQUE (CEA)

Held in Manchester, England. Organised and sponsored by BHRA, The Fluid Engineering Centre and co-sponsored by the British Valve Manufacturers Association and the Institution of Mechanical Engineers Process Industries Division

who carried out all the tests. In the first step of this program, FRAMATOME investigated different solutions as:

- increasing mass or inertia of the internals of the valves;
- use of assist devices (pneumatic and magnetic);
- hydraulic and dry friction damping of the stem of the valve.

Beside these technological investigations, FRAMATOME has developed a mathematical model (see appendix) which confirmed the general trends obtained with the different test configurations (1).

Finally, it has been judged that the hydraulic damping of the valve was the best solution to eliminate valve instabilities. Consequently, FRAMATOME has developed a damping device specially suited for use with spring loaded safety valves operated in liquid service.

2. Control of the stability of spring loaded safety valves

2.1 Spring loaded safety valve operation

Mechanical or spring loaded safety valves are designed to open at a defined set pressure, to discharge a defined flowrate at the relieving pressure and to ensure tightness at the reseating pressure after operation.

These safety valve characteristics must be precisely set with reproducibility. Moreover reseating pressure is often required to be very close to the set pressure.

Usual values of these parameters for mechanical safety valves are as follows:

- relieving pressure: 110 % of set pressure;
- reseating pressure: above 92 % of set pressure.

To obtain these characteristics full lift pressure relief valves are used. They are usually provided with a bell shaped flow deflector around the disc and sometimes with an adjustable nozzle ring to allow changes in relieving and reseating pressures. These values are not independent: a lowering of the relieving pressure (or overpressure) induces a lowering of the reseating presure and an increase of the blowdown.

2.2 Means for safety valve stability improvement

As described in a previous paper (1) instabilities can be explained by different mechanisms and occur for different operating conditions. Here we want to prevent high frequency oscillations which result from an interaction between disc lift and pressure propagation through the line. Indeed this kind of instability is very dangerous as it can be very quickly destructive for the valve (stem binding) and for the line (supports damage). Field experience, loop testing and analysis have shown that to cope with this type of instability there were different approaches which are discussed below:

2.2.1 Valve/Pipe decoupling

From computations (1) and experiments realized by FRAMATOME and the COMMISSARIAT A L'ENERGIE ATOMIQUE we can conclude that valve position modification or valve moving mass modification are rather ineffective when the valve cannot be installed very close to the tank to be protected.

2.2.2 Increase in blowdown

This can be done only with valves fitted with rings. The efficiency of this solution is limited as it does not suppress the instability in all cases and so as to obtain good stability, large blowdowns are required. Such blowdowns are often not acceptable because reseating pressure has to be greater than normal operating pressure and because for economical reasons the margin between operating pressure and safety valve set pressure is as small as allowed by normal operating transients.

2.2.3 Assist device

Tests realized by CROSBY and FRAMATOME with a pneumatic assist device on a pressurizer safety valve fitted with a cold loop seal showed a significant improvement of the stability during discharge of the loop seal(3).

The development of this system was not carried out as two major tasks remained to be done:

- qualification to accidental conditions;
- guarantee of absolute reliability as the system was very sensitive to the adjustment of the assist device.

2.2.4 Dry friction damping

Dry damping is always present in a safety valve as there is some friction between the stem and the guides. This effect, which is minimized by the use of adequate materials as it decreases the accuracy of the setting, can be sufficient to prevent oscillations in non-critical situations. This has dramatically been observed on a PWR pressurizer valve during loop seal discharge tests on ELECTRICITE DE FRANCE(EDF)'s Indira test loop: the improvement in the guiding made by FRAMATOME to prevent galling (use of stellite) led to the worse observed oscillations.

Some manufacturers provide a controlled dry friction damping for valves susceptible to chatter. This solution which does not change the overall dimensions of the safety valve has the drawback of increasing the blowdown, and having an efficiency limited to no extreme cases. But above all, it does not have a total reliability owing to the poor control of the value of the friction coefficient and of its evolution with time.

2.2.5 Hydraulic damping

Hydraulic damping is already used to ensure the stability of some critical pneumatically operated control valves. Some manufacturers incorporate a hydraulic snubber as an integral part of the actuator to provide trouble free operation.

The application of hydraulic damping to safety valves was, to our knowledge, first done by EDF in 1980 to solve the problems of the heat removal system (RHR) safety valves. In spite of a successful test on loop, the RHR safety valves fitted with hydraulic snubbers are being replaced by EDF by pilot operated valves. Indeed, due to excessive friction of dynamic seals and leakage problems, the plant operators found it too difficult to maintain the required tightness and set pressure accuracy, using the standard maintenance procedures. Nevertheless as the principle of safety valve hydraulic damping was very attractive, as confirmed by computation simulations done in FRAMATOME (1) and in USA (2) we decided, with the help of the COMMISSARIAT A L'ENERGIE ATOMIQUE to carry out the achievement of a successful hydraulic damper which could be retrofitted to existing mechanical safety valves. This solution could also possibly be applied to the pilots of pilot operated safety valves which are prone to the same instability problems as their sisters of larger size.

3. Design of a hydraulic damping device for a safety valve

3.1. Advantages of hydraulic damping

From field experience, loop tests and computational studies it can be concluded that hydraulic damping is the best-fitted means to improve the stability of spring loaded safety valves.

Indeed it permits:

- an easy and precise control and adjustment of the damping coefficient;
- a stabilizing effectiveness increasing with the amplitude of the perturbation;
- an absence of influence on transient maximum overpressure with a damping

system only actuated during valve closure and allowing free opening. Indeed numerical simulations and testing have shown that this kind of damping is sufficient to stabilize safety valves;
- an absence of influence on opening characteristics of the safety valve: set pressure, accumulation and opening time;
- a small increase in closing time which does not result in a significant increase in blowdown with usual pressure rates.

3.2. Description of the damping device

FRAMATOME has developed a damping device (fig. 2) specially designed for mechanical safety valves. It is based on hydraulic damping and employs a single acting cylinder.

The device is fitted on the top of the safety valve (fig. 3) where it is connected to the stem. This disposition allows an easy maintenance and the backfitting of existing safety valves.

The device which is completely sealed can be used in very severe ambient conditions and in particular can be installed inside the nuclear reactor containment. The materials are compatible with the most severe ambient conditions in the nuclear plants.

Bellows, rather than dynamic seals, have been employed to ensure rod leak tightness. Indeed efficient dynamic seals induce high friction forces depending on the valve movement. Conversely, the bellows acts as a spring ensuring the reproducibility of the valve performance and a control of the additional force applied which can be taken into account in the valve setting. However we chose a bellows of very low stiffness, to avoid the replacement of the original spring while keeping the same safety valve design performance.

Damping is realized by a transfer of flow through a sharp edge orifice. This design has been prefered because its damping effect is independent of the viscosity of the fluid and therefore independent of fluid temperature.The section of this orifice can be modified by a set screw for damping adjustment. A lock-out prevents accidental adjustment in the field. A large cross-section check-valve allows the free opening of the safety valve. This check-valve which has a high frequency response (more than 200 Hz) has been especially designed for use with high pressure safety valves. As a matter of fact high pressure safety valves have high frequency characteristics (up to 150 Hz) due to the use of strong springs.

Damping technology experience shows that, even with double-ended piston rods, cylinders must have accomodation for changes in fluid volume which appear with temperature variation. This has been realized by using a bellows. The visual control of the position of the end of the bellows allows to check easily the presence of the correct volume of hydraulic fluid and of the operability of the damping system.

4. Experiments

4.1 Loop tests

Our damping device has been validated by tests with a 4"M6" spring loaded safety valve. Its efficiency has been assessed by comparison with previous tests without a damper and by comparison with tests with willfully insufficient damping adjustment.

The tests have been carried out on the CLAUDIA test loop which is a facility used by the COMMISSARIAT A L'ENERGIE ATOMIQUE for the improvement and qualification of safety and check valves.

The valve is posisioned at the end of a pipe connected to the tank (fig. 4). The tests consist of an ambient temperature water release through the valve with controlled pressure time rate and flowrate by pump discharge valve operation. It was attempted to increase the flowrate step by step but, due to the coupling between the loop and the valve characteristics, we observed low frequency instability. In fact, high frequency instability mainly occurs experimentally either at low flowrate or with these low frequency lift oscillations.

Previous tests without damper revealed high frequency instabilities with strong pressure pulsations and disc lift oscillations (fig. 5). The experiments with the damper were realized initially with a sufficient damping adjustment (fig. 6). Then the area of the damper's orifice had been increased step by step until high frequency instability appeared (fig. 7). Once again strong pressure pulsations and disc lift oscillations were present showing the significant improvement obtained with this hydraulic damping device when the damping adjustment is sufficient. The constancy of the pressure parameters such as set and reseating pressure over a wide range of damping adjustment and the small difference between set and reseating pressure shows that the damping device eliminates the high frequency instabilities without adversely affecting valve operation.

4.2 Damping characterization

Originally a spring loading characterization procedure was employed. For this procedure we used the safety valve spring. The test consisted in compressing and then suddenly releasing the spring connected to the damper. We obtained the damping coefficient from the measurement of the displacement amplitude decrease as a function of the time.

We now use a simpler procedure which consists of recording the speed of the rod when it is dropped with added weights connected to it. This procedure is better fitted than spring loading characterization for single way dampers and non linear damping.

The tests realized with different weights confirm that the damping force is a quadratic function of the speed. This could be expected as the damping is realized by a turbulent flow through a sharp edge orifice. Figure 8 shows the experimental results obtained with a previous design.

The test results are introduced in our computational model to confirm the validity of the selected damping adjustment (if a quadratic damping is needed the damping coefficient can be introduced as a function of the velocity).

5. Conclusion

Among the different ways to cope with the high frequency instabilities of the spring loaded safety valves, hydraulic damping appeared the most attractive.

The damping device developed by FRAMATOME confirmed the effectiveness of this principle. Indeed this device which benefits both from hydraulic damping and safety valve behaviour knowledge suppresses the high frequency instabilities without adversely affecting valve operation.

Future work will consist in testing this device on high pressure steam valves fitted with a loop seal in order to qualify this device in all possible fluid conditions.

6. Acknowledgements

The authors wish to aknowledge the "COMMISSARIAT A L'ENERGIE ATOMIQUE" and particularly the DRE (Water Reactors Department) team in CADARACHE near AIX-EN-PROVENCE, who carried out all the tests realized for this study.

7. References

1. Coppolani,P. Henry,J.M. Caumette,P. Huet,J.L. and Lott,M.: "Stability of self actuating safety valves in liquid service" SMIRT IX Conference (Lausanne, Switzerland: Aug. 17-21, 1987) Paper F1 PP3-16.
2. Singh, A.: "An analytical study of the dynamics and stability of a spring loaded safety valve" Nuclear Engineering and Design, 72, 1982, PP 197-204.
3. EPRI/C.E. PWR Safety Valve Test Report - NP-2770LD Vol. 9 - Test Results for Crosby/Framatome Safety Valve With Assist Device.

Appendix: Valve instability transient simulation model

Nomenclature

A	pipe inside area
C	valve damping coefficient
Cd	valve discharge coefficient
Co	sound velocity
D	nozzle diameter
Fh	hydraulic force
K	spring stiffness
ℓ	valve lift
M	valve moving mass
P	relative pressure
Pt	set pressure
Q	volumetric flowrate
S	nozzle area
Z	pipe abcissa
ϕ	pipe inside diameter
λ	friction factor
ρ	specific mass
β	hydraulic stiffness
m.o.c	method of characteristics

To investigate the stability of the pipe/valve system FRAMATOME has used a monodimensional transient model of the line with resolution of the equations of flow by m.o.c.. The boundary conditions are assigned pressure at the inlet of the line connected to the tank and pressure/flow function at the valve computed from the disc lift. The motion of the disc submitted to a hydraulic force Fh is determined by mass/stiffness model with or without damping. Several hydraulic stiffness depending on valve lift can be introduced in the hydraulic force model. For the input of this hydraulic force model a special experimental procedure has been employed (1).

m.o.c

$$\begin{cases} \begin{cases} \dfrac{1}{A}\dfrac{dQ}{dt} + \dfrac{dP}{\rho C_0 dt} + \lambda \dfrac{Q|Q|}{2\phi A^2} = 0 \\ \dfrac{dZ}{dt} = Co \end{cases} \\ \begin{cases} \dfrac{1}{A}\dfrac{dQ}{dt} - \dfrac{dP}{\rho C_0 dt} + \lambda \dfrac{Q|Q|}{2\phi A^2} = 0 \\ \dfrac{dZ}{dt} = -Co \end{cases} \end{cases}$$

valve flow $\quad Q = Cd\,\pi\,D\,\ell\sqrt{\dfrac{2P}{\rho}}$

valve movement $\quad M\dfrac{d^2\ell}{dt^2} + C\dfrac{d\ell}{dt} + K\ell = Fh - Pt.S$

hydraulic force $\quad Fh = PS\,(1+\beta\ell)$

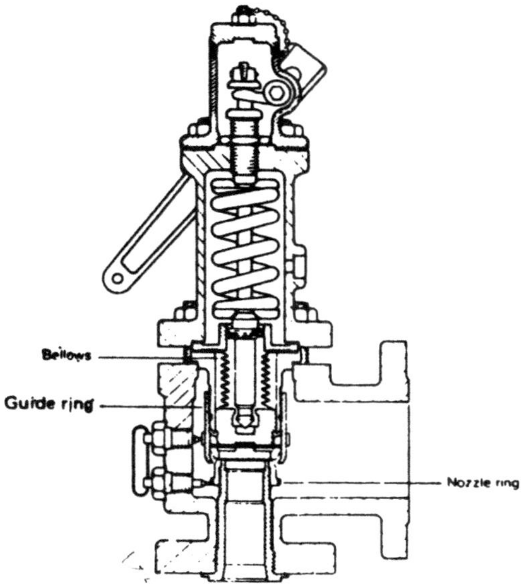

FIGURE 1 : CROSS SECTION OF A TYPICAL SAFETY VALVE

1 BODY FILLED WITH HYDRAULIC FLUID

2, 3 PISTON ROD

4 ADJUSTABLE ORIFICE

5 CHECK VALVE

6 BELLOWS FOR VOLUME CHANGES ACCOMODATION

7 VOLUME CONTROL INDICATOR

8 SEAL BELLOWS

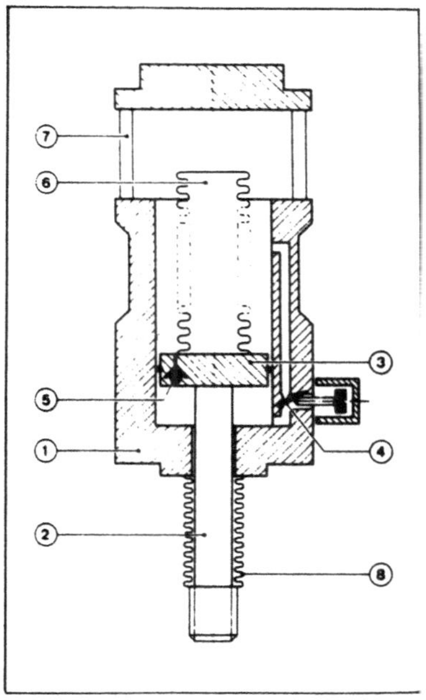

FIGURE 2 : SCHEMATIC DIAGRAM OF THE HYDRAULIC DAMPER

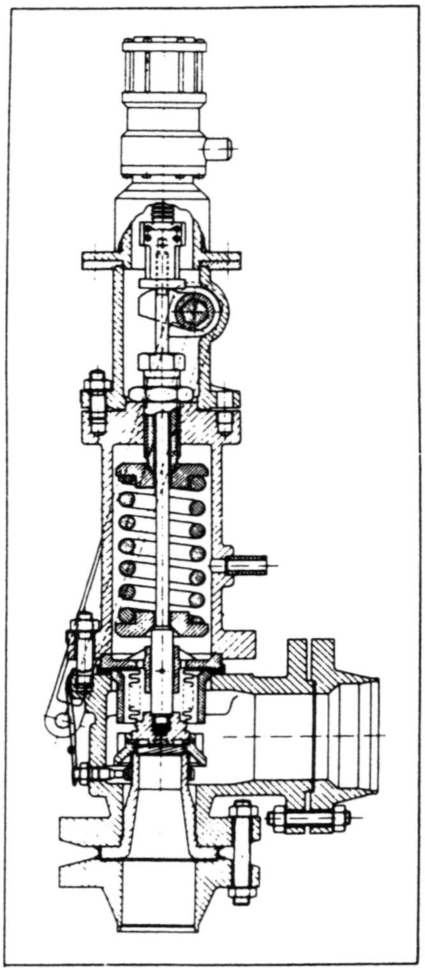

FIGURE 3 : MOUNTING OF THE HYDRAULIC DAMPER ON A SAFETY VALVE

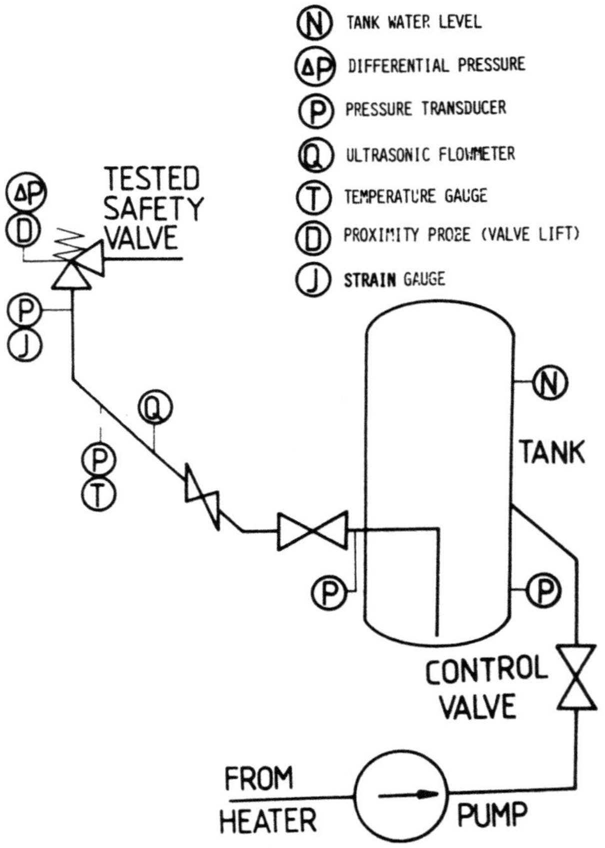

FIGURE 4 : CLAUDIA TEST LOOP

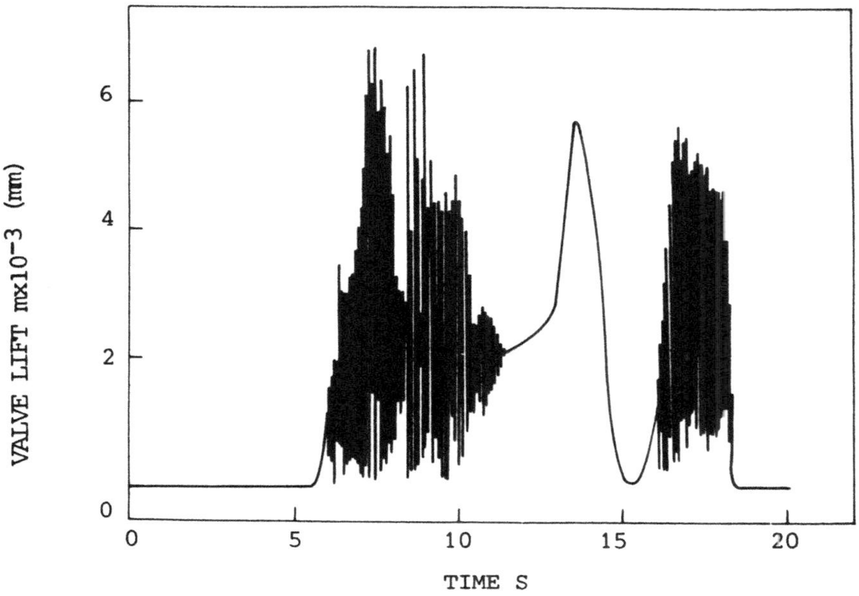

(5a)

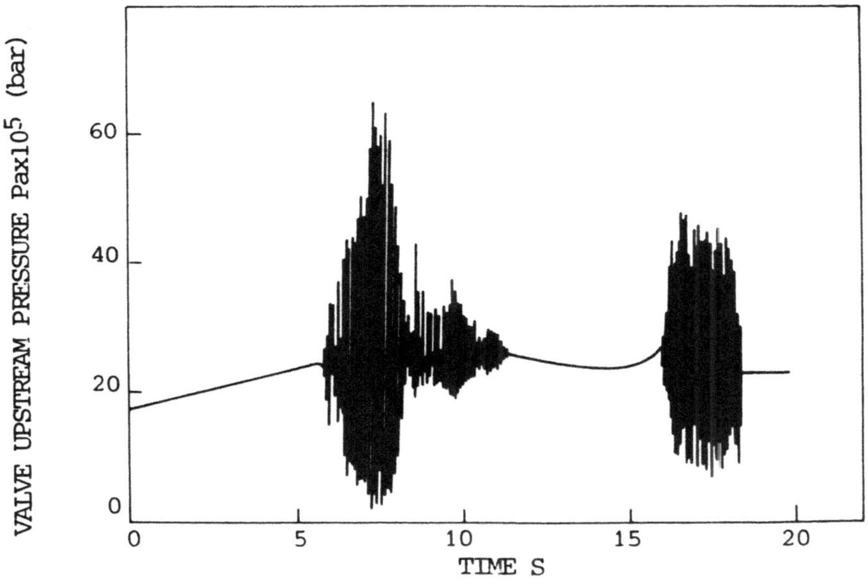

(5b)

FIGURE 5 : SAFETY VALVE TEST RESULTS WITHOUT DAMPER

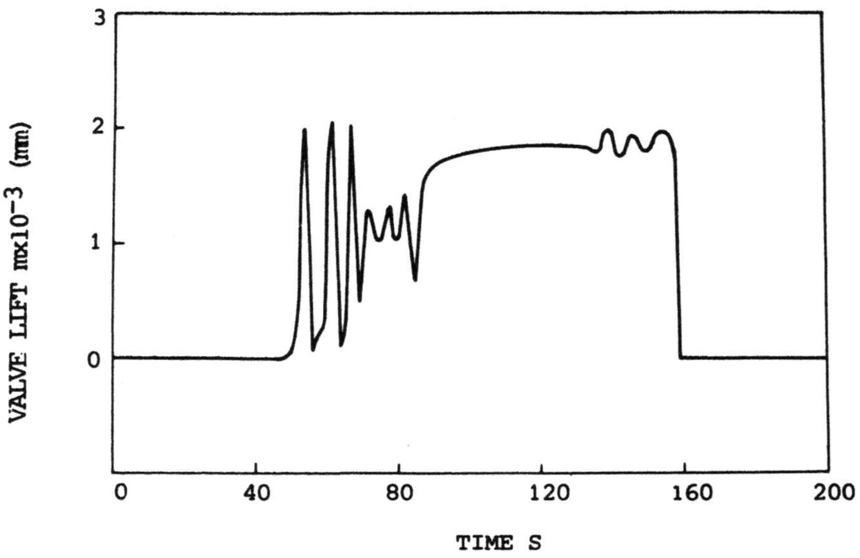

(6 a)

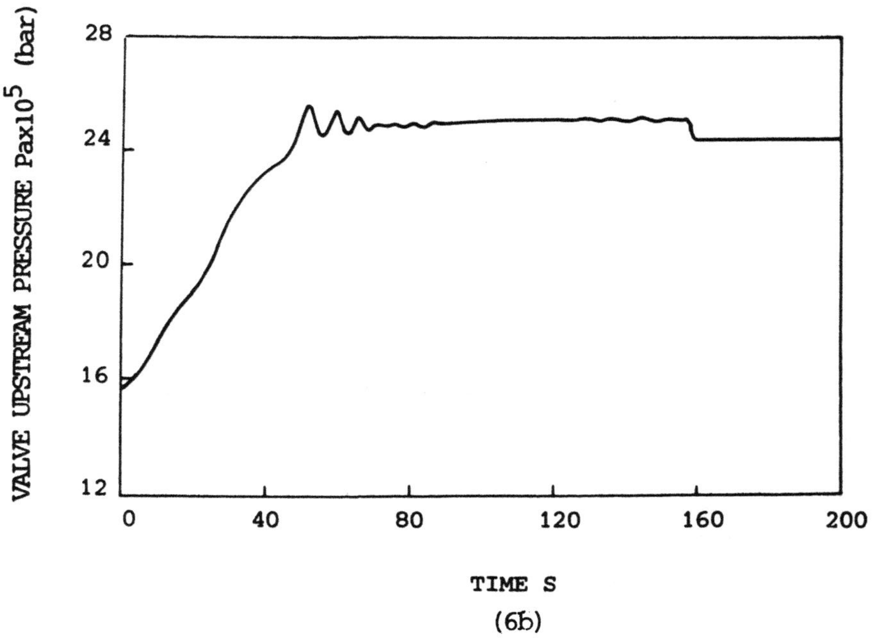

(6b)

FIGURE 6 : SAFETY VALVE TEST RESULTS WITH DAMPER

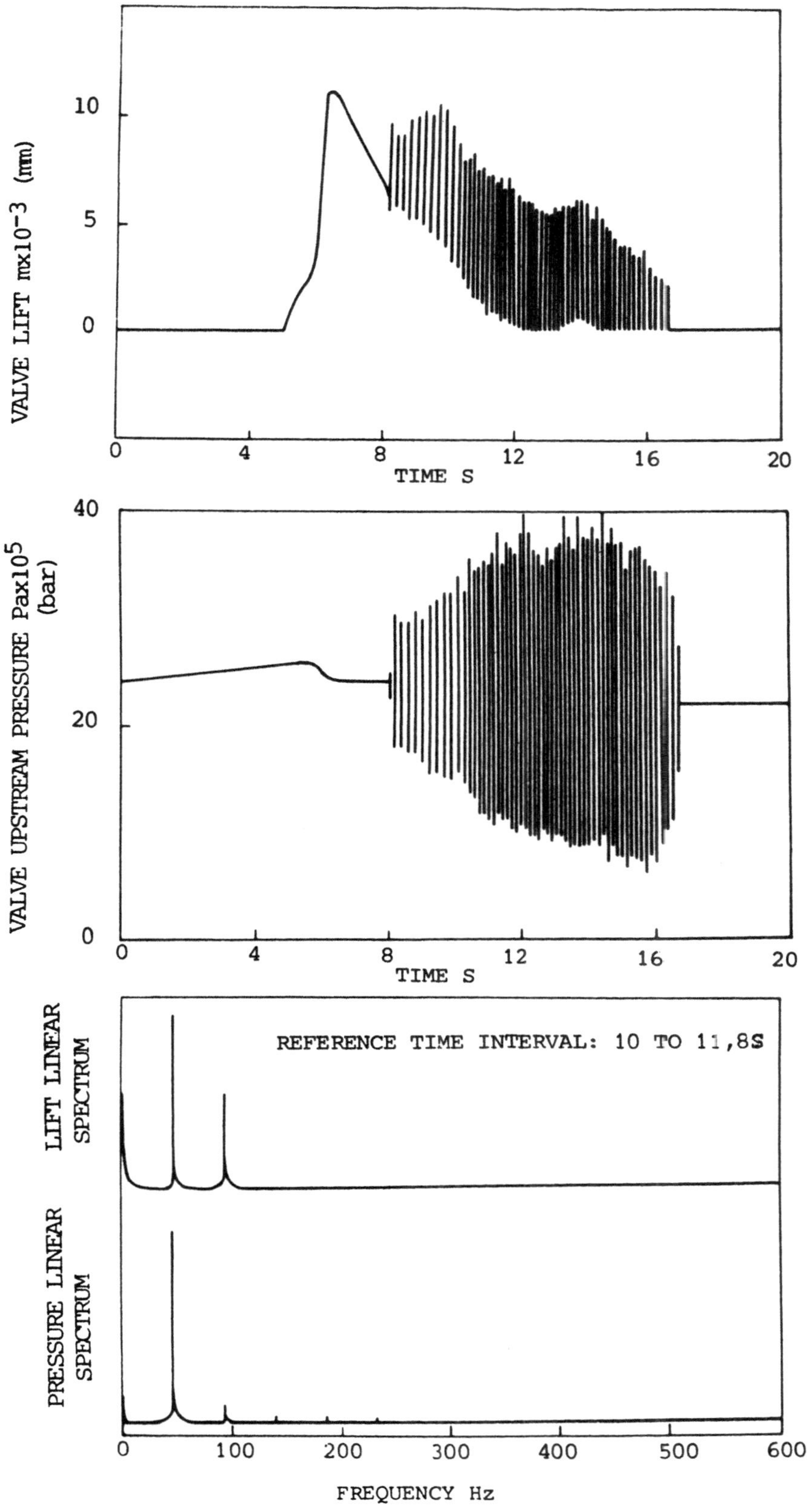

FIGURE 7 : TEST RESULTS WITH INSUFFICIENT DAMPING

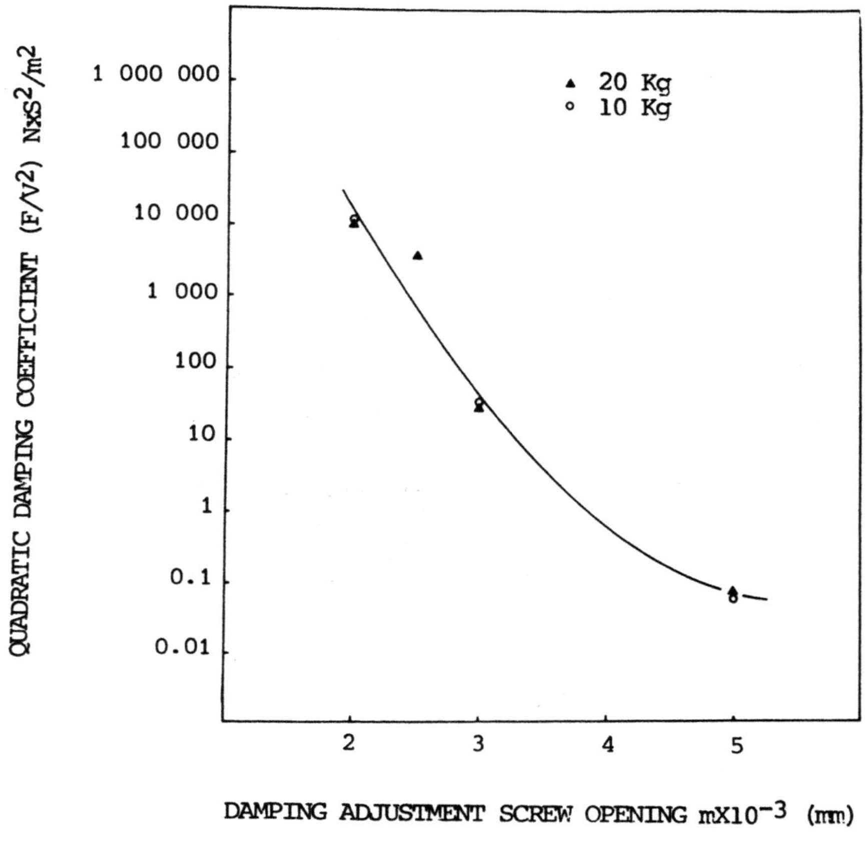

FIGURE 8 : DAMPING CHARACTERIZATION RESULTS

2nd International Conference on

Developments in Valves and Actuators for Fluid Control

Manchester, England: 28-30 March 1988

PAPER E1

THE DEVELOPMENT AND APPLICATIONS OF DUPLEX STAINLESS STEEL FOR CONTROL VALVE EQUIPMENT.

BY: David R Brammer, Wm Cook PLC (Catton & Co. Ltd., Division)
Peter J Hughes, Fisher Controls Ltd., England.

For many years wrought form of Duplex stainless steel has been used widely in many applications from Oil Refineries to various types of Acid Plants. The potential for corrosion resistant materials has widened to the Offshore Oil & Gas Industry. This new offshore metallurgical technology has had a dramatic impact on types of materials used in the design of valves, pumps, etc.

While it is acceptable onshore to use traditional materials with corrosion allowances or inhibitors, it is not acceptable offshore. This is due to the increased weight (for every 1 tonne topside it requires approximately 12 tonnes of steel structure), therefore, traditional materials are uneconomic. In consequence, a cast version of a Duplex stainless steel, with near identical properties to the wrought material, was needed. For sometime, one major company (Sandvik) has co-operated with specific foundries to develop a cast equivalent of their trade name SAF 2205, which has the chemical composition of UNS S31803 (1) /DIN 1.4462 (2). The cast material has near identical properties to the wrought material manufactured by Sandvik. Both the foundries and Sandvik have developed the expertise together to ensure this compatibility. This arrangement has allowed the foundries to be licensed by Sandvik to use their tradename.

THE DEVELOPMENT OF THE CAST DUPLEX STAINLESS STEEL, UNS S31803/DIN 1.4462

A. Historical Review and Technical identity

The term "stainless steel" covers a great variety of engineering alloys, those in commercial use can be categorised broadly in three different groups.

Firstly, there are those which can be hardened and tempered in much the same way as carbon steels, these are usually known as "Hardenable stainless steel" or sometimes "Martensitic stainless steels". Secondly there is a group known as "Ferritic stainless steels" which are really high chromium irons, having low carbon contents, which cannot be hardened to any useful extent. Finally, there is a third group which differs markedly from the other two in terms of mechanical properties. These are usually termed "Austenitic stainless steels".

Held in Manchester, England. Organised and sponsored by BHRA, The Fluid Engineering Centre and co-sponsored by the British Valve Manufacturers Association and the Institution of Mechanical Engineers Process Industries Division

The earliest discovery of a stainless steel is ascribed to Mr. Harry Brearley in 1913, and was of the martensitic type, i.e. (12 - 14% chromium), and due to the 1914/18 war its commercial development was somewhat retarded. The "ferritic" stainless group was a natural development of this early discovery due to progressive increases in chromium additions.

The third group or the "Austenitic stainless steels" were discovered in Germany around 1912/13 and were researched there. The introduction of these steels into the U.K. took place around 1923 under licence from Krupp and were taken up in the U.S.A. at a somewhat later date.

B. Origin of Duplex Stainless Steel

In the austenitic group of stainless steels the effect of (a) chromium and (b) nickel is quite distinct. Nickel increases the temperature range in which gamma iron or austenite is stable, whereas chromium behaves in an opposite manner. Pure austenite or gamma iron can hold in solution no more than 12% chromium, but when austenite contains 8% nickel and a small amount of carbon, usually present in these steels, it can dissolve up to 19% chromium. If more than this is added, with the nickel remaining at 8%, the austenite does not become unstable (as happens at lower chromium percentages which do not contain enough nickel) but a part of it changes completely to ferrite (alpha iron), the remainder being still stable austenite. Actually two different solid solutions of chromium and nickel in iron are formed, one containing more chromium and less nickel than the average which crystallises in the alpha form, i.e. ferrite, the other in the gamma form, i.e. austenite.

It is from this phenomenon that the "Duplex stainless steels" are derived. In most cases the austenitic stainless steels are a form of duplex stainless, usually containing some delta ferrite to improve weldability, or to some extent ambient temperature yield or ultimate strength.

The balance of composition in a duplex steel is controlled by balancing the austenite forming elements against those which preferentially form ferrite, as per the Schaeffler diagram, as modified by Delong & Schoeffer for nitrogen content (see Appendix 1, Figs. 1 & 2) which is detailed in ASTM A800/A800M in 1984. (3)

One of the first Duplex type stainless steels was developed around 1935 under designation AISI 329 (4) with a nominal composition of 0.08% carbon, 25.0% chromium and 4% nickel. Little further happened until 1963 when a copper grade was developed under designation CD4MCu, (5) and since this time many other alloys have been registered in both wrought and cast forms. (See Appendix 1.)

The originally developed steels suffered from poor ductility and weldability which was restrictive since they exhibited high yield strengths together with excellent resistance to chloride and acid corrosive attack. However, the more recently developed duplex steels have favoured a much lower carbon content (controlled by AOD), (6) together with molybdenum and nitrogen additions, and in some cases copper additions. The nitrogen in particular assists in controlling the chemical balance between austenite formers and ferrite formers and hence the ferrite/austenite ratio.

Duplex stainless steels are normally accepted as stainless steels which contain between 35/75% ferrite with the remainder being austenite. This differentiates them from the more common austenitic stainless steels which may contain up to some 25% of delta ferrite.

The later developments with duplex stainless steels coupled with the needs of industry, particularly in the pursuit of offshore oil, has led recently to a much wider use of duplex stainless steels, since their corrosion resistance has remained prominent and the welding difficulties have now been substantially overcome.

C. Requirements of Duplex Stainless Steels

Due to the recent new and demanding applications particularly in the offshore oil and gas industries, the duplex steels have found continuing applications for pumps, valves, vessels for processing and pipelines for transportation of wet sour gases, and oil. Also their use, in the fields of desalination, urea, pulp and paper plants, waste water treatment, the food industry, the brewing industry, sugar refining and the storage and transportation of phosphoric acid and many other chemicals, is increasing.

Providing the necessary precautions are observed the steels are readily weldable and suitable welding consumables are available for MMA (Manual Metal ARC), TIG (Tungsten Inert Gas), MIG (Metal Inert Gas) and submerged arc processes. Thus joint welding (i.e. fabrication) as well as cladding or repair are all quite acceptable.

Several duplex stainless steels are available in plate, sheet, strip, tube, and forgings from the wrought alloy with plate available up to 1.0 inch thick. Some specific specifications are also available as castings.

It was at this stage Wm. Cook PLC with foresight undertook a limited market appraisal of the industry's requirements and found that a common requirement for duplex was for an alloy designated UNS S31803 which was also registered under DIN 1.4462. Subsequent discussion with customers led to the identification of a need for castings in this type of alloy in many applications, particularly for small pressure components. Thus after discussions with Sandvik Steel in Sweden a technical agreement was signed to develop a cast version of the alloy with the foundry group supplying the foundry technology and Sandvik the melting and metallurgical expertise, and testing the ensuing product.

D. Foundry Development

A somewhat protracted period followed during which test material and trial components were manufactured and submitted to the necessary testing programmes for mechanical properties, corrosion testing, and metallurgical evaluation. The trial pieces were also subjected to full sample evaluation including radiography, dye penetrant examination, and dimensional conformity. During this period it became obvious that very tight technical procedures had to be observed every time, if injurious defects were to be avoided during the manufacture and processing of components. Once the balance of composition had been established the match with wrought data, and mechanical properties, was quite pleasing with only marginal differences being observed. Again in terms of corrosion testing all the tests that were undertaken gave satisfactory results and these included the accepted practices of ASTM A262 - 85a (7) and also the NACE test TM-01-77 (8) conducted at ambient temperature and atmospheric pressure.

It is true to say that duplex stainless steels are somewhat difficult to produce in the cast form for a number of reasons.

Solidification of a stainless steel may commence with the precipitation of either austenite or ferrite as indicated by the phase diagram. At high levels of nickel, solidification commences with austenite, which results in a coarse grained structure with any impurities rejected to the last liquid to solidify giving segregation which results in "hot tearing". At high levels of chromium, solidification starts and continues with ferrite giving a coarse embrittled structure unsuited to castings. At a certain balance of chromium and nickel, ferrite is precipitated first but later, austenite is formed from the liquid or by solid transformation so that an austenitic or austeno-ferritic structure results, leading to some grain refinement.

Similarly since most of the impurities have a higher solubility in ferrite and their diffusion rate is higher in this phase, the risk of their concentration at the grainboundaries is minimised. However, some segregation does occur with differing compositions for ferrite and austenite. Carbon must be kept as low as possible and for all these reasons it is essential that electric melted steel is AOD refined. Small quantities of the alloy are produced by re-melting AOD pig, but all revert material is returned for AOD refining before re-use. Sulphur is also controlled in this way. Other features to be accommodated in foundry practice are the fluidity of the steel, its compatibility with moulding media, venting of moulds, and risering to allow adequate escape for gases and the careful introduction of nitrogen into the steel.

Minor modifications to the chemistry of the steel had to be made for a number of reasons and, of course, each of these can affect the austenite-ferrite ratio to some extent.

Tight technical procedures were found to be necessary in both the heat treatment areas, where solution temperature is critical, and welding areas, where weld procedure specifications, control of consumables, and qualified operators are essential, and a strict discipline which ensures that all welds are re-solution treated to restore their corrosion resistance, which may be impaired during the weld deposition. To sum up, although there are many difficulties associated with the production of cast duplex stainless steels which require a well-ordered discipline in manufacture, these can be monitored by the existence of a meaningful quality assurance approach which will ensure at each stage of the procedural flowline, the necessary observance of recorded detail has taken place. This can be administered under any formal framework, i.e. BS5750 (9), AQUAP/4 (10) or other company led schemes, but it is still essential to build a thorough understanding between customer and supplier of the potential problems and the manner in which they are controlled.

Specifications

The designation UNS S31803 is not a specification as such, and is merely a designation or registration of an alloy against a number in the Unified Numbering System which is a joint scheme administered by the Society of Automotive Engineers and the American Society for Testing and Materials, and even proprietary alloys are listed under this scheme without specification references. Similarly the DIN 1.4662 is merely a registration of a composition under German National Standards Deutsches Institut fur Normung eV) and may not cross reference a given material specification. The only parameters fixed by the above numbers, are the chemical composition.

On the other hand there are a number of engineering specifications now in circulation where some clauses of the specification are either impractical or impossible to fulfil and, it would seem, have been written by personnel who have neither used nor manufactured a duplex stainless steel. The lesson to be learnt, we believe, in the present circumstances should be to fully discuss requirements with the supplier who has the experience of what can and cannot be achieved in practical terms.

With regard to NACE approved materials, according to the NACE specification MR-01-75 (11) and its amendments up to February 1986, only wrought materials in duplex stainless steel are approved under Werkstoff No. 1.4462 with the addition of one French national standard for cast duplex stainless steel to 6CNDU 20.08 M N.F. A320-55 (12).

DESIGN, APPLICATION AND PROCEDURES USED FOR MANUFACTURE OF EQUIPMENT USING DUPLEX STAINLESS STEEL - UNS S31803 / DIN 1.4462

The availability of wrought material was not sufficient to allow economical manufacture of valves. With the advent of the cast version of UNS S31803/ DIN 1.4462, the Industry was enabled to seriously consider this option.

In the early days when the offshore end users and contractors started to request the cast material, no international standards covering manufacture, inspection, testing and supply of Duplex stainless steel had been written. This was a major concern to Fisher, as a valve manufacturer and therefore, we decided to write our own specification in conjunction with the suppliers who have specialist knowledge of this material. We have also conducted our own testing programme to satisfy ourselves that the material we opted for meets our quality assurance standards. The Fisher Controls, USA, laboratory has carried out various tests which are detailed in Appendix 2.

The first design decision we had to make was whether to reduce the wall thicknesses of our equipment due to the increased strength of this material. Our conclusion was not to change our patterns but to emphasise the increased safety aspect, which is of prime importance offshore. On this basis we decided to use the pressure/temperature curve from ASTM B16.34-1981 Table F4-A (13) in the absence of any other recognised international standards for duplex stainless steel. However we have only used the working temperature range of - 30°C to 260°C (see Appendix 4), since temperatures outside of this range need careful checking to ensure retention of physical and mechanical properties. We know from tests already carried out that the material strength is not impaired down to - 100°C and up to +300°C. Secondly, this additional wall thickness would enable the end user to reduce maintenance costs by permitting a greater weight loss through corrosion.

Another major engineering decision was needed. Valves and other process equipment have to be manufactured not only from castings, but from wrought bar, tube, etc., so the welding of this material was looked at in detail both from the foundry repair point of view and in the fabrication of equipment from components.

It was necessary to review various methods and standards to find the optimum procedure that suited fabrication of our equipment, hence we decided to use ASME IX for weld procedures and qualification test which, for the UK, is certified by Lloyd's (see Appendix 3). It is essential that all parts to be welded together are compatible.

As the material's prime uses are for services subjected to sulfide stress cracking (SSC) and chloride stress corrosion cracking (SCC), it is essential that the properties of the parent material are carried through the heat affected zone (HAZ) and the weld itself.

Sandvik recommend a maximum heat input of 1.5kJ/mm and an interpass temperature of 150°C (302°F) maximum. The H.A.Z. and weld metal should be free from any grain boundary carbides and intermetallic phases, also the maximum hardness must be 28 HRC.

To maintain the Ferrite/Austenite balance plus parent metal properties in the weld, composition of the filler material should be of the type 22% Cr, 6-9% Ni, 3% Mo. Coated or non-coated rods can be used, however, each type/manufacturer of filler material should be tested to ensure that the Ferrite content in the weld metal is between 25% and 60%. The determination shall be made to ASTM E562 (14). For results see Appendix 3.

Our qualified welders had no difficulty in welding Duplex stainless steel, in fact they consider it to be easier to weld than some grades of commercially available 316 stainless steel (15).

In order to fabricate, it is necessary to machine both cast and wrought material, we have found that the material allows us to use the same tools we use for 316 Stainless Steel. However, to produce finishes normally associated with valve equipment, the feed rates are reduced by 10 to 15% to those used on 316 Stainless Steel.

The reduced feed rate is needed when undercutting for preparation prior to applying the buffer zone and hard facing to Duplex stainless steel. The final machining of the hard facing surface is the same whether base material is Duplex or 316 Stainless Steel.

The extensive range of valves used in the Oil and Gas production Industry is required not only to have the capability to withstand SSC and SCC attack, but also the highly erosive fluids which need controlling. Therefore, to overcome these problems, valve internals need hardened trims which are not susceptible to these adverse condition. Again, no standards had been written on the application of hard surfacing to Duplex stainless steel. We, therefore, undertook a test programme, organised by our French factory, to establish a procedure for the application of Cobalt Chrome Alloy to Duplex stainless steel (UNS S31803/DIN 1.4462).

The main problem encountered with hard facing of Duplex stainless steel is that of the base metal becoming very brittle in the heat affected zone and the possible occurence of intergranular corrosion.

Four methods were evaluated:

Method 1

Duplex stainless steel with CoCr Alloy.

Method 2

Duplex stainless steel with AISI 309 (16) buffer zone plus CoCr Alloy.

Method 3

Duplex stainless steel with AISI 309 buffer zone plus CoCr Alloy. The completed part is then solution annealed at 1100°C and air cooled.

Method 4

Duplex stainless steel with Haynes 25 (UNS R30605) plus CoCr Alloy.

The results of our tests are given below, however, photographic documentation is in Appendix 5 with cover sheet.

Method 1 (Exhibit 1)

The fusion zone and heat affected zone showed a lot of intergranular carbides and dilution which cause problems of corrosion attack and brittleness. Therefore, we rejected this method as being unacceptable due to the reduced corrosion properties between base metal and hard facing material.

Method 2 (Exhibit 2)

Interface between the Duplex stainless steel and AISI 309 showed small dilution and no intergranular carbides within the fusion zone and heat affected zone. A large amount of carbides (picture 4) can be seen in the interface between AISI 309 and Cobalt Chrome Alloy. As AISI 309 is not subject to brittleness, this method is suitable for hard surfacing. However, it is necessary to restrict heat input to 8 to 12 kJ/cm with a minimum preheat temperature of 250°C to reduce the dilution and therefore, the risks of corrosion.

Method 3 (Exhibit 3)

To remove the carbides within the interface between the AISI 309 and Cobalt Chrome Alloy, we tried heat treating the finished part (solution annealing at 1100°C, air cooled). The annealing dissolved the carbides in the heat affected zone, but not those in the fusion zone.

This method was an improvement over method 2, however, the possibilities of deformation which may occur during the heat treatment process, increase the failure rate of the finished part.

Method 4 (Exhibit 4)

This method produced the best results. By substituting AISI 309 by Haynes 25 (UNS R30605) as the buffer zone, there were no problems of heat input and only a small dilution.

We have now adopted this method in preference to method 2. Costs are similar.

APPLICATIONS

This paper, so far, has defined the technical parameters of the Duplex material. The use of both cast and wrought versions in our equipment has always been defined by the contractor/end user because we have relied upon their expertise to define the process fluid, and to specify whether duplex material is needed. However, our experience in valve design dictated when hard faced trim materials were needed to overcome the erosive action of the process fluid. This joint approach to the applications, allowed correct types of equipment to be used both in design and materials. We still review very closely each application to ensure compatibility of material, product and any contractor/end user specification.

The range of equipment supplied so far spans globe control valves 1" to 16" with pressure ratings up to ANSI 1500 lb, including noise reduction trim; butterfly valves; level controllers and safety relief valves.

Not all safety relief valves supplied were manufactured completely in Duplex stainless steel. To reduce costs, a large proportion had Duplex nozzle, disc, disc holder and blowdown ring, with carbon steel or stainless steel body and bonnet. The reason being that many safety systems have no pressure or fluid upstream of the safety relief valve, hence no major corrosion problem. The body and bonnet on a full nozzle type Farris safety relief valve does not come into contact with 'live' process fluid. 'O' ring seals can be used in the disc assembly to remove any chance of leakage, during normal plant operation.

QUALITY ASSURANCE

Due to the lack of fully defined international standards for the cast UNS S31803/DIN 1.4462 alloy designation, we have spent a considerable amount of time reviewing suppliers and procedures for manufacture to ensure that the customer receives a product that has all the anti-corrosive material properties and the high quality product that he expects.

With all materials, especially Duplex stainless steel, it is essential that the raw material arriving at manufacturer is correct. Duplex stainless steel is more susceptible than most to changes in metallurgical controls. Therefore the quality procedures at foundry need particular reviewing by manufacturer. Fisher use BS5750 part 2 as a minimum for the UK or other similar Quality Standards outside UK as a basis. We also have our own standards to supplement national standards. When reviewing foundries we evaluate their metallurgical and engineering facilities very closely, as an essential part of reviewing the total Quality System.

All this is necessary to give us confidence that the correct metallurgical controls have been effected. Any permissible weld repairs to castings have to be performed with correct matching electrodes, reheat treated and documented. (We use ASTM A703 supplementary requirements S20 for major repairs (18)).

Once we are satisfied that the foundry has adequately controlled its own operations our own quality procedures (based upon BS5750) guide the raw material through various stages of manufacture.

It is essential that closely controlled quality procedures are agreed with foundry. Without this dedication a customer could receive a product which may have inadequate properties, for example incorrect heat treatment will not give adequate ferrite content and microstructure thereby affecting the anti-corrosive properties of UNS S31803/DIN 1.4462, and the mechanical properties which can be achieved.

CONCLUSION

We hope this paper has demonstrated that simply requesting materials to UNS S31803 or DIN 1.4462 will not guarantee receipt of a material with the optimum properties. We have found that by working closely with material suppliers, it is possible to formulate a specification covering material composition, manufacture and testing for both wrought and cast versions giving near identical properties to each other. Your customer/end user can thus evaluate whether your specification is suitable for his specific applications. Additionally, Good Quality systems and procedures ensure that both material and product will give the end user trouble-free service.

Since the initial introduction of cast material, interest in the use of this material has continually expanded. For the future, improvements in manufacture and its properties will continue.

LIST OF REFERENCES

1 UNS 31803. Designation of duplex composition in the "Metals and Alloys Unified Numbering System - ASTM and SAE Joint publication.

2 DIN W/No. 1.4462. Designation of given duplex composition identified in the German national standards.

3 ASTM A800/A800M-84

Standard practice for steel castings, austentic alloy estimating ferrite content thereof. American Society for Testing and Materials.

4 AISI 329

Specification for Ferritic- Austenitic Cr-Nickel Stainless Steel, American Iron and Steel Institute.

5 ASTM A296 Grade CD4MCu replaced by ASTM A744/A744M-84 Standard Specification for castings, iron chromium nickel, nickel-base corrosion resistant, for severe service, American Society for Testing and Materials.

6 A.O.D. Argon, Oxygen, Decarburisation - Patented process (Union Carbide).

7 ASTM A262-85a. Standard practices for detecting susceptibility to intergranual attack in Austenitic Stainless Steels.

8 TM - 01 - 77. Testing of materials for resistance to Sulphide Stress Cracking at ambient temperatures NACE document.

9 BS 5750. British Standard on quality systems (1987)

10 AQUAP/4. NATO Quality Assurance system for manufacture and inspection used by Ministry of Defence.

11 NACE MR-01-75. Material Requirements - Sulphide Stress Cracking Resistant Metallic Materials for Oil Field Equipment (Jan.1984) from National Association of Corrosion Engineers (USA)

12 6CNDU 20.08 MNF A320-55. Designation of a Cast Duplex Stainless Steel as a French national standard mentioned in NACE MR-01-75 (1984).

13 ANSI B16.34 - 1981

Specification for valves flanged and butt welding ends.

14 ASTM E562 - 1983

The standard practice for determining volume fraction by systematic manual point count.

15 316 Stainless Steel

Specification for Austenitic Cr-Nickel Stainless Steel.

16 AISI 309

Specification for Austenitic Cr-Nickel Stainless Steel. American Iron & Steel Institute.

17 ASTM A 370

Specification for mechanical testing of steel products.

18 ASTM A 703 Supplementary Requirement S20

Specification for steel castings for pressure containing parts.

ACKNOWLEDGEMENTS

SAF 2205 - Tradename of AB Sandvik Steel, Sandviken, Sweden.

HAYNES 25 - Tradename of Haynes Stellite Co.

Constitution Diagram for stainless steel weld metal (Schaeffler diagram) by Anton L. Schaeffler.

Modification to Schaeffler diagram to include nitrogen by De Long.

APPENDIX 1

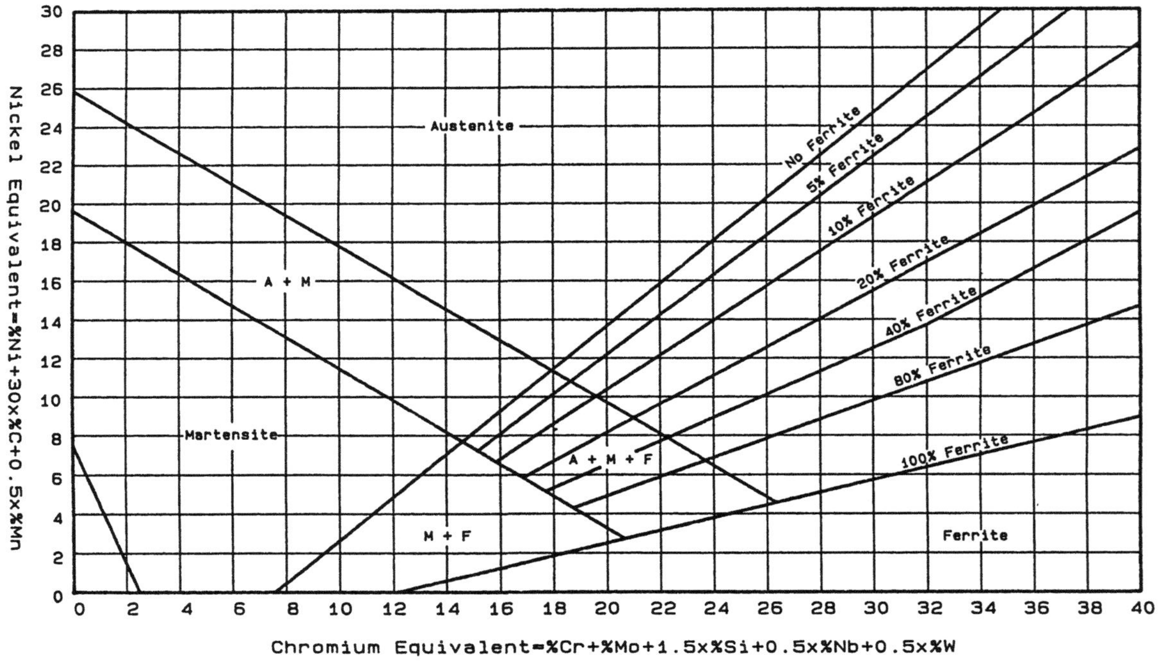

A = Austenite F = Ferrite M = Martensite

Fig 1 CONSTITUTION DIAGRAM FOR STAINLESS STEEL WELD METAL By Anton L. Schaeffler

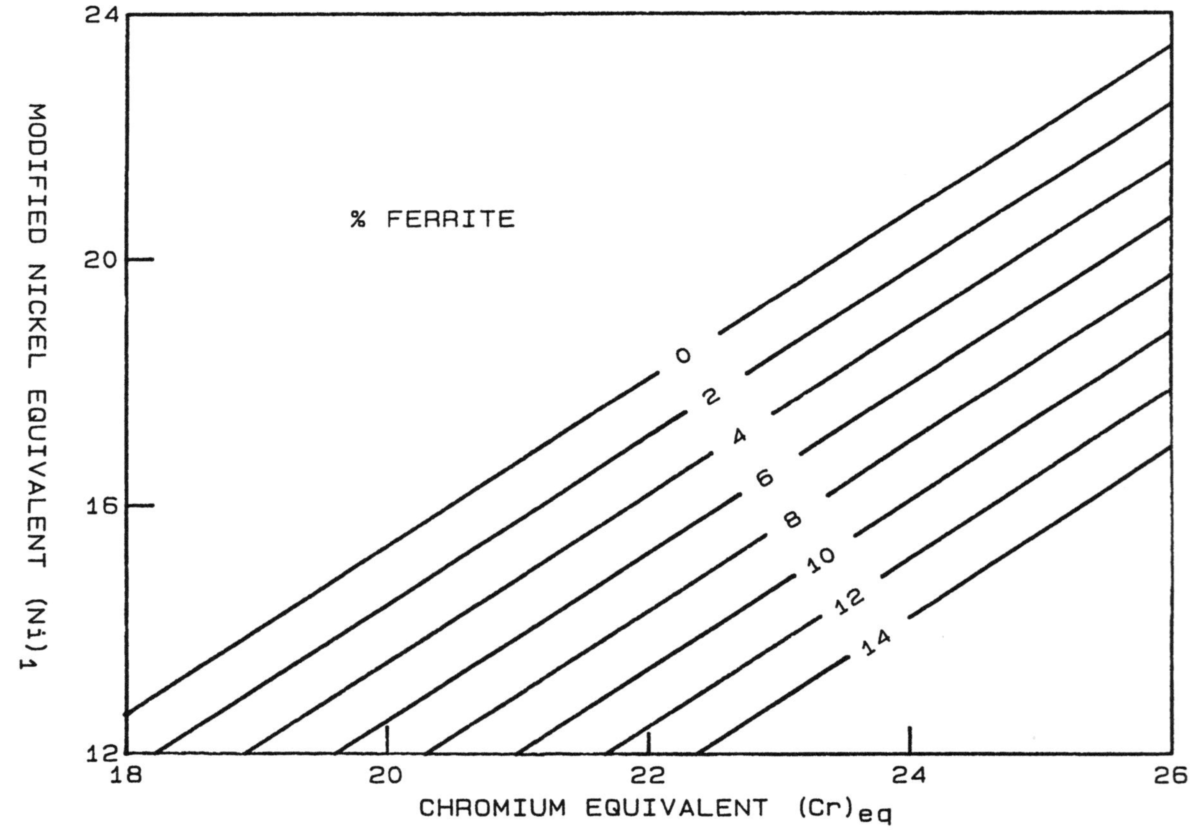

Fig.2. -- Modification of the Schaeffler diagram by DeLong to include a factor for nitrogen [$(Ni)_1 = (Ni)_{eq} + 30 \times \%N$].

APPENDIX 1

TABLE 1 The Composition of Some Modern Duplex Cast Steels

Steel Designation	NOMINAL OR AVERAGE COMPOSITION C(max)	Cr	Ni	Mo	Cu	N	W	Product Form
SAF 2205	0.02	22	5.5	3	--	0.10	-	Wrought & Cast
FERRALIUM 255	0.04	26	6	3	1-4	0.10	-	Wrought & Cast
FERRALIUM 288	0.05	28	7	3	0.75/1.75	0.12	-	Wrought & Cast
ZERON 25	0.03	25	7	2	--	-	-	Cast
ZERON 100	0.03	25	7	4	0.3	0.25	1.0	Cast
FMN	0.04	25	5	2.2	--	0.20	-	Cast
URANUS 50	0.07	21	7	2.5	1.5	0.15	-	Wrought & Cast
ASTM A 296 CD4 M Cu	0.05	26	5	2	3	-	-	Wrought & Cast
PARALLOY 3FL	0.03	25	5	2	3	-	-	Cast

APPENDIX 2.

Results of tests conducted by Fisher Controls, U.S.A.

Test bars taken from cast SAF 2205 manufactured by Catton & Co. Ltd. (division of Wm Cook plc).

1. Tensile Test

SAMPLES	A	B
Ultimate tensile strength - N/mm^2	687	688
Yield strength (0.2% offset) - N/mm^2	492	510
Elongation % in 50mm minimum - %	29	30
Reduction in area - %	43	44

2. Charpy Impact Test - to ASTM A370 (17)

SAMPLES	A	B	C
Temperature ^{o}C	-50	-73.3	-195.5
Value - Joules / cm^2	118	72	6.8

3. Weld Test - 2 Samples

U-Bend - Both good

Ultimate Tensile Strength - 704 N/mm^2

Yield strength (0.2% offset) - 577 N/mm^2

Elongation % in 50mm Min. - 30%

Reduction in area - 42%

4. Test to ASTM A262 Practice E (Strauss' Test)

Test piece prepared, tested and passed in accordance with above standard.

5. Test to ASTM A262 Practice C (Huey Test)

Two test pieces prepared and subjected to five periods each of 48 hours duration.

Sample 'A' average corrosion rate = 7.68 mils per year

Sample 'B' average corrosion rate = 7.36 mils per year

Average corrosion rate for both samples = 7.53 mils per year.

APPENDIX 3

Test results taken from Lloyd's WPQ Cert. No. LON 606183/1

DESTRUCTIVE TESTS		UTS (N/mm^2)	YIELD STRESS (N/mm^2)	LOCATION OF FRACTURE
SAMPLE	1	736	570	) BROKE OUTSIDE
	2	737	572	) WELD.

CHARPY IMPACT TEST OF WELD AT -50°C		
SAMPLE	1	62 JOULES
	2	66 JOULES
	3	87 JOULES

HARDNESS TEST - ROCKWELL 'B' SCALE *			
SAMPLE	1	2	3
PARENT METAL	97	99	98
WELD	97	99	99
H.A.Z.	97	99	99

* NOTE: 100 HRB = 20 HRC

1. Test Pieces Used - 4.5 inch OD x 0.5 inch thick tube manufactured by Sandvik, reference SAF 2205 (ASTM A790 UNS S31803).

2. Position of Test Pieces - Tube rotating about horizontal axis.

APPENDIX 4

Pressure / Temperature Ratings

Standard Class valves - Flanged and buttwelding end.

ANSI WORKING PRESSURES/ TEMPERATURE	150 lbf/in^2 bar	300 lbf/in^2 bar	400 lbf/in^2 bar	600 lbf/in^2 bar	900 lbf/in^2 bar	1500 lbf/in^2 bar	2500 $l/f/in^2$ bar
-20° to 100°F -30° to 38°C	290 20	750 52	1000 69	1500 103	2250 155	3750 259	6250 431
200°F 93°C	260 18	750 52	1000 69	1500 103	2250 155	3750 259	6250 431
300°F 149°C	230 16	730 50	970 67	1455 100	2185 151	3640 251	6070 419
400°F 204°C	200 14	705 49	940 65	1410 97	2115 146	3530 243	5880 405
500°F 260°C	170 12	665 46	885 61	1330 92	1995 138	3325 229	5540 382

APPENDIX 5 EXHIBIT 1

COVER SHEET

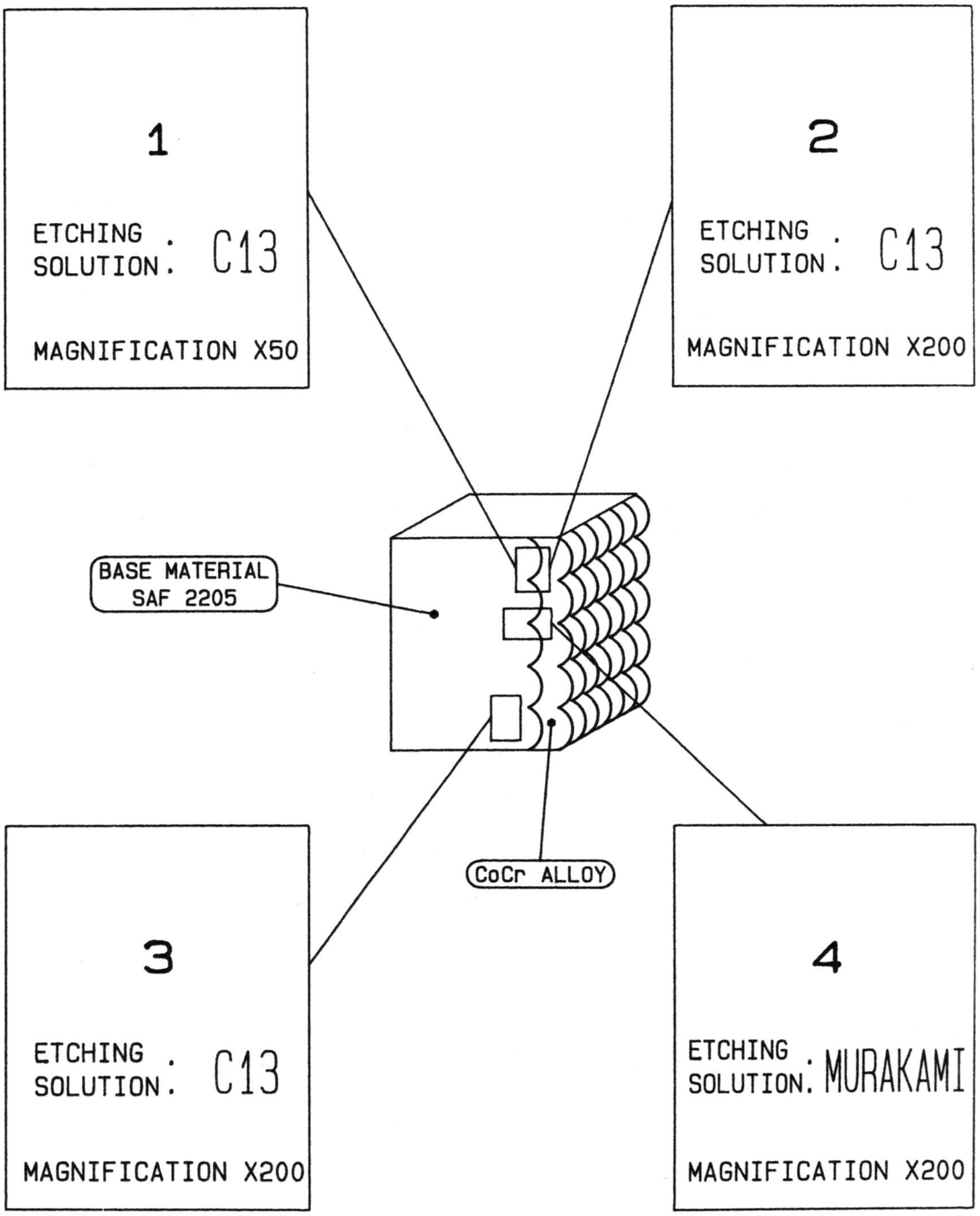

APPENDIX 5

EXHIBIT 1

BASE MATERIAL

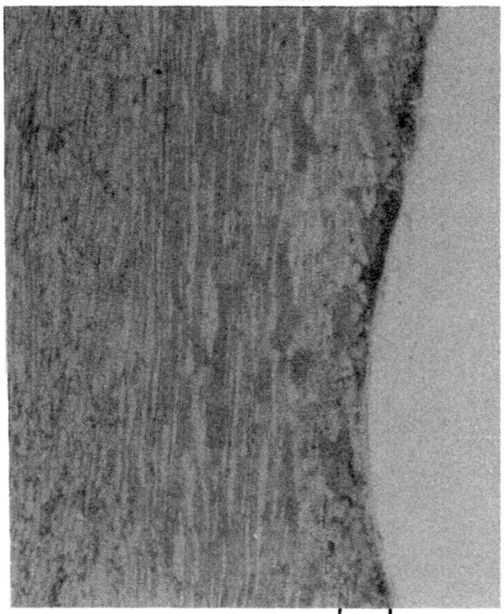

x50

FUSION ZONE

CARBIDES IN FUSION ZONE

BASE MATERIAL

CoCr ALLOY

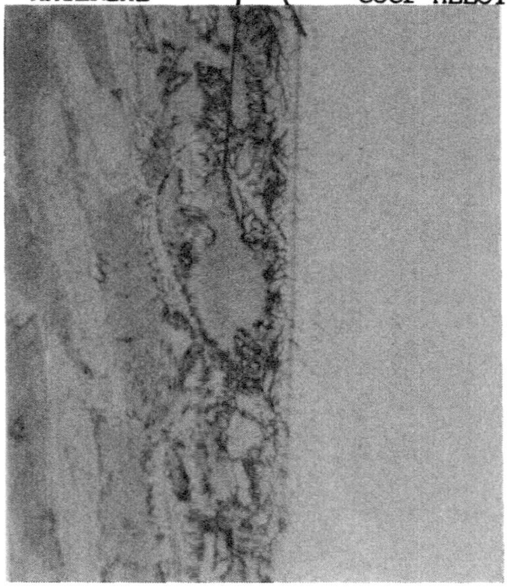

x200

BASE MATERIAL x200

INTERGRANULAR CARBIDES

BASE MATERIAL

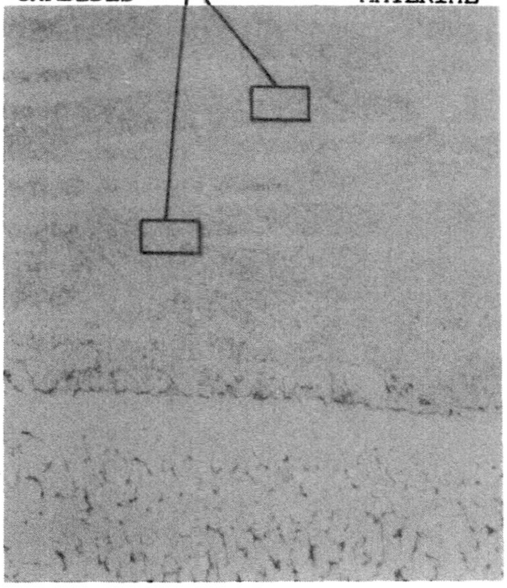

CoCr ALLOY x200

APPENDIX 5 EXHIBIT 2

COVER SHEET

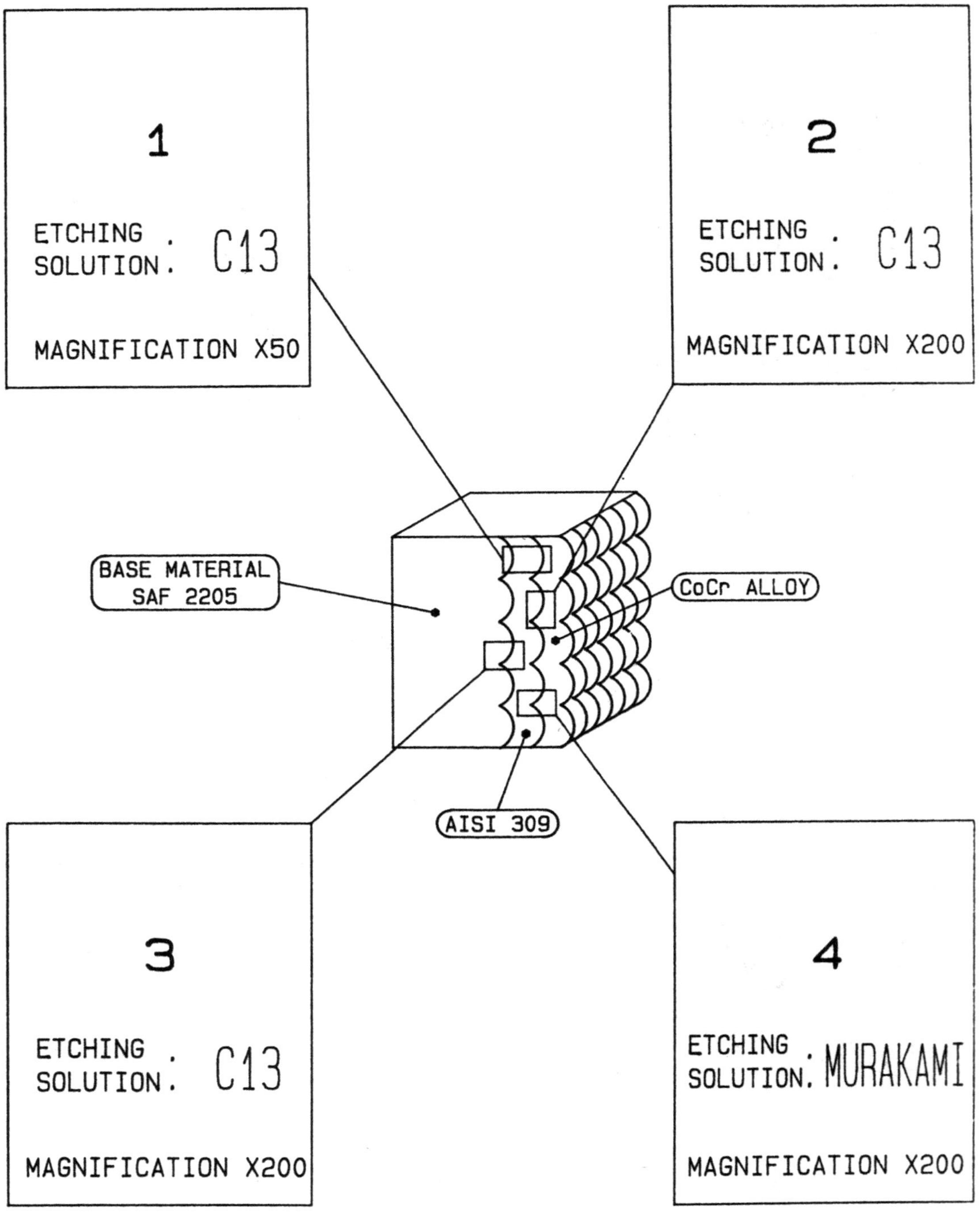

EXHIBIT 2

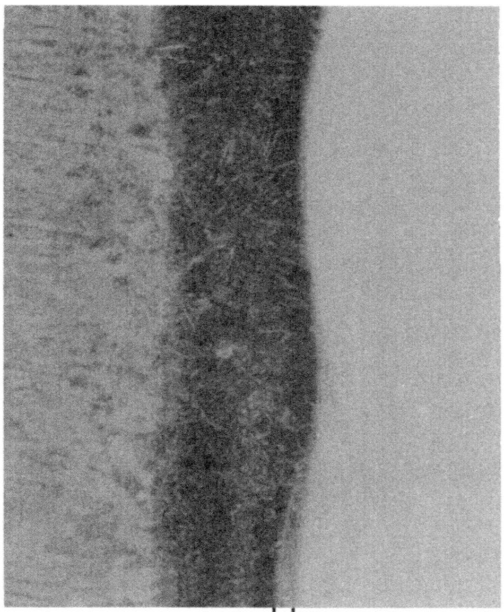

FUSION ZONE x50

CARBIDES IN FUSION ZONE

AISI 309 CoCr ALLOY

x200

BASE MATERIAL

AISI 309 x200

AISI 309 INTERGRANULAR CARBIDES

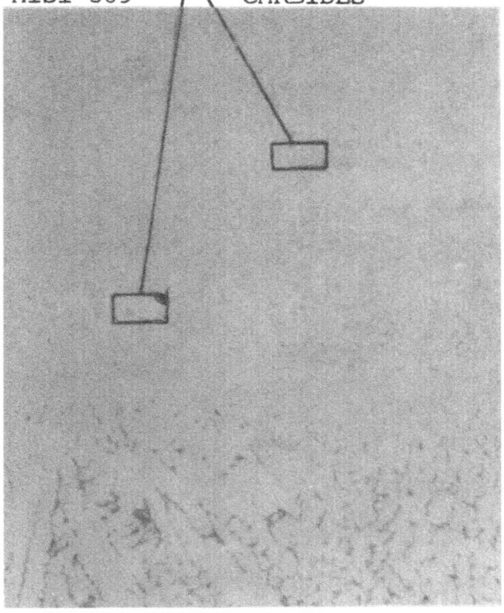

CoCr ALLOY x200

APPENDIX 5

EXHIBIT 3

COVER SHEET

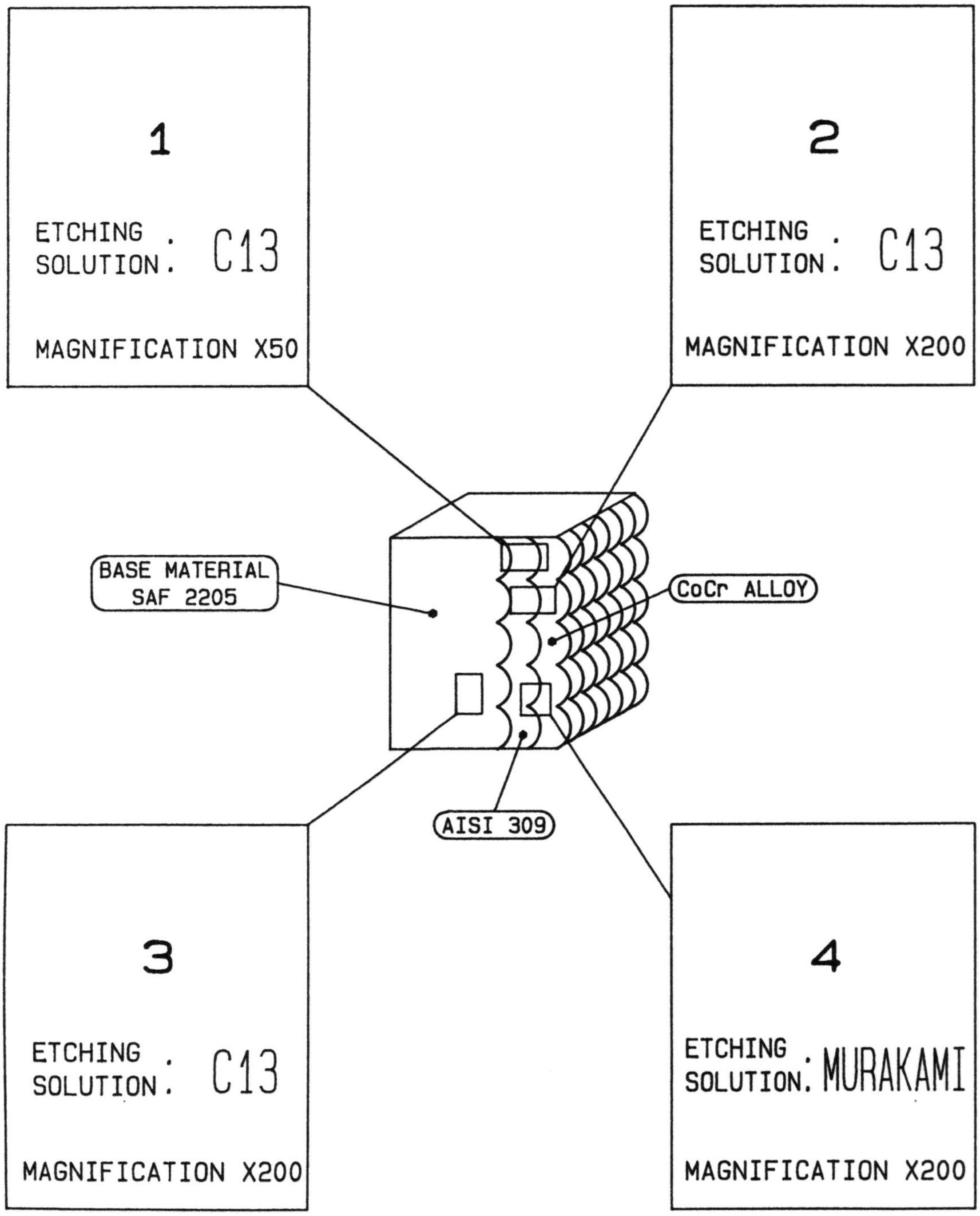

APPENDIX 5

EXHIBIT 3

BASE AISI 309 CoCr ALLOY

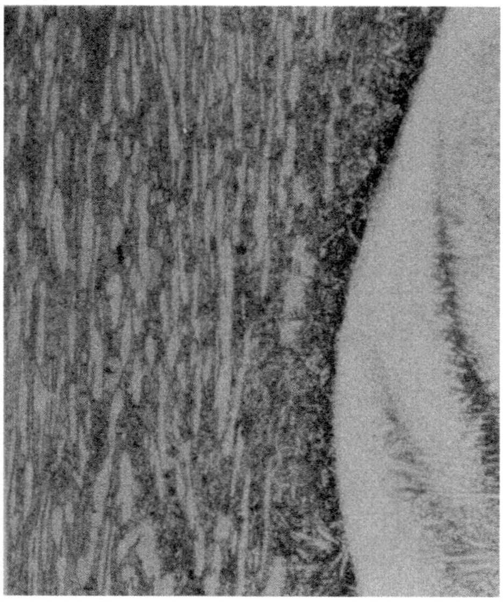

x50

CARBIDES IN FUSION ZONE

AISI 309 CoCr ALLOY

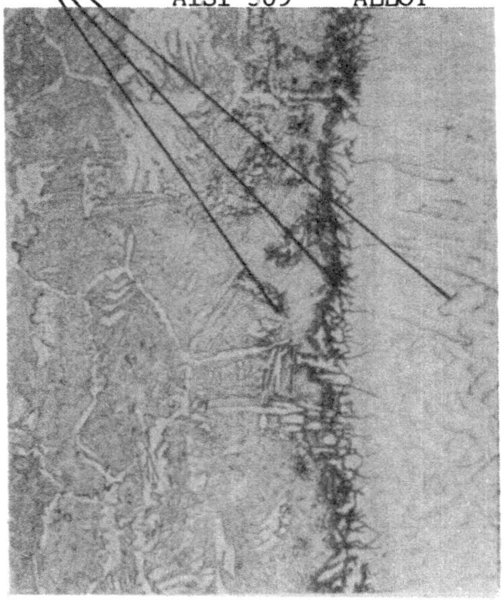

x200

BASE MATERIAL x200

NO INTERGRANULAR CARBIDES

AISI 309

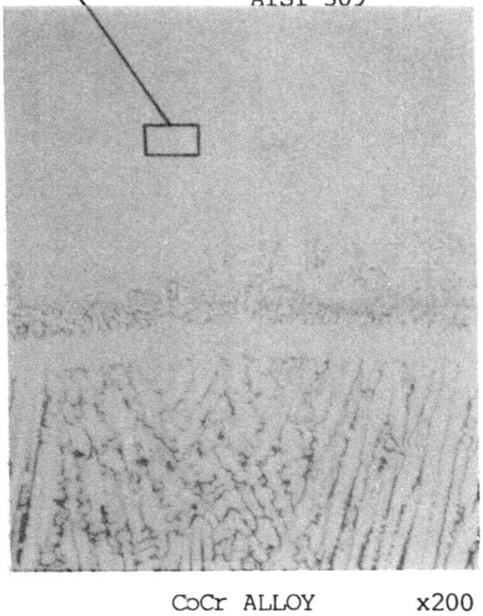

CoCr ALLOY x200

APPENDIX 5

EXHIBIT 4

COVER SHEET

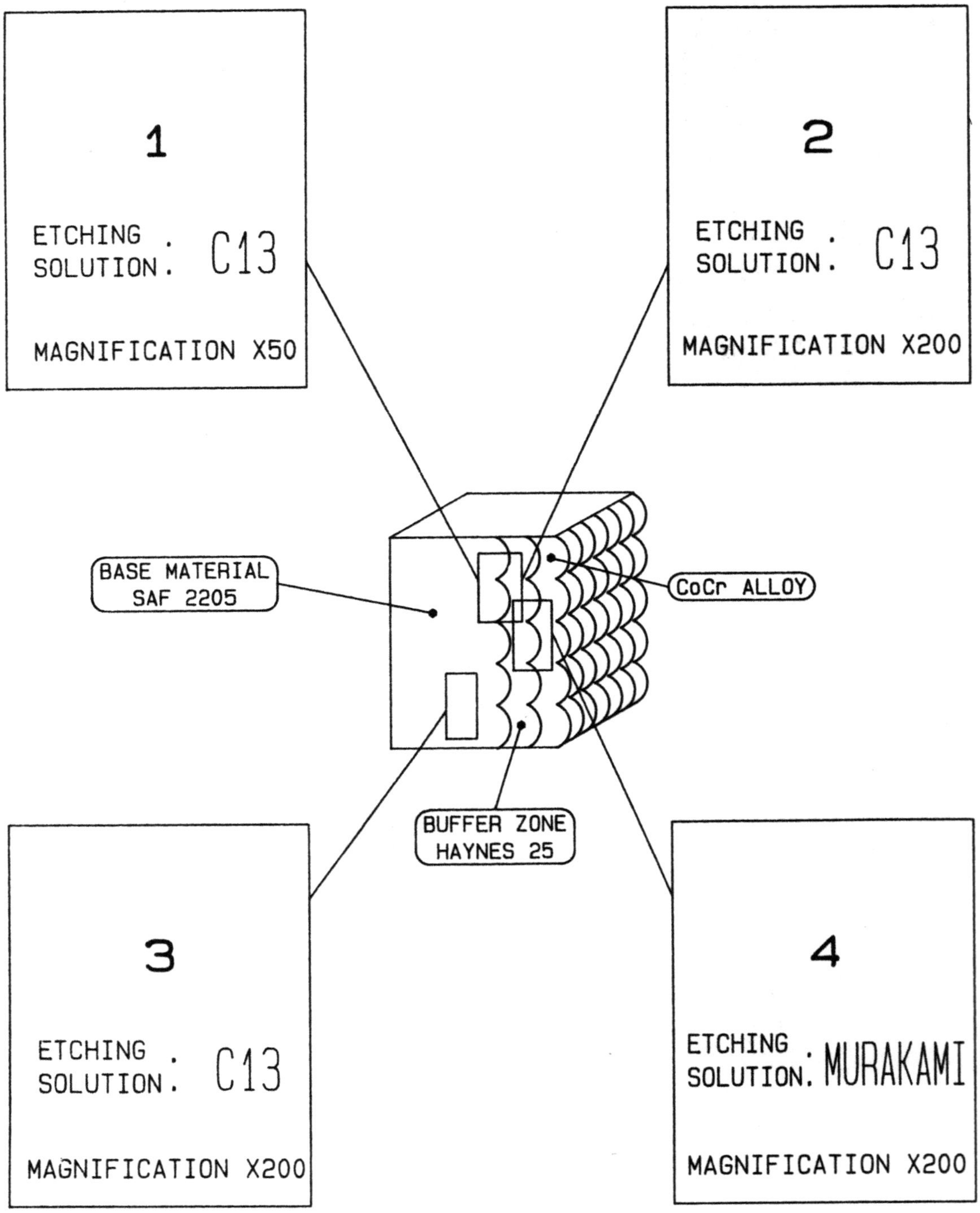

APPENDIX 5

EXHIBIT 4

BASE MATERIAL HAYNES 25

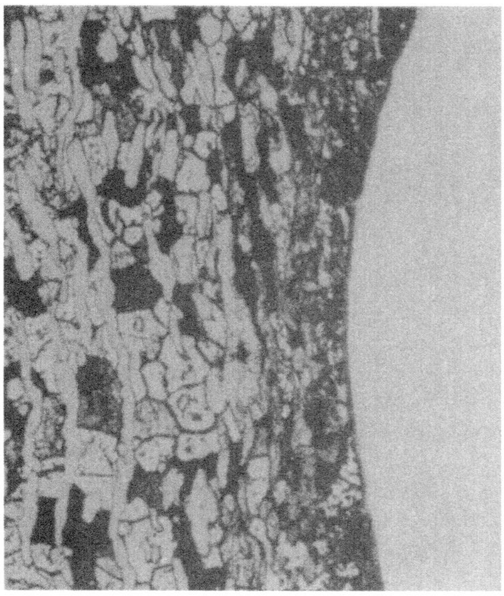

x50

BASE MATERIAL HAYNES 25

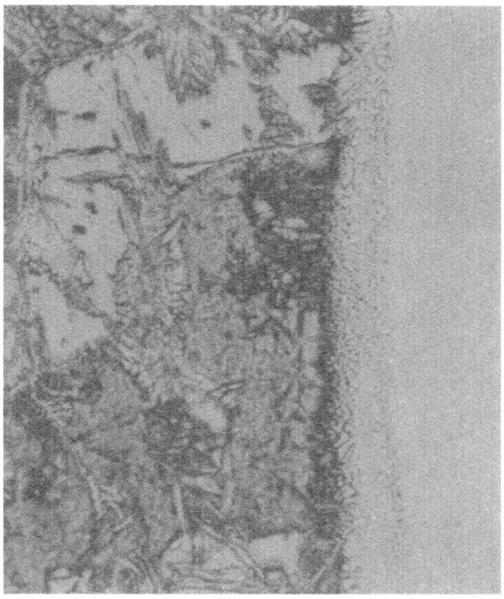

x200

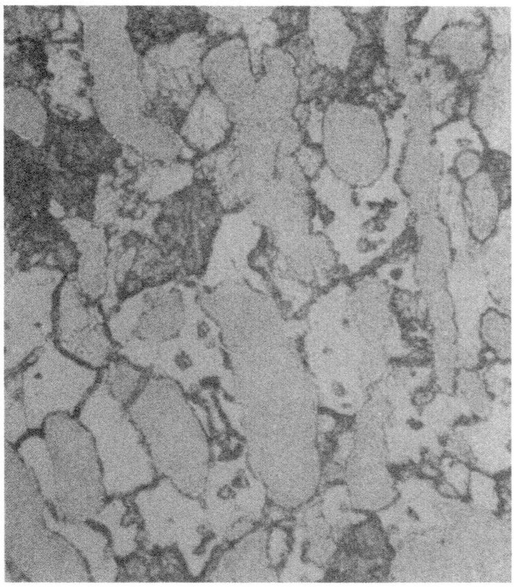

BASE MATERIAL x200

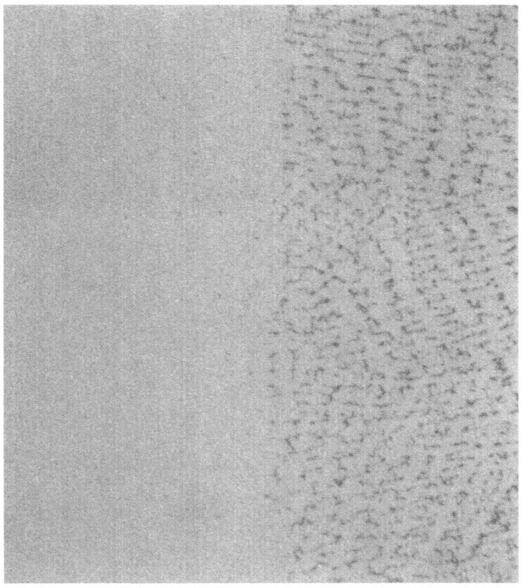

HAYNES 25 CoCr ALLOY x200

2nd International Conference on

Developments in Valves and Actuators for Fluid Control

Manchester, England: 28-30 March 1988

PAPER E2

NEW AND IMPROVED, HIGH NICKEL ALLOY CASTINGS

James L. Gossett
Fisher Controls International, Inc.
R.A. Engel Technical Center
Marshalltown, Iowa, USA

Summary

High nickel alloy castings have long been a chronic problem for the process control industry due to poor casting quality, late deliveries and poor corrosion resistance compared to the wrought equivalents. Detailed specifications, new casting alloys and advanced processing techniques have been combined to solve these problems. ASTM and ASME specifications and/or the common tradenames, Hastelloy C, Alloy 20 and Monel, are not sufficient for ordering high integrity, reliable castings.

1. Introduction

Most control valves are produced from castings. This is dictated by the intricate passages required for precise flow control. The most commonly used valve body materials are WCB cast carbon steel and CF8M cast stainless steel (SST). Carbon steel is typically used for non-corrosive applications such as steam, dry gases and organic materials. The SST is used for more severe applications such as dilute acids, caustics and cryogenic service.

Nickel base alloys or alloys with high nickel contents are selected for the most corrosive applications such as hot, concentrated acids and caustics. Due to the critical nature of these applications and the hazards involved, maximum integrity along with timely delivery is required. Several corrosion failures have recently been encountered in environments where the respective alloys are commonly used. While the cast components were severely corroded, the wrought valve components were unattacked. This is contrary to the perceptions held by most in the industry, that the wrought and cast materials are equivalent. Most available corrosion data was developed for wrought alloys.

Unfortunately, the high nickel alloys are much more difficult to cast than the common carbon steels and SST's. The high nickel alloys are also prone to hot tears, cracking, porosity and/or gassing. These defects may appear at any

Held in Manchester, England. Organised and sponsored by BHRA, The Fluid Engineering Centre and co-sponsored by the British Valve Manufacturers Association and the Institution of Mechanical Engineers Process Industries Division

stage; shake-out, heat treating, machining or final pressure testing. Castings are commonly poured several times to produce and install a serviceable product. This makes late deliveries commonplace.

There are a number of commonly cast high nickel alloys. The compositions, cast designations, wrought equivalents and corresponding tradenames of the alloys discussed in this paper are listed in Tables 1 and 2. Although many users purchase cast products to the common tradenames, Hastelloy, Alloy 20 and Monel, the results are potentially dangerous. There are some 15 different Hastelloy grades. The corrosion properties of the different grades vary tremendously depending upon the application. Specifying "Hastelloy" when a C type alloy is needed, can have disastrous results. Significant differences will be shown among the cast "Hastelloy" C grades.

2. Common Problems

2.1 Foundry

Orders for high nickel alloy cast products are generally for small quantities, and short lead times are often required. These factors often result in several actions which have detrimental effects on delivery and quality. Foundries prefer to wait until they accumulate enough orders to melt and pour a full heat which extends delivery times. With the expense of these alloys and the inherent characteristics which limit yields to 30% or less, this is understandable. Smaller than optimum heat sizes also aggravate gas pickup and make compositional control difficult.

If delivery constraints force a foundry to use existing steel or SST pattern equipment to pour any of the high nickel alloys, the results are often unsatisfactory. Properties such as shrinkage rates, solidification characteristics, thermal conductivity, solidification range and gas evolution tendencies during solidification, vary tremendously among the high nickel alloys, carbon steel and SST. More time may be lost repairing or repouring castings than would have been required to make dedicated patterns with special rigging.

2.2 Weld Repairs

The nickel-copper alloy M35-1 (UNS N04400 or cast Monel 400) is a standard material for oxygen service. While wrought N04400 and the other common wrought high nickel alloys are routinely welded and some even hardfaced, cast alloy weld repairs have always been difficult; M35-1 is the most difficult alloy.

Many castings have been scrapped because minor porosity or sand inclusions could not be successfully repaired. Even though the defects are easily removed and the cavity successfully welded, cracks appear in the heat affected zone (HAZ), Figure 1. These cracks must then be ground out and welded. All too often, more HAZ cracks will then appear. Each round of defect removal and weld repair, enlarges the affected area until defeat is accepted and a new casting ordered.

2.3 Corrosion Resistance of CW12MW

Hastelloy C is considered the standard material of construction for sodium hypochlorite and chlorine dioxide services (1,2,3,4). In actuality Hastelloy C is an obsolete material. The most commonly used wrought alloy is N10276 which differs in composition from the original Hastelloy C which is no longer produced. The most commonly supplied casting alloy is CW12MW. Most equipment is merely ordered as "Hastelloy C" with no thought of whether the component parts are cast or wrought. This leads to unexpected problems.

High nickel alloy castings have exhibited corrosion resistance inferior to their wrought counterparts in field applications. In sodium hypochlorite and chlorine dioxide services for times as short as 2 months, cast CW12MW bodies, bonnets and cages were severely pitted. The crevice areas were the most severely attacked, Figure 2. The wrought plug, stem and seat ring were in like new condition and fully functional. The cast parts were not suitable for service.

The CW12MW castings were thoroughly analyzed and found to comply with the compositional requirements of ASTM A494 as ordered. Metallographic examination did not reveal grain boundary carbide precipitates or any other microstructural defects which would indicate improper heat treatment. The microstructure did contain an interdendritic second phase which was preferentially attacked, Figures 3 and 4. This second phase was corroded at a higher rate than the matrix material. Samples of the castings were solution heat treated for various times and temperatures as given in Table 3. The castings could not be homogenized nor the second phase dissolved.

Boiling laboratory corrosion tests were run in 10% hydrochloric acid, ferric sulfate-sulfuric acid per ASTM G28 and 3% sulfuric acid. The procedures used are given in Appendix 1. The values in Table 3 are averages from several components and duplicate samples. The corrosion rates for CW12MW were inconsistent and 3 to 6 times those obtained for the wrought N10276. In 10% HCl, CW12MW was 17 mm/y (660 mpy) compared to 5.6 mm/y (220 mpy) for N10276; in the G28 test, 22.9 mm/y (900 mpy) and 5.8 mm/y (230 mpy); and in 3% sulfuric acid, 0.86 mm/y (34 mpy) and 0.15 mm/y (6 mpy), respectively. These results were totally unexpected based on past experience and lack of customer feedback.

The solution annealing heat treatments did not produce consistent improvements in corrosion resistance. In many of the laboratory corrosion tests, the cast samples suffered interdendritic attack like that found in the field. The attack on unwelded N10276 was uniform in all cases with no intergranular or localized attack. As the second phase content of the CW12MW was not noticeably changed by heat treatment, it appeared that the second phase was responsible for the CW12MW accelerated attack.

F. G. Hodge (5,6) identified the second phase as an intermetallic μ phase. A μ phase has a rhombohedral crystal structure with 13 atoms per cell. This μ phase is enriched with Mo compared with the bulk composition. His work also confirmed the interdendritic corrosion problem. In reducing solutions, such as HCl, the matrix-intermetallic interface is attacked. In oxidizing solutions, such as the ASTM G28, ferric sulfate-sulfuric acid test, the μ phase suffers accelerated attack.

Hodge noted that the as-cast structure also contained M_6C carbides. Thermal treatments may dissolve the carbides but not the μ phase; therefore, reducing the carbon content would not significantly improve the corrosion resistance of properly heat treated CW12MW. The amount and distribution of the μ phase is a function of casting size, location, cooling rate and other factors, therefore, all castings do not behave in a uniform manner. Our tests clearly showed this phenomenon.

2.4 Corrosion Resistance of CN7M

CN7M (cast UNS N08020, alloy 20Cb3) was originally developed for, and is the standard material of construction for sulfuric acid service (7). In the field cast CN7M has shown unsatisfactory corrosion resistance compared to wrought UNS N08020 in 3% sulfuric acid at 49-82°C (120-180°F), 98% sulfuric acid at 113°C (235°F), 30% sulfuric acid at 27°C (80°F) and 40-50% nitric acid at 93°C (200°F). Several bodies and investment cast CN7M cages were severely corroded along the grain boundaries, Figure 5. In some cases the attack was so severe that the grain boundaries were completely corroded away. Only the mechanical interlocking of the grains was holding the castings together. This, of course, could lead to a catastrophic failure under impact loading such as, a water hammer. Such failures are unknown to date.

The castings were in compliance with the appropriate ASME SA351 and ASTM A743 specifications. The impurity levels were very low for carbon, sulfur, phosphorus and silicon. Metallographic examination did not find any second phases, carbides or other grain boundary precipitates. None of the characteristics of stress corrosion cracking (SCC) were found. Surface carbon pick-up was not suspected because sand, shell and investment castings had been attacked.

Laboratory corrosion tests were run in boiling 10% hydrochloric acid, boiling ferric sulfate-sulfuric acid per ASTM G28, boiling 3% sulfuric acid and boiling 65% nitric acid per ASTM A262 Practice C, see Appendix 1. In 10% HCl the corrosion rates are 36 mm/y (1400 mpy) for CN7M and 33 mm/y (1300 mpy) for N08020, Table 4. The data scatter is large for duplicate samples from the same casting and for samples from different castings. On rare occasions the rates may even be lower than the wrought alloy.

In the G28 test CN7M was 0.61 mm/y (24 mpy) compared to 0.28 mm/y (11 mpy) for N08020; in 3% sulfuric acid, 0.69 mm/y (27 mpy) and 0.17 mm/y (6.5 mpy); and in the A262 nitric acid test, 0.38 mm/y (15 mpy) and 0.20 mm/y (8 mpy) respectively. In these tests, the rates for the cast CN7M were generally higher than the wrought N08020 but not in all cases. The intergranular corrosion seen in the field was not reproduced in the laboratory.

Another typical CN7M problem was through wall leaks after a short time in service. It is impossible to determine whether the leaks are from hot tears which did not leak during the hydrotest or are from intergranular corrosion in service. After a body has been in corrosive service, the evidence as to the cause is destroyed.

3. Problem Solving

Following a thorough analysis of the problem, extensive testing was done working closely with several foundries. Several items were identified to produce solutions to the high nickel alloy casting problem.

3.1 Foundry Consolidation

The consolidation of high nickel alloy casting efforts into a minimum number of foundries, produces several benefits. Heats of a given alloy are accumulated faster which decreases lead-times. Each foundry's learning curve is accelerated as heats are poured more often. The benefits of larger heats are also gained.

Duplicate patterns should always be made to provide dedicated equipment for the high nickel alloys. Compromises in pattern equipment are totally unacceptable. As mentioned earlier, properties such as shrinkage rates, solidification characteristics, thermal conductivity, solidification range and gas evolution tendencies during solidification, vary tremendously among the high nickel alloys, carbon steel and SST. Different gate and riser arrangements are required for the high nickel alloys. Many shrinkage and porosity problems will be eliminated with the proper rigging.

With two sets of pattern equipment, the steel and SST castings can be produced at a high volume foundry while the high nickel alloys are produced at a speciality foundry. This also helps concentrate the high nickel alloy casting work at the minimum number of foundries.

3.2 Foundry Qualification

The wrought alloys UNS N10276, N08020 and N04400 are routinely welded. Weld procedures are qualified to codes and standards which require a 180° weld bend test. The weld bend test has been successfully applied as a foundry qualification test. Each foundry is required to pass this test for each high nickel casting alloy. If the foundry has a process (chemical control, gas pick-up, raw material, etc.) which can pass a 180° bend test in the as-welded condition, good casting integrity and weldability are assured.

The bend test in ASTM A494 per Figure 1(b) is used, Figure 6 (8). The 25 x 85 x 155 mm (1 x 3 1/4 x 6 inch) test plates are welded with matching composition filler material per normal weld procedures. The 10 mm (3/8 inches) bars are bent over a 38 mm (1 1/2 inches) mandrel and examined per ASTM A494 Paragraph S2.1.2. No cracks exceeding 3.2 mm (1/8 inch) are permitted with the exception of edge cracks, weld slag inclusions or casting defects. Orders should not be placed with foundries which fail this test. Retests are permitted as the foundry process is optimized.

Excellent correlation has been found between casting integrity and weld bend test performance. Weld repairs required in the foundry or in-house are no more difficult than for the wrought alloys. Fabrication welding, such as welding on forged flanges, can now be considered.

3.3 Improved Alloys and Compositions

a) CW2M

The development of alloy CW2M, a new Hastelloy C type casting alloy, by Haynes International (formerly the Stellite Division of Cabot) solved the inherent problems with the old CW12MW alloy. The CW2M composition differs from CW12MW in eliminating the tungsten (3-4% vs. zero) and reducing the maximum carbon content from 0.12% to 0.02%, Table 2. These changes prevent μ phase formation. The typical elongation is increased to 40-60% compared to 6-10% for CW12MW. The minimum elongation specified in ASTM A494 is increased from 4% to 20%.

Figures 3, 7 and 8 show the dramatic difference in the microstructure and the corrosion resistance of the two alloys. The interdendritic attack is eliminated in CW2M. The corrosion rate in boiling 10% HCl drops to 4.4 mm/y (175 mpy) from 17 mm/y (660 mpy) for CW12MW. This is in the same range as wrought N10276 at 5.6 mm/y (220 mpy). The G28 corrosion rates are 4.8 mm/y (190 mpy) for CW2M, 5.8 mm/y (230 mpy) for N10276 and 22.9 mm/y (900 mpy) for CW12MW. The difference is also clearly seen between CW2M and CW12MW in 3% sulfuric acid with 0.08 mm/y (3 mpy) vs. 0.86 mm/y (34 mpy). The average CW2M corrosion rates in the three environments were all lower than for wrought N10276. Forty-five heats of CW2M have been shipped without any reoccurrence of the interdendritic corrosion problem in the field. The 1 to 25 kg castings have been in service for up to 18 months without any failures.

b) CN7M

The development of the Argon-Oxygen-Decarburization (AOD) process has produced great improvements in the corrosion resistance of cast alloy CN7M. In 3% sulfuric acid, conventionally produced CN7M was failing in the field in a few weeks or months. AOD refined CN7M has performed without a single failure in the same environment for 2 years. After shipping 107 CN7M AOD refined heats over 18 months, no failures have been recorded of the 1 to 25 kg castings.

AOD is a ladle refining process. The metal is first melted in an electric arc or induction furnace as is the normal practice. When the melt reaches the proper temperature, it is transferred to the refractory lined AOD vessel. Submerged tuyeres permit the injection of argon, oxygen, nitrogen and combinations thereof. The process gives simultaneous decarburization (C <0.015%), deoxidation, desulfurization (S <0.01%), degassing and reduction of metallic oxides from the slag to produce a cleaner, more uniform product. Improvements in impact toughness, weldability, cleanliness, compositional control and nitrogen control are well documented.

The improvements in AOD CN7M are not detectable by metallography or other standard techniques. As past failures were intergranular, grain boundary impurities were suspected as the cause. With AOD processing and the resulting improvements in purity, the grain boundary impurity levels are reduced. Excellent performance in field applications indicate that AOD processing is an effective solution.

CN7M laboratory corrosion tests do not show the improvements clearly seen in the field. The intergranular attack seen in the field has never been duplicated in the laboratory. Due to the scatter in the laboratory corrosion rates, the average rates in Table 4 do not show the superiority of AOD refined CN7M compared to conventionally processed CN7M.

In 10% HCl the alloy 20 corrosion rates in the cast and the wrought conditions are all about 35 mm/y (1350 mpy). In the G28 test, AOD refined CN7M records slightly better results than N08020 of 0.20 mm/y (8 mpy) vs. 0.28 mm/y (11 mpy). In 3% sulfuric acid, both forms of CN7M exhibit corrosion rates of about 0.7 mm/y (27.5 mpy) compared to 0.17 mm/y (6.5 mpy) for N08020. In nitric acid, AOD processed CN7M is 0.23 mm/y (9 mpy), conventional 0.38 mm/y (15 mpy) and N08020 0.20 mm/y (8 mpy).

c) M35-1

Alloy M35-1 is the least weldable of the three high nickel casting alloys. Imposing the weld bend test, requires each foundry to work diligently to develop its process. Close control of foundry practices is required and is more important than compositional modifications. AOD refined or virgin materials and dedicated pattern equipment are required. Other process improvements are considered proprietary by each foundry and cannot be discussed here. Over 100 weldable heats have been produced in casting sizes of 1 to 50 kg.

3.4 Specifications

Ordering castings to the applicable ASTM and ASME standards and/or wrought tradenames, does not guarantee the quality and purity of material necessary to obtain these improvements. Detailed specifications which address the following items have been developed.

a) Raw material

The CN7M specification requires AOD refined material or AOD refined ingots for remelting. The CW2M and M35-1 specifications require virgin melt stock, AOD refined material or AOD refined ingots. No revert, risers or scrap material is permitted. Risers, gates, etc. can only be used if again AOD refined. This does add to the cost but should be done until experience proves otherwise.

b) Composition

Several elements are restricted to lower maximums than those listed Table 2 from ASTM and ASME: CN7M; 0.03% carbon, 1.0% silicon, 0.03% phosphorus and 0.015% sulfur: CW2M; 0.015% sulfur and M35-1; 0.18% carbon.

c) Pouring Temperature

Castings are poured at the lowest possible temperature to produce the smallest grain size possible. So many factors influence the minimum pouring temperature at each foundry, that a temperature cannot be specified. Small grain size increases relative grain boundary area. This distributes low melting impurities and carbides over a larger area, reducing their detrimental effects and improving resistance to hot tears and intergranular corrosion.

d) Weld Repairs

Weld repairs are tightly controlled. Minimum heat input, stringer beads and interpass temperatures are specified. All defects must be excavated by grinding. Air-arc gouging is prohibited. Liquid penetrant (LP) inspection is required before welding to verify complete defect removal. The LP is repeated after welding. Older alloy 20 and Hastelloy C grades require solution annealing after welding for good corrosion resistance. These grades do not require post weld heat treatment as shown by preliminary corrosion tests; AWS grade 320LR for CN7M, NiCrMo-7 or NiCrMo-10 for CW2M and NiCu-7 for M35-1.

e) Solution Annealing Temperature

The solution annealing temperature is specified for each alloy as ASTM and ASME specifications generally neglect this important item. CN7M is solution annealed at 1120°C (2050°F) and CW2M at 1230°C (2250°F). M35-1 is used as-cast.

f) Dedicated Pattern Equipment

Dedicated pattern equipment for the high nickel casting alloys is recommended. This is coupled with radiographic qualification for each new pattern-alloy combination. This permits the earliest possible detection of any pattern or rigging problems. When production begins, spot checks are used to detect foundry process changes which may occur over a period of time.

g) Radiographic Examination

Although 100% radiographic examination of high nickel alloy castings is commonly used, the difficulty in radiographing these alloys compared to steel and SST, makes the expense difficult to justify. The large grain sizes and other characteristics of the nickel alloys makes radiography relatively ineffective. Radiographic examination at the foundry puts more burden of repairing castings on the foundry; however, with the new alloys and weldable material, defects found during machining are easily repaired.

h) Weld Bend Test

Finally the same weld bend test required for foundry qualification is required for each production heat. This insures that the same quality, controls and standards shown at the time of qualification are maintained. Even with the tighter compositional limits and other requirements of our specifications, weldability and casting integrity may vary. The weldability test provides indirect control over the parameters which have not been defined. It has proven most effective.

4. Conclusions

In conclusion, it is not necessary to tolerate the historic problems of the high nickel alloy cast components. Improvements in alloy chemistry and foundry melting techniques can solve many of the problems and, when coupled with stringent specifications and foundry qualification, produce high quality castings with the same excellent, consistent corrosion resistance as the "equivalent" wrought alloys. Merely imposing strict LP and radiographic requirements will not eliminate the chronic quality problems.

5. Acknowledgments

The assistance of R.P. Western was invaluable in developing the weld bend test as a foundry qualification method. Western and D.E. Campbell performed unending corrosion tests and accumulated the large volume of data. I would also like to thank Fisher Controls for the time and support need to perform this work and write this paper.

6. References

1. "Corrosion Resistance of Hastelloy Alloys", Cabot Corporation, Kokomo, IN, 1978, pp. 16,17,48,49,80.

2. "Jessop JS Alloy 276 Data Sheet", Jessop Steel Corporation, Washington, PA, 1986, p.1.

3. "Resistance of Nickel and High Nickel Alloys to Corrosion by Hydrochloric Acid, Hydrogen Chloride and Chlorine", International Nickel Company, Inc., New York, NY, 1975, pp.29-31.

4. Passinen, K. and Ahlers, P.E., "Corrosion of Acid-Proof Steel in ClO_2 Solutions", Paperi ja Puu, 48, 1966, pp. 549-560. English Translation by The Finnish Pulp and Paper Research Institute.

5. Hodge, F.G., "Hastelloy Alloy C-4C, An Improved Nickel Base Casting Alloy for the CPI", presented London, England, March, 1981.

6. Hodge, F.G., "Cast Alloy Resists Corrosion by Hot Acids and Oxidizing Agents", Industrial Research & Development, July, 1983.

7. Fontana, M.G. and Greene, N.D., "Corrosion Engineering", New York, NY, McGraw-Hill, 1967, pp.228-230.

8. "Ferrous Castings; Ferroalloys; Shipbuilding", Volume 01.02, American Society for Testing and Materials, Philadelphia, PA, 1984, pp.336-345.

9. "Metal Corrosion, Erosion and Wear", Volume 03.02, American Society for Testing and Materials, Philadelphia, PA, 1984, pp.160-165.

10. "Metal Corrosion, Erosion and Wear", Volume 03.02, American Society for Testing and Materials, Philadelphia, PA, 1984, pp.1-27.

11. "Metal Corrosion, Erosion and Wear", Volume 03.02, American Society for Testing and Materials, Philadelphia, PA, 1984, pp.207-214.

Table 1: This is a listing of the high nickel casting alloys discussed in this paper with ASTM cast designations, tradenames of the wrought "equivalents" and generic designations.

ASTM Designations	Wrought Tradenames	Generic Designations	UNS Number for Wrought Alloy
CN7M	Carpenter 20Cb3	Alloy 20	N08020
CW12MW	Obsolete Hastelloy C	Alloy C	N10002
---	Hastelloy C276	Alloy C276	N10276
CW2M	New Hastelloy C, C-4C	Alloy C	--
M35-1	Monel 400	Alloy 400	N04400

Table 2: Compositions of the high nickel casting alloys and wrought "equivalents".

Elements	CN7M	N08020	CW12MW	CW2M	N10276	M35-1	N04400
	Grades and Composition, %						
C, max.	0.07	0.07	0.12	0.02	0.02	0.35	0.3
Mn, max.	1.5	2.0	1.0	1.0	1.0	1.5	2.0
Si, max.	1.5	2.0	1.0	0.8	0.08	1.25	0.5
P, max.	0.04	0.045	0.03	0.03	0.04	0.03	---
S, max.	0.04	0.035	0.04	0.03	0.03	0.03	0.024
Cu	3-4	3-4	---	---	---	26-33	28-34
Mo	2-3	2-3	16-18	15-17.5	15-17	---	---
Fe	bal	bal	4.5-7.5	2 max	4-7	3.5 max	2.5
Ni	27.5-30.5	32-38	bal	bal	bal	bal	63 min
Cr	19-22	19-21	15.5-17.5	15-17.5	14.5-16.5	---	---
Cb	---	8 X %C	---	---	---	---	---
W	---	---	3.75-5.25	1 max	3-4.5	---	---
V	---	---	.2-.4	---	0.35 max	---	---

Table 3; Average boiling acid corrosion rates in mm per year (mils per year) for cast CW12MW, UNS N10276 (Hastelloy C276) and cast CW2M.

Grade	Condition	10% HCl	G28	3% H_2SO_4
CW12MW	as-received	17.0(660)*	22.9(900)*	.86(34)*
	1177°C (2150°F)-1 hr.	9.4(370)	--	--
	1177°C (2150°F)-2 hr.	--	32.5(1280)	--
	1177°C (2150°F)-6 hr.	21.6(850)	34.5(1360)	--
	1232°C (2250°F)-1 hr.	9.9(390)	--	--
	1232°C (2250°F)-4 hr.	--	21.3(840)	--
	1232°C (2250°F)-18 hr.	9.1(360)	--	--
	1260°C (2300°F)-4 hr.	--	17.0(670)	--
N10276	as-received	5.6(220)	5.8(230)	.15(6)
CW2M	as-received	4.4(175)*	4.8(190)*	.08(3)*

* These corrosion rates are averages for several samples and duplicate tests.

Table 4; Average boiling acid corrosion rates in mm per year (mils per year) for cast CN7M and UNS N08020 (Alloy 20Cb3).

Grade	Condition	10% HCl	G28	3% H_2SO_4	A262-C
CN7M	Conventional	36(1400)*	.61(24)*	.69(27)*	.38(15)*
N08020	Wrought	33(1300)	.28(11)	.17(6.5)	.20(8)
CN7M	AOD Refined	33(1300)*	.20(8)*	.71(28)*	.23(9)*

* These corrosion rates are averages for several samples and duplicate tests.

Appendix 1:

Laboratory corrosion tests were run in boiling ferric sulfate-sulfuric acid per ASTM G28 (9) and boiling 65 weight per cent nitric acid per ASTM A262 practice C (10). All requirements of the standards were followed regarding sample preparation, test time, solution concentration, solution changes, etc. The 24 hour test time for the N10276 alloy in ASTM G28 was used for the CW12MW and CW2M tests. A 120 hour test period was used for CN7M and N08020. Five-48 hour test periods were used for all A262 Practice C tests.

Boiling 10 weight per cent hydrochloric acid and boiling 3 weight per cent sulfuric acid tests were run following the same procedures and test methods. A modified Allihn condenser without a drip tip was used as described in ASTM G36 (11). One-48 hour test period was used for the HCl test and 2-168 hour periods for the sulfuric acid test.

The corrosion test samples were taken from the cast test plates used for the weldability studies. These were generally cast in air set sand molds, appropriately heat treated in air, water quenched and grit blasted to remove the oxide scale. Corrosion test samples were cut so no remaining surface was within 3 mm (1/8 inch) of any cast surface. Typical sample sizes were 6 x 25 x 25 mm (1/4 x 1 x 1 inch).

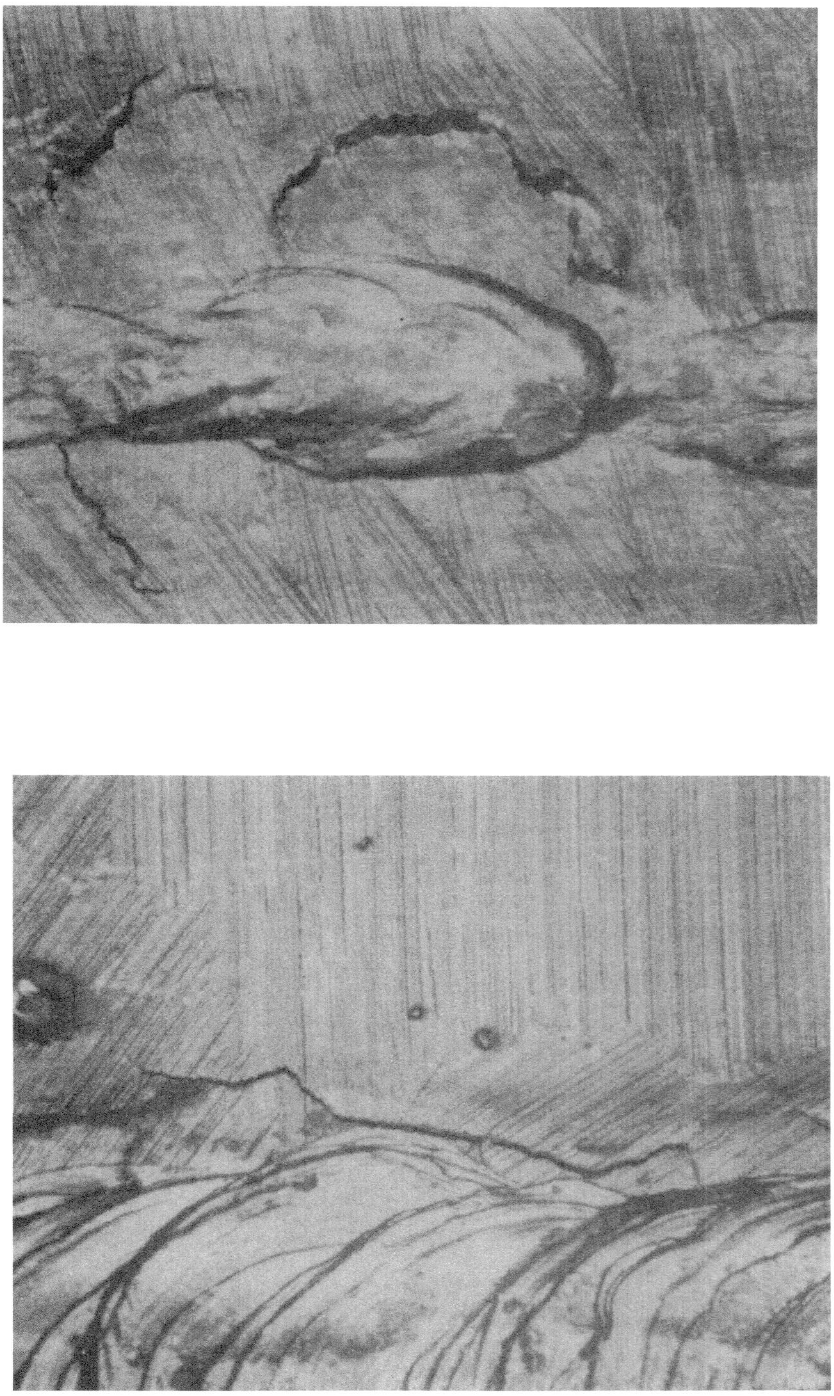

Figure 1: This is a typical example of base metal cracking in the weld heat affected zone of a cast high nickel alloy.

Figure 2: Unexpectedly severe corrosion has been experienced on CW12MW when the wrought N10276 components were not attacked.

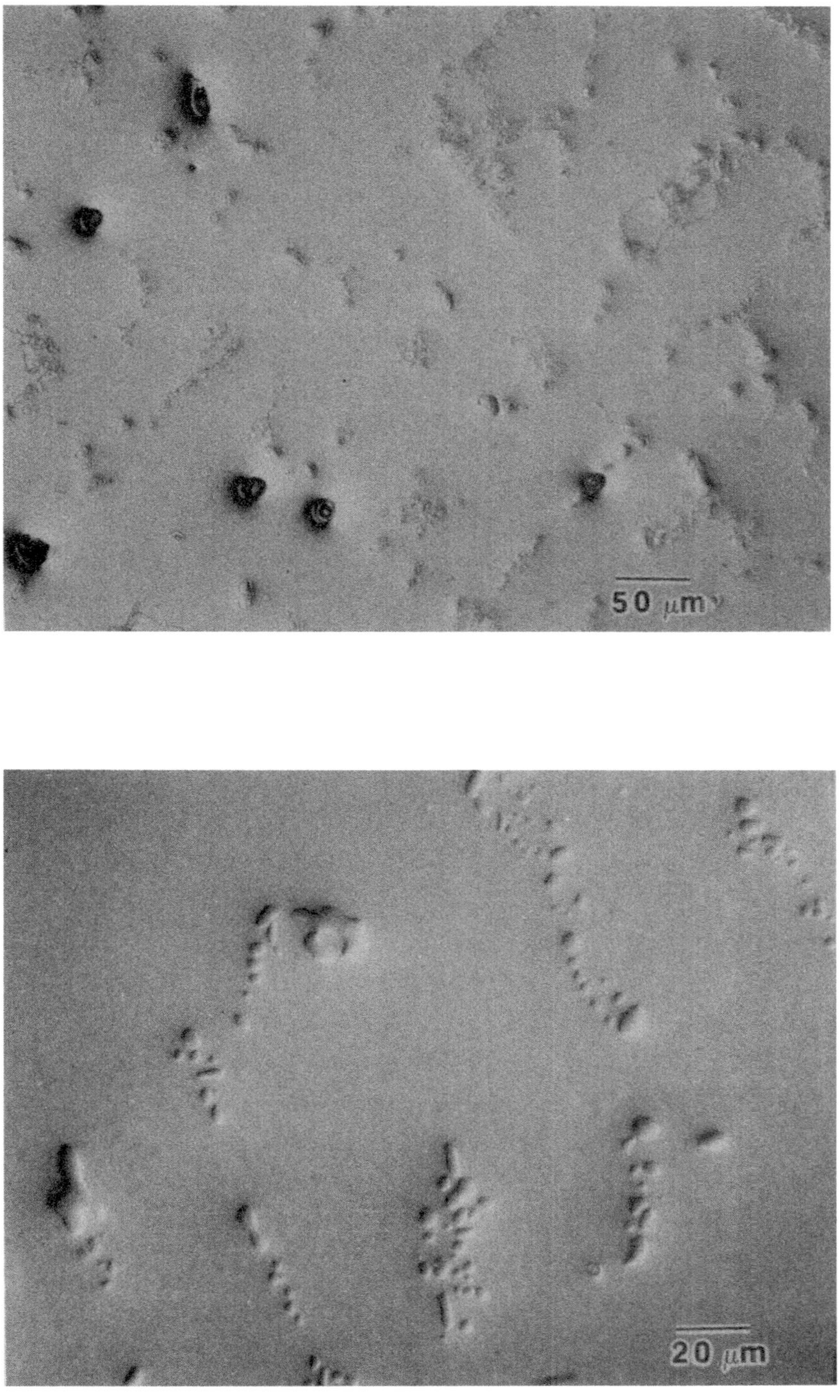

Figure 3: Typical CW12MW microstructure showing the large amounts of the interdendritic second phase.

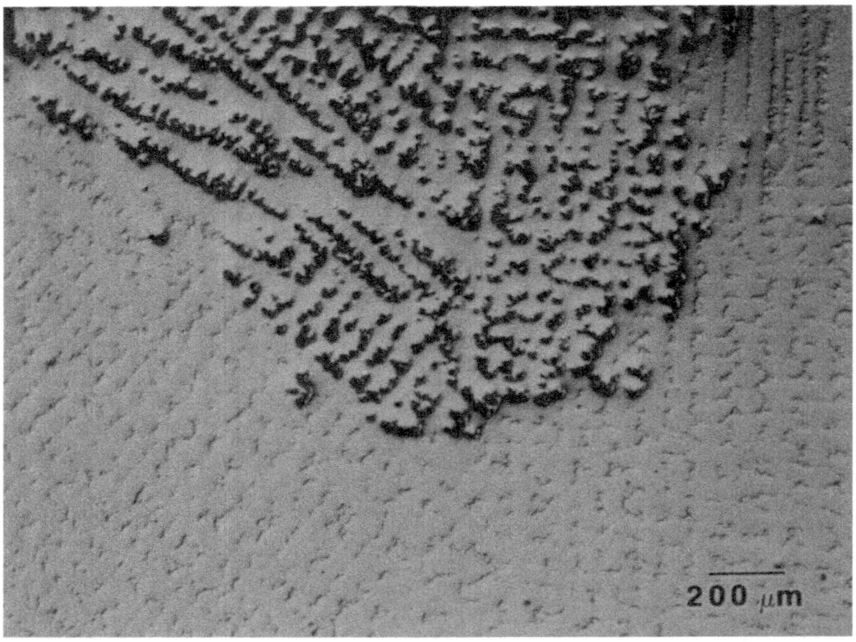

Figure 4: CW12MW showing the preferential corrosion of the second phase.

Figure 5: CN7M cages and bodies have suffered severe intergranular corrosion.

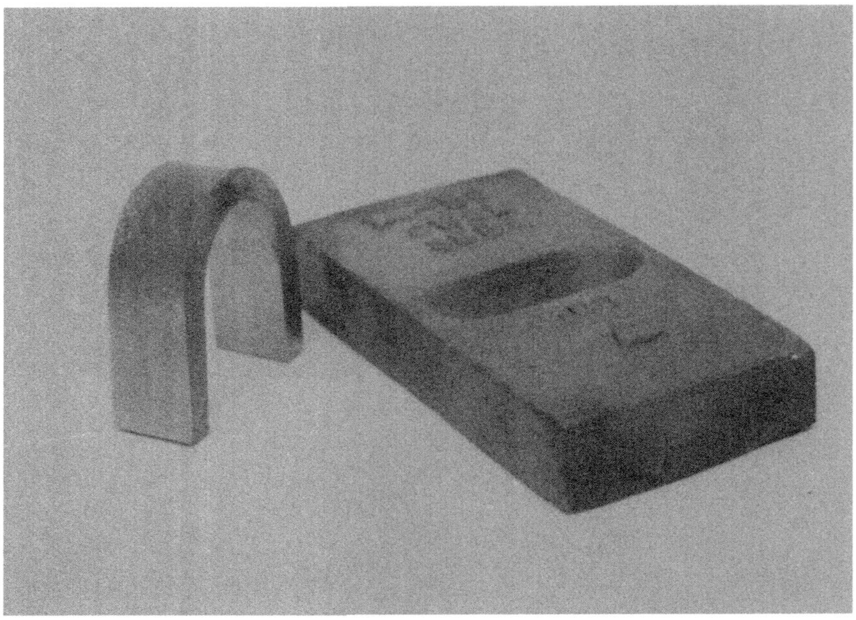

Figure 6: All heats of cast high nickel alloys are qualified by welding a test plate and bend testing.

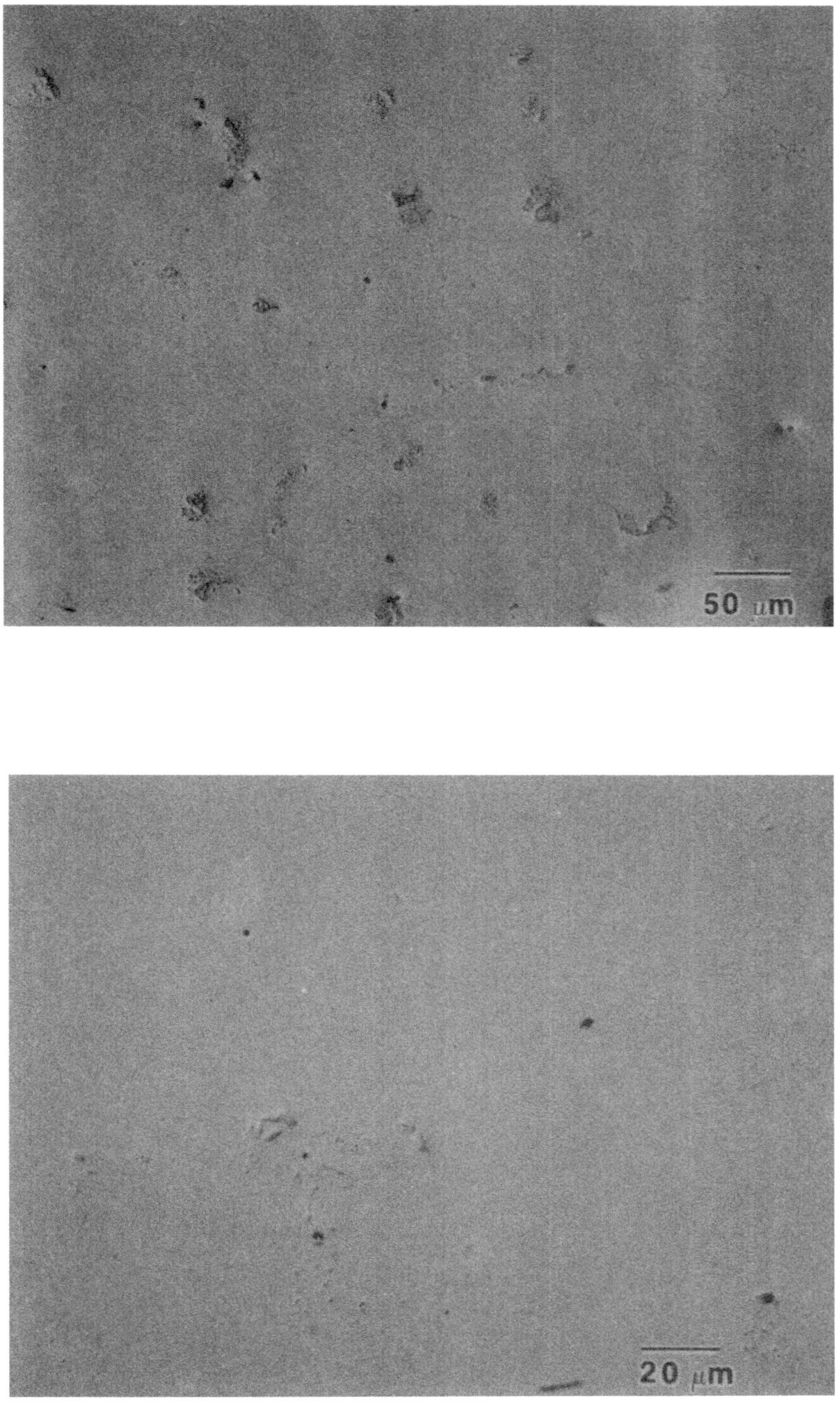

Figure 7: Typical CW2M microstructure with only a minimum amount of the second phase. The corrosion resistance is greatly improved.

Figure 8: Typical G28 corrosion test results on CW2M (left) and CW12MW (right). The lower photo shows the CW12MW at 50X.

2nd International Conference on

Developments in Valves and Actuators for Fluid Control

Manchester, England: 28-30 March 1988

PAPER E3

DEVELOPMENTS IN SOFT SEATED TRIMS FOR CONTROL VALVES

By: Walter.W.Mott
Technical Director
Copes-Vulcan Limited

BVMA Technical Committee Chairman

Summary

The long standing and high cost problems of leaking control valves in severe duty liquid applications continues to cause concern, both in terms of losses due to energy wastage and more importantly, damage and failure of valve trims due to wire drawing and cavitation. Failure of valves in this manner can result in costly down time.

Metal seated valves have been found to be unsuitable for these applications and early attempts at soft seated valve trims were only partially effective due to material limitations and the orientation of the soft seats in the valve trims, resulting in water washing of the soft seat during normal flow conditions. Developments have continued in this area of valve design and this paper discusses one of the latest designs of soft seated trim which maintains zero [bubble tight] leakage throughout the service life of the trim. This trim is particularly useful for feed pump recirculation duties and the paper also addresses this application as a follow on to the paper on feed pump recirculation presented at the last BHRA Conference.

1. Introduction

Leakage through control valves, particularly on liquid applications is a subject that has attracted a lot of attention as pressure drops across control valves have increased due to the higher ratings of modern power plants. Leakage through liquid control valves is costly in terms of lost energy and in certain applications will result in wire drawing and cavitation occurring, which will very rapidly destroy the valve seat.

A particular application where no leakage can be tolerated is that of feed pump recirculation where the valve is required to remain shut for most of its life, with an extremely high pressure drop across the valve seat. Any

Held in Manchester, England. Organised and sponsored by BHRA, The Fluid Engineering Centre and co-sponsored by the British Valve Manufacturers Association and the Institution of Mechanical Engineers Process Industries Division

damage to the valve seat in this application means that a leak path is created, and fluid passes through this leak path all the time the valve is in the closed position.

When any damage occurs to the seating surface of a valve trim a leak path is created through which the fluid is forced at high velocity due to the pressure differential which occurs across the valve seat. Even very small imperfections in the valve seating surfaces allow this action to occur, which results in a stream of high velocity fluid passing through a restriction, with resultant wear due to the abrasive action of the high velocity jet. This action is known as wire drawing and can quickly result in total destruction of the valves shut off capability.

Considerable losses in energy can be incurred on modern plants due to leakage through a single feed pump recirculating valve with some estimates putting the value as high as $100,000 per year and this figure will be considerably higher if one takes into account down time as a result of early valve failure caused by this leakage resulting in cavitation damage to the valve trim.

Various designs of soft seated trims have been developed and have proven to be successful to varying degrees and this paper studies some of the latest methods being employed to achieve tight shut-off.

2. Background

Two basis approaches have been employed in valve seating - metal to metal and soft seat design.

In the metal seat design, as shown in Figure 1, a differential angle is used between the plug and seat to achieve a line contact surface between the seats.

In this manner the actuator thrust is developed over a fairly narrow band, thus focusing the seating stress over a relatively small area. This concept is illustrated in Figure 2, from which it will be seen that the inherent surface microstructure is such that various peaks and valleys occur between the seating joint.

In order for the seat to provide zero leakage, the actuator thrust must be sufficient to yield the peaks in the seat surfaces, such that perfect contact is achieved for 360 degrees around the plug's contact area.

This kind of contact, although possible to obtain, remains fairly inconsistent and repeatability of shut-off is very questionable.

In the event that leakage does occur with this arrangement it is easily seen that cavitation or wire drawing across the valve seat will quickly follow. Once leakage commences it will increase quite rapidly, particularly on valves handling severe pressure differentials.

By using stellite, or hardened metal seats, resistance to wire drawing is increased, yet in services where shut off differentials exist in excess of 125 bar, there is no existing seat material that can survive the continuous leakage of a metal seat design.

A further problem with a metal to metal seat design is that nearly perfect plug to seat alignment is essential in order to provide zero leakage. This is almost impossible to achieve and maintain due to normal machining tolerances.

Clearly metal to metal seats are widely used on a large number of applications with a considerable amount of success. It is, however, in the area of high pressure drop liquid applications where most problems occur with metal to metal seats, because it is in this field of application that Class VI leakage to ANSI B16.104 is normally required, and it is impossible to achieve this shut-off class with metal to metal seats. [See Appendix 'A' for acceptable leakage rates, as called up by this standard.] It is this problem that has caused the valve industry to look closer at the use of resilient seating materials, and to develop different forms of soft seating valve trims which will achieve Class VI shut-off and also give a satisfactory life under arduous pressure drop conditions.

3. Early Designs of Soft Seats

With a soft seat design, a resilient seating material is used to mate with the seat on the plug, as illustrated in Figure 3. When the valve approaches the closed position the smooth surface of the plug seat is embedded into the resilient seat material, which deforms to embrace the plug surface, ensuring that perfect plug to seat conformance is obtained, with reasonable actuating thrusts on the valve stem. In theory only enough stress to prevent leakage need be applied to the soft seat, as it is essential that the resilient seat be able to support the actuator load for extended durations, particularly if the valve is going to operate in the closed position for long periods of time.

With this design repeatability is assured as the soft seat surface is not permanently deformed and perfect plug to seat conformance can be repeated. In addition, seat to plug alignment is not as critical as with a metal seated trim, since the resilient seat will correct for minor variations in alignment.

There are several drawbacks with this type of soft seat. Firstly, it is difficult to limit the actuator thrust with the valve on its seat without providing a mechanical stop, which would effectively remove the seating force from the resilient seat. There is a danger of permanent deformation of the elastomer, particularly if the valve is held in the closed position for long periods of time, with the full actuator thrust deforming the seat, and after extended periods of operation the natural resilience of the material may be affected.

Secondly, with this design the soft seat surface is the initial throttling point in the valve during the opening cycle, thus the resilient seating material will see the high pressure drop, and resultant high velocity which occurs when the valve begins to open, and in consequence undergoes the same wire drawing effect as the metal seated design. The result of this is that the high velocities present during the initial opening may quickly have an adverse effect upon the soft seating material.

Earlier soft seat designs of this type would only survive a few opening and closing cycles under the conditions prevailing in modern plants. These designs can still be found giving satisfactory service in valve sizes 50 mm or less, and with pressure drops not exceeding 175 bar, but for sizes and operating conditions in excess of these parameters problems can be expected. The larger seating areas are subject to wire drawing and a further problem is that the seat insert tends to loosen and disintegrate, probably due to the turbulence caused by high velocity fluids and larger mass flow rates.

4. Self Energising Soft Seat Designs

With improvements in materials and sealing technology for dynamic seals, attempts were made to use these items in valve trims, and indeed they are being used very successfully for pressure balancing and for soft seated designs under certain conditions.

A typical trim design using a self energising seal is shown in Figure 4, which illustrates a balanced single seat valve trim with a soft seat concept.

In this design, the metal seat in the cage has been replaced by a flexible seal incorporating a reinforcement spring. The orientation of the seal in the cage is such that when pressure is applied to the internal surface it will expand and make contact with the cage and the bottom of the plug. In this way a self energising seal is formed using the fluid pressure on the inlet side of the valve.

It will be seen from Figure 4 that a spring loaded 'U' cup seal is provided in the valve seating area in order to give Class VI shut off. The seal used in this design incorporates a spring device inserted in the 'U' shaped section of the seal, in order to force open the legs of the 'U' to come in to contact with the sealing surfaces of the plug and the cage. This provides the initial sealing force and also ensures that when the valve is open the lip of the seal is moved out of the flow path of the fluid.

Although the energising spring provides the initial contact load, the main sealing force is achieved by the inlet fluid pressure which enters the 'U' section of the seal and deforms this section to make positive contact between the seal and the contact surfaces, thus providing a positive shut off.

A mechanical stop is incorporated in the trim in the form of a landing in the cage bore to ensure that the seal is not over compressed and in the closed position the plug rests on this stop with the bottom of the plug slightly compressing the edge of the seal. This mechanical stop acts as a back-up in the event that the soft seat fails.

This type of trim has proved to be very successful in valve sizes up to 75 mm with pressure drops of up to 300 bar, although initially there were problems with the concept due to excessive clearances between the plug and cage, resulting in the seal being deformed. In addition, the high fluid turbulence occurring in the area of the seat just at the time the valve opens, can cause the lip of the seal to vibrate, resulting in fatigue of the seal material.

These problems were overcome on smaller valves by providing a metal lip in the seat ring, as shown in Figure 5, to protect the lip of the seal and this design is now used successfully within the sizes and pressure limitations referred to above.

5. Latest Developments

In the area of feed pump recirculating valves there has been a greater awareness of the cost penalty for leakage through recirculating valves, and with the advent of fully modulating leak off systems, which in themselves give considerable savings in energy costs, the subject of valve leakage has been highlighted.

Valve designers have become more aware of the problems and have realised that to solve these successfully it is necessary to move the area of throttling away from the soft seating area. Figure 6 illustrates a design that uses this concept.

In this design the valve plug assembly consists of an outer primary plug, an inner throttle plug, and a series of springs and shoulder bolts for maintaining the inner plug. When the valve is off its seat the inner plug is extended from the outer plug by virtue of the springs in the assembly and thus provides an initial metal contact with the metal portion of the seat ring, as the valve closes.

The seat ring itself is provided with a resilient seat insert, which is removed from the throttling area of the fluid as the valve initially closes.

In operation when the valve plug assembly is away from its seat, throttling occurs through the ports in the cage. As the valve moves towards the closed position the initial shut-off is provided by the inner plug making contact with the inner surface of the seat ring. This provides a metal to metal shut-off and reduces the flow to almost zero.

The actuator continues to move in a downward direction, compressing the springs between the inner and outer plug, thus allowing the seating surface of the outer plug to come into contact with the resilient insert in the valve seat ring, thus giving tight seating, enabling the Class VI shut-off requirement of ANSI B16.104 to be achieved.

In the opening mode the initial movement of the actuator will raise the outer plug separating the seat on this plug from the resilient seat insert, allowing the metal seat on the inner plug to provide the shut-off by virtue of its contact with the metal portion of the seat ring. Further movement of the actuator will cause the inner plug to move away from the seat ring, thus allowing flow to commence. In this manner a two-stage seat approach is employed and the soft resilient seat is protected from the effect of wire drawing or turbulence due to the reaction of the flowing medium.

In designing this type of soft seat trim, use is made of the technological developments occurring in thermoplastic polymers which have been developed to such an extent that they have mechanical properties approaching those of aluminium and yet still give the kind of resilience which is vital in this application, typical mechanical properties of one such material are given in Appenxix 'B'.

6. Entrained Solids

It is prudent at this time to mention one other problem that faces valve designers when considering liquid flow applications, and that is the subject of entrained solids in fluid flow paths. Although this is a problem on all valve designs, it is particularly relevant to valves incorporating soft seats.

If solids are present in the fluid there is always the risk that some of these will become trapped between the plug and seat preventing positive seating, causing a leak path to occur, with resultant erosion due to the high velocity fluid action.

With the two-stage soft seat trim referred to above this is less critical as the trapping is likely to occur between the metal to metal seat of the inner plug, and some leakage through this is acceptable, due to the backup of the soft seat on the outer plug.

Most soft seat designs are able to deal with solids of 1.5 mm diameter if the concentration is low. Metal seats on the other hand will encounter problems if the entrained solid is harder than the plug or seat ring, thus impairing full valve closure. Normally one relies upon the suction strainer installed with the pump. The main drawback with this approach is that pipe scale and welding slag can occur in the piping downstream of this strainer, and this will eventually find its way into the valve. This occurs most usually during the start-up period when most damage is seen on valve seats.

It is possible to guard against this by the use of an inlet filter element which can be attached to the valve seat, and thus become an integral part of the valve trim. This concept is shown in Figure 7, from which it will be seen that a strainer element is located below the valve seat ring, and is held in position by a series of socket headed shoulder screws. This element will trap particles of a size that would cause damage to the valve seating by eliminating the larger debris from the throat of the valve. Smaller debris passing through the filter will also pass through the valve trim, but as mentioned above, soft seats are able to deal with particles up to 1.5 mm diameter.

The design of this filter element is such that it can easily be removed for cleaning, but more important on critical applications such as feed pump recirculating flow, where the minimum flow must be maintained at all times, it is possible to design the filter element with a built in relief feature. The design of the filter is such that in the event of the amount of trapped solids beginning to build up to such an extent that the pressure drop across the filter is likely to affect the flow through the valve the filter will collapse. This will therefore preserve the valve's basic function, which is to always ensure a minimum flow through the pump.

7. Conclusion

As operating parameters increase in severity, the challenge to the valve designer becomes greater and it is encouraging to note that greater use is being made of advancing material technology, and a greater awareness of the phenomena causing valve problems is allowing manufacturers to design valve trims capable of meeting today's challenges.

Figure 8 shows one of the latest designs of valve used on feed pump recirculating duty. This valve embodies an anticavitation trim that uses the pressure profile concept to ensure that at no stage during the fluid's path through the valve trim is a pressure drop encountered that will give rise to cavitation. This trim also incorporates the latest two stage seat concept of a metal seat backed up by a soft seat to give Class VI shut off, and also incorporates a filter element to minimise the risk of damage due to the ingress of foreign matter.

This particular valve is now being used successfully on a large number of applications world wide for feed pump recirculation control, and is allowing the modulating concept to be adopted with considerable savings in energy being achieved.

8. References

ANSI B16.104 — American National Standard for Leak Control Valve Seat Leakage

Mathew L. Freeman — ISA Handbook of Control Valves Second Edition

Dale S.Tripp — Boiler Feed Pump Recirculation Valve & System Requirements

APPENDIX 'A'

Acceptable maximum leakage rates through valve trims per ANSI B.16.104

[Test medium air or nitrogen gas at 10-52 deg C pressure of test medium shall be rated differential or 3.5 bar, whichever is least].

Nominal Port Diameter

Millimeters	ml per Minute	Bubbles per Minute
25	0.15	1
38	0.30	2
51	0.45	3
64	0.60	4
76	0.90	6
102	1.70	11
152	4.00	27
203	6.75	45

<u>APPENDIX 'B'</u>

Typical physical properties of one resilient soft seat material

Tensile Strength [at yield] N/MM2	105.2
Tensile Strength [at yield] %	6
Elongation [at break] %	35
Flexural Modulus N/MM2	4,399
Shear Strength N/MM2	82.7
IZOD Impact [unnotched]	No Break
Heat Distortion, deg C @ 18.2 Bar	182
Coefficient of Thermal Expansion mm/mm deg C x 10 -5	150 Deg C-2.61 150 Deg C-5.99
Water Absorption % 24 hours	.14
Specific Gravity	1.30
Compressive Strengt N/MM2 [ULT]	118.0
Co-Efficient of Friction	.34

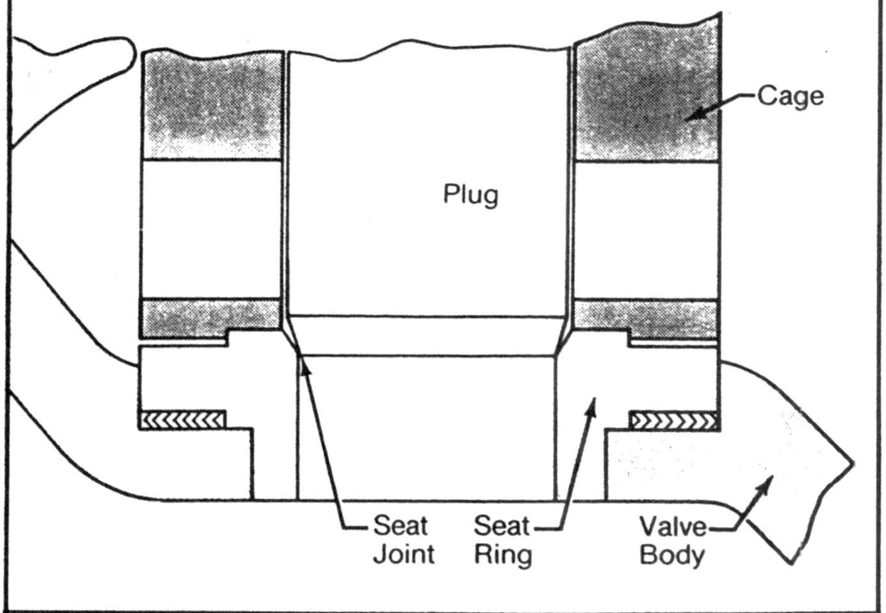

Figure 1: Metal Seat Concept

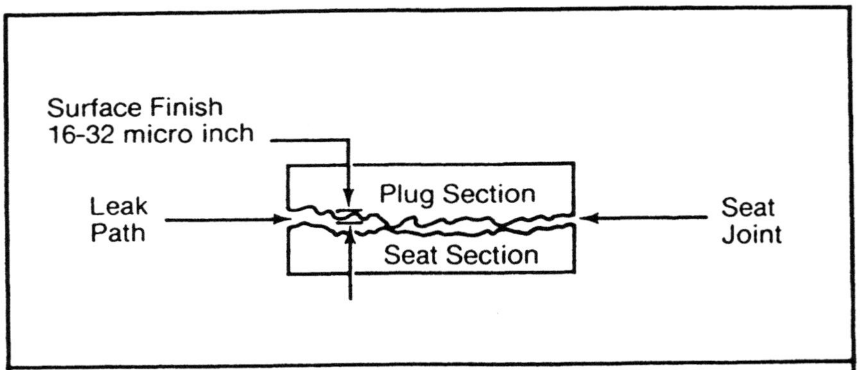

Figure 2 Metal Seat Micro-Structure

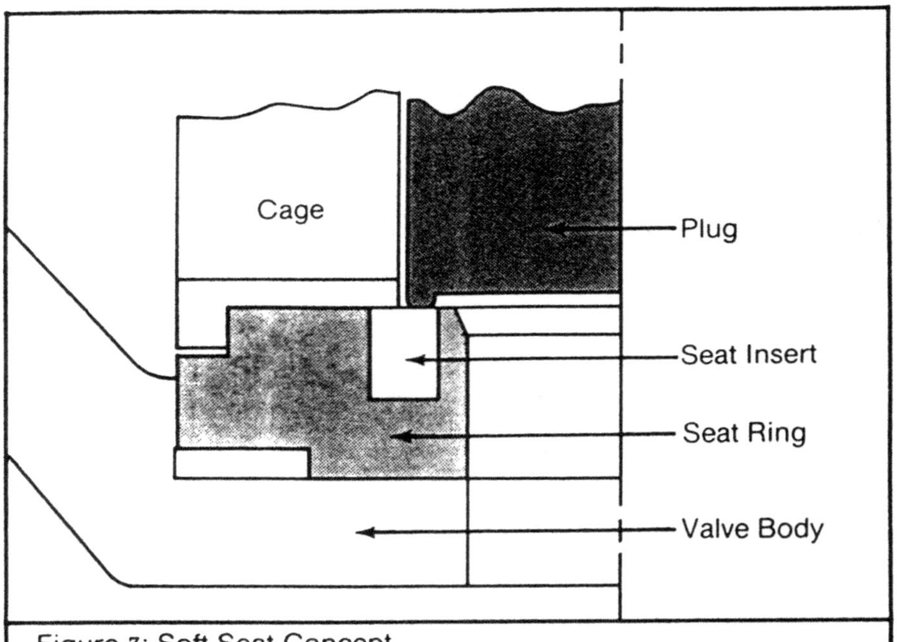

Figure 3: Soft Seat Concept

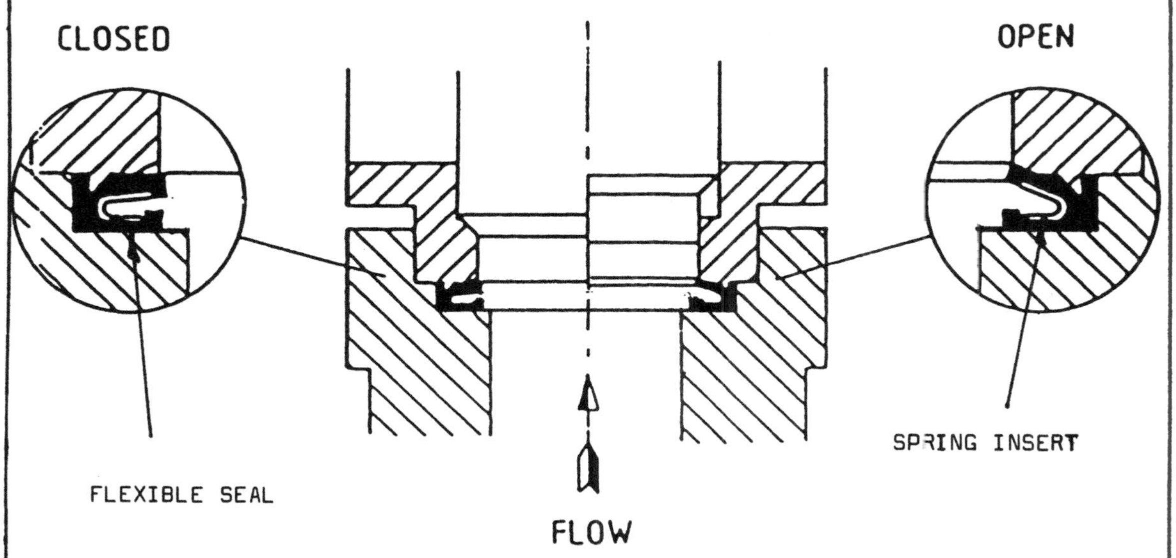

FIGURE 4
SOFT SEAT SEAL DESIGN
USING SELF ENERGISING SEAL

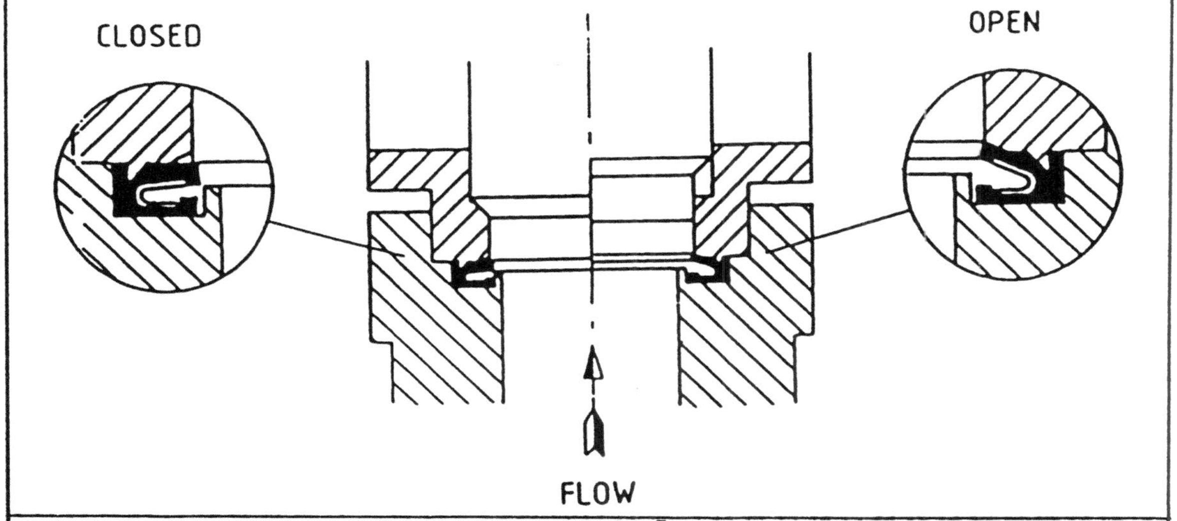

FIGURE 5
SOFT SEAT SEAL DESIGN
USING SELF ENERGISING SEAL WITH LIP

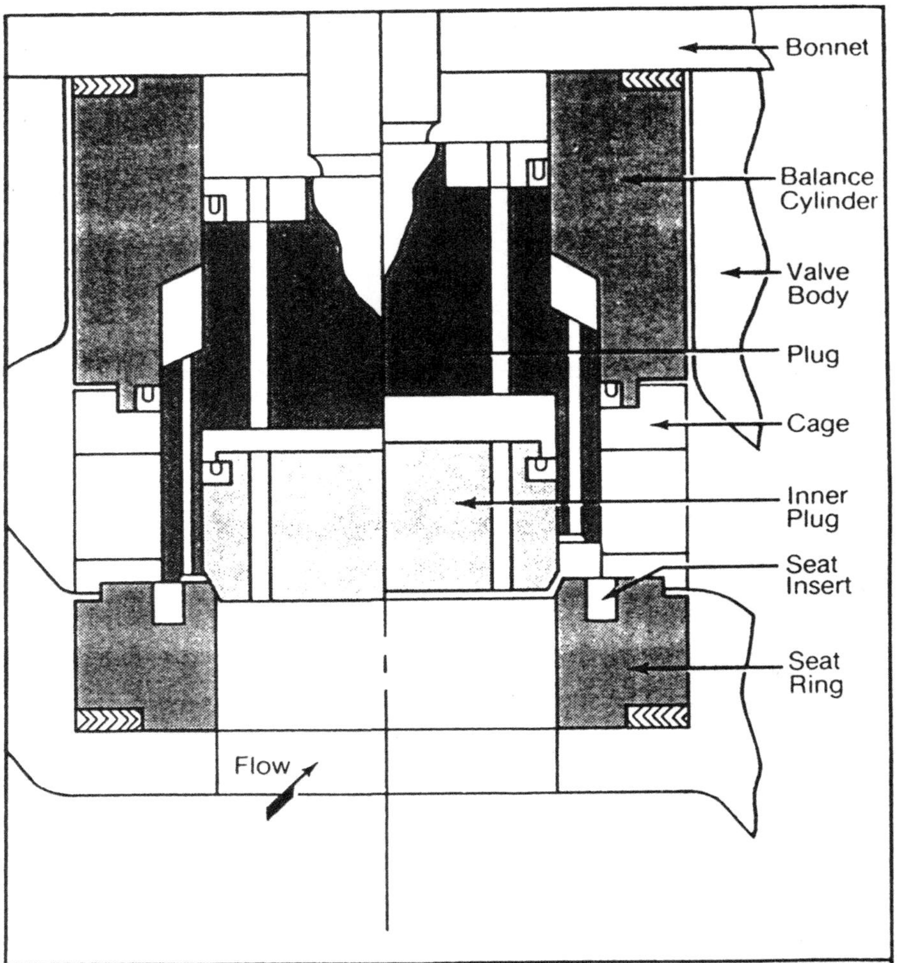

Figure 6 : Two-Stage Soft Seat Trim

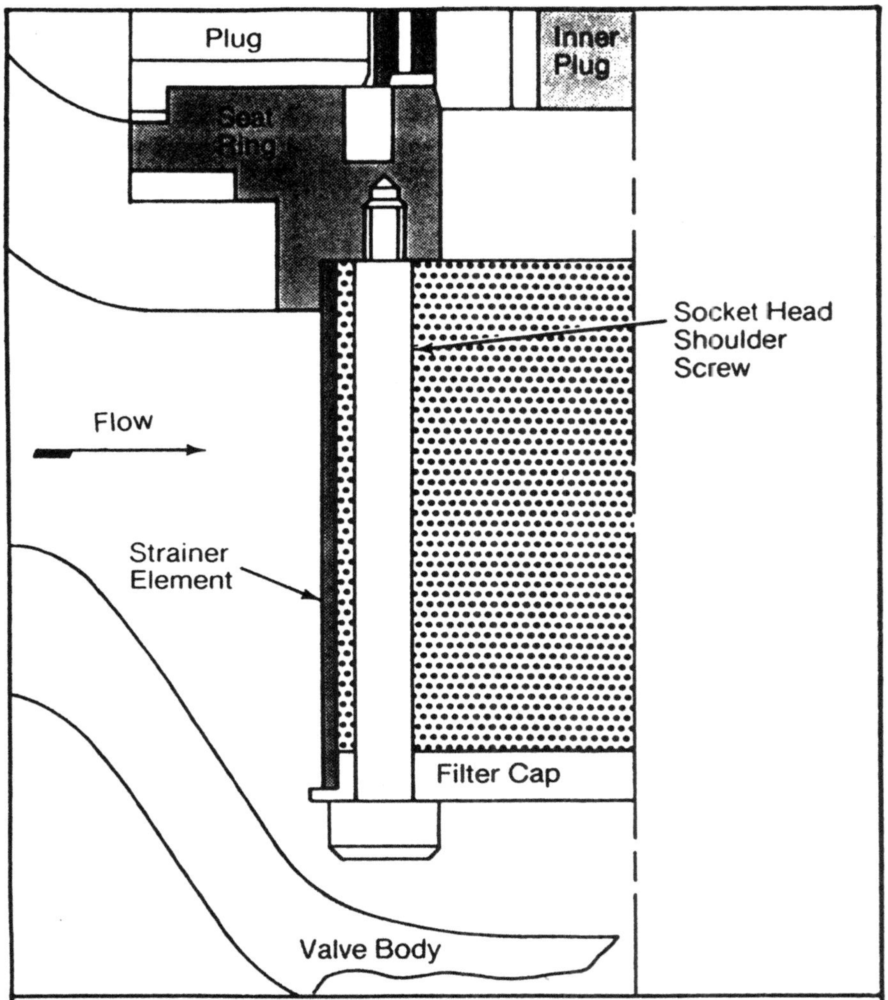

Figure 7 : Inlet Strainer

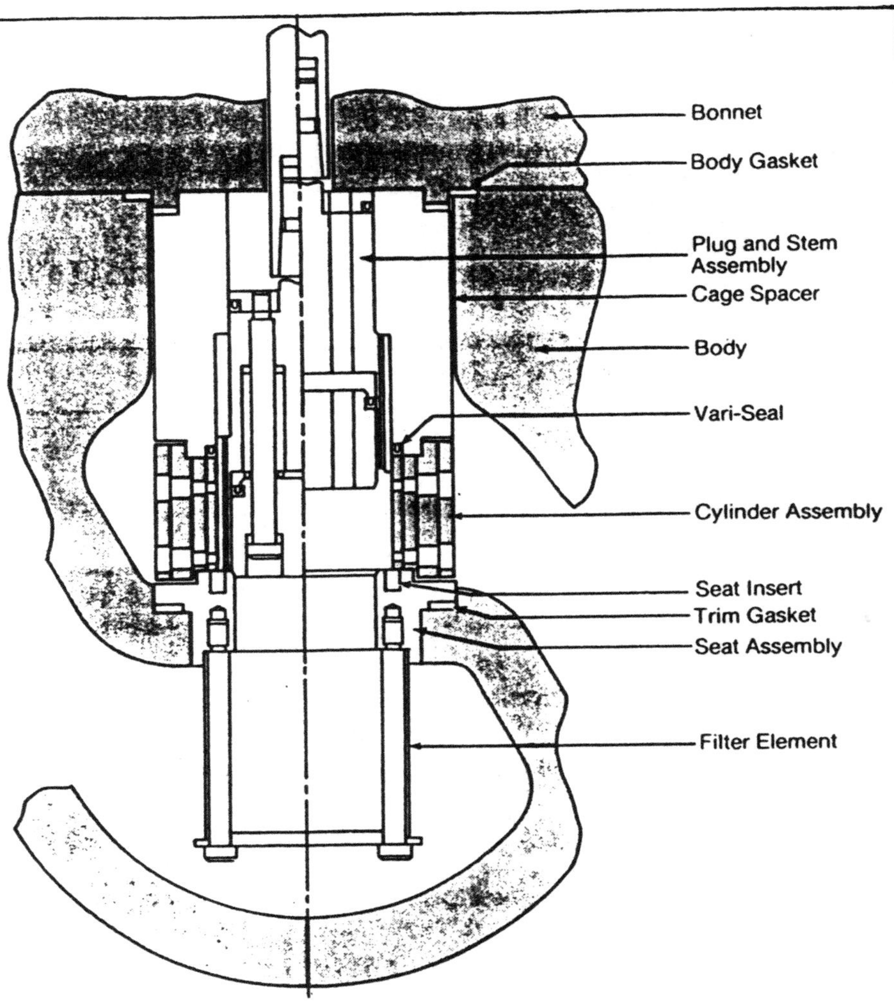

Figure 8. Recirculation Valve

2nd International Conference on

Developments in Valves and Actuators for Fluid Control

Manchester, England: 28-30 March 1988

PAPER E4

ENGINEERING THERMOPLASTICS IN ACTUATOR CONSTRUCTION

P A Williams, B.Sc, Prod.Eng., T. Eng, A.M.P.R.I,
Project Manager - Diaphragm Valve Products
Saunders Valve Company Limited,
Grange Road,
Cwmbran
Gwent
NP443XX

SYNOPSIS

The need to be competitive in industry and to remain so, has led mechanical engineers to be increasingly aware of alternative materials. The relatively new advancements in polymer technology has opened many potential doors to the innovative engineer/ designer. Therefore, it is essential that the material supplier works closely with the engineering team, to ensure that the correct polymer has been chosen for the application, and a good working relationship must be built between the two. This will eventually lead to a better understanding between the engineer and material technologist and can only enhance the status of both.

The growth of engineering plastics has primarily been achieved in the replacement of metals which, coupled with the fact plastics are easier to process and assemble, major factors generally resulting in lower part costs.

Held in Manchester, England. Organised and sponsored by BHRA, The Fluid Engineering Centre and co-sponsored by the British Valve Manufacturers Association and the Institution of Mechanical Engineers Process Industries Division

Cost effectiveness, functionability and aesthetics play a major role in the search for alternative materials and the engineer/designer must be aware of production efficiencies before design freedom can be given.

The following paper outlines the choice of an Engineering Plastic capable of elevated temperatures, working primarily as a replacement to metal components.

INTRODUCTION

Saunders Valve Company Limited have been in existence for over 50 years manufacturing the weir type diaphragm valve (Fig.1),and is even now still generally considered as an engineering biased company converting metals into finished product. Although primarily this is true, a large proportion of its generated income is associated with the conversion of Rubber and Plastics into finished components.

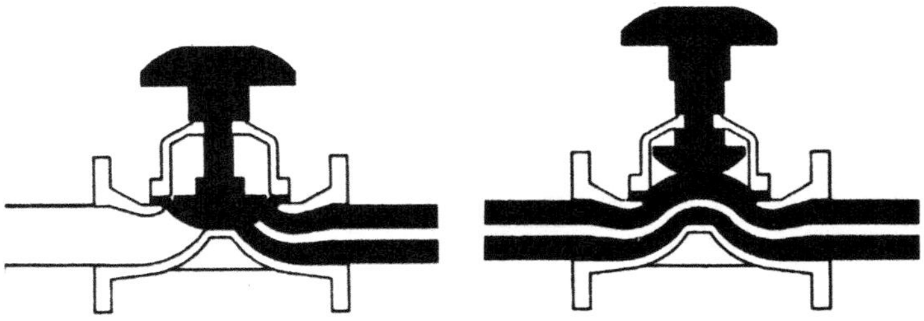

FIG.1

This factor has generated considerable expertise in the use of the more exotic materials, e.g PFA, Perfluoroalcoxy, PVDF Polyvinylidene fluoride (Solef), ETFE Ethylenetetrafluoroethylene (Tefzel) to name but a few. These materials are used to line the valve bodies (Fig 2) to enhance the chemical resistance of the unit.

It is therefore understandable that engineering plastics are often considered as metal replacements. In fact, many standard products now use plastic components as direct replacements to previously metal parts.

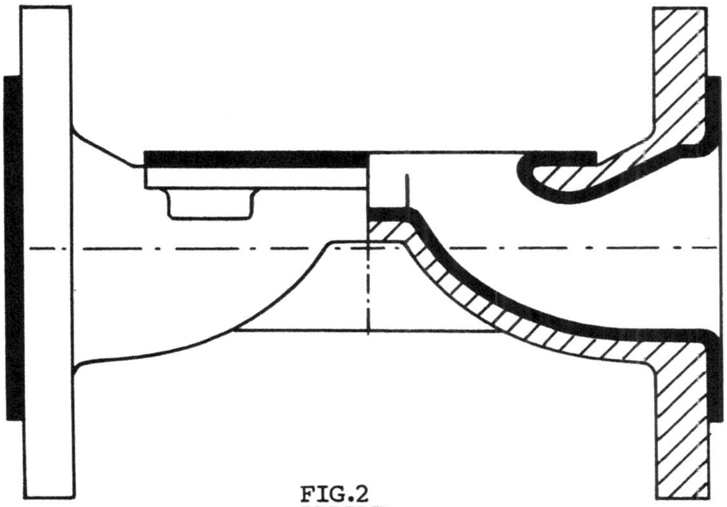

FIG.2

PRODUCT IDENTIFICATION

As an alternative to manual operation, Saunders weir type valves may be fitted with power operated bonnet assemblies commonly known as actuators (Fig 3).This enables the valve to be remotely controlled from a central station.

Two power sources may generally be used :

1] Fluid - Compressed air, water or hydraulic medium.

2] Electric - Motors with gear or belt drives.

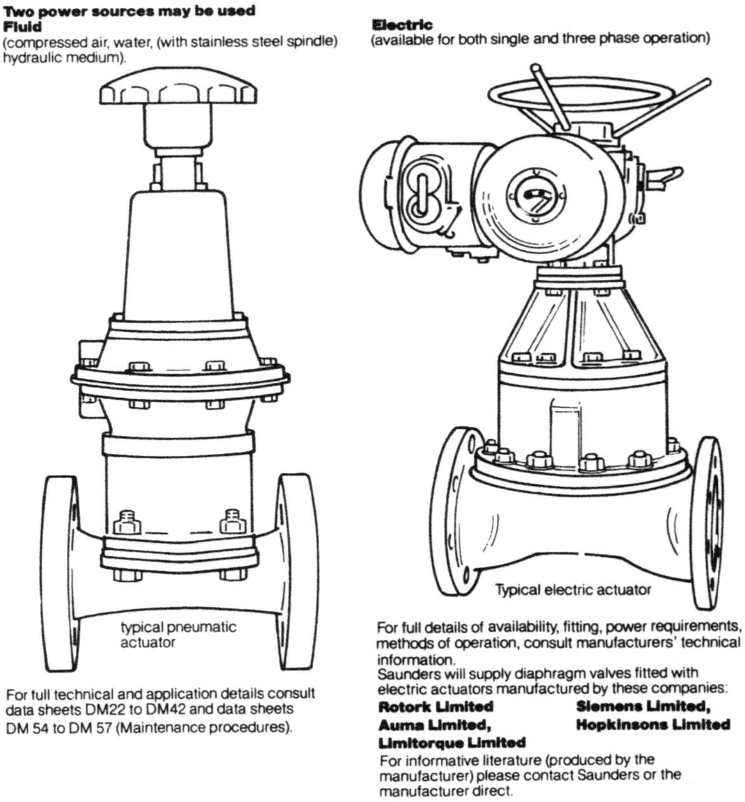

FIG. 3

Analysis of the actuator market can be likened to a triangle (Fig.4). Where the market is divided into three sectors. Fine control at the pinnicle, automatic process control in the middle with simple ON-OFF duties at the bottom.

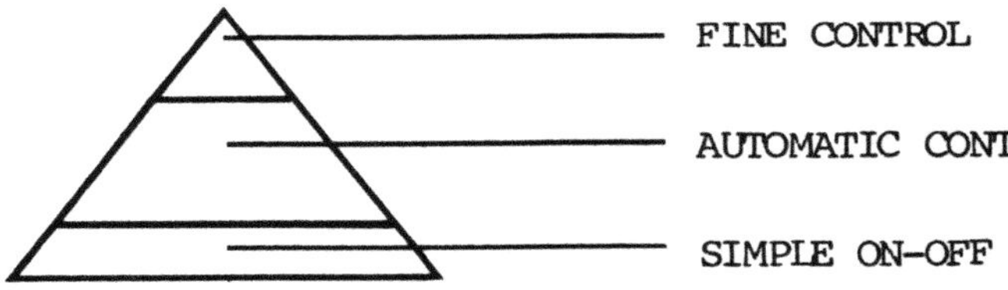

FIG.4

It was impossible to cover the entire range with the standard ES actuator, which is of metal construction and diaphragm operated. (Fig.5)

Operating Pressure (maximum)

All ES actuators are designed for 6 bar maximum. For pressures in excess refer to Saunders.

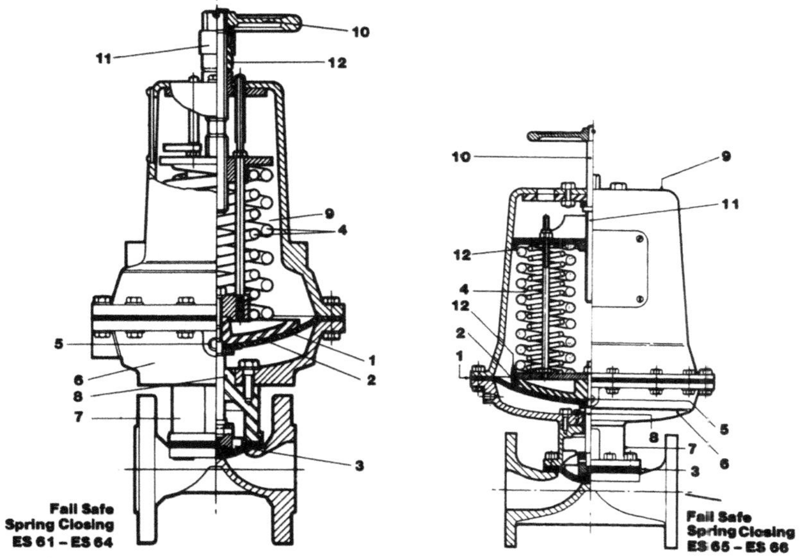

Item	Component ES 61 — ES 64
1	OPERATING DIAPHRAGM
2	DIAPHRAGM PLATE
3	'A' TYPE LINE DIAPHRAGM
4	SPRINGS
5	OPERATING FLUID INLET
6	OPERATING CYLINDER
7	BONNET
8	'O' RING
9	SPRING COVER
10	MANUAL OVERRIDE
11	VISUAL INDICATOR
12	LOCKING BUSH

Item	Component ES 65 — ES 66
1	OPERATING DIAPHRAGM
2	DIAPHRAGM PLATE
3	'A' TYPE LINE DIAPHRAGM
4	SPRINGS
5	OPERATING FLUID INLET
6	OPERATING CYLINDER
7	BONNET
8	'O' RING
9	SPRING COVER
10	MANUAL OVERRIDE
11	SPRING ADJUSTING SCREW
12	UPPER & LOWER SPRING PLATES

FIG.5

It was considered necessary, to produce a product that more readily meets the simple ON-OFF category as the existing range of actuators meets the automatic control parameters.

The initial brief was therefore an actuator with the following features :-

PROJECT BRIEF

SUITABLE FOR :	'A' TYPE WEIR VALVE
SIZE RANGE :	A008, A010, A015, A020, A025, A040 & A050
ACTION :	ON/OFF
OPERATION :	PNEUMATIC
OPERATING PRESSURE :	6 BAR (AIR)
LINE PRESSURE:	5.5 BAR

MAX AMBIENT TEMP : 95°C

MAX STERILISATION TEMP (STEAM) - 150°C

MODE : SPRING CLOSE *

MUST BE:-

1] Compact Unit

2] Lightweight

3] Low cost

4] Integral Bonnet

5] Microswitch adaption

6] Steam and Chemical sterilisation.

In considering the above, it was considered that a piston operated actuator fitted both the requirement of compact design and would also meet the operating parameters identified.

The spring closing actuator must also be able to be dismantled safely, without special tools, giving it a feature which is not normally available in the marketplace.

* FAIL SAFE SPRING CLOSING

Fail safe spring closing actuators close the valve against the line pressure in the event of failure (or designed withdrawal) of air or hydraulic pressure to the operating head. This mode is used when the valve is normally in the closed position (to avoid the need for constant application of operating pressure).

DESIGN

The Design Department produced a prototype based on the aforementioned criteria, making the design as small as physically possible, but without considering the material of construction and manufacturing processes.

Once the concept had been envisaged considerable emphasis was placed on the construction and process route.

To achieve the low cost, and lightweight factors, the obvious choice was the use of an engineering polmer, which, was capable of being near or nett shape formed from an injection moulding machine, thus possibly eliminating the costly post machining operations associated with the following processes.

1] Sand Casting
2] Investment Casting
3] High and Low Pressure die casting
4] Stamping

MATERIAL SELECTION

The polymer chosen for this application had to be capable of meeting the following parameters :

a] Steam Sterilisation at 150°C for 30 minutes.

b] Chemical sterilisation - Caustic Hydroxide and dilute nitric acid

c] Resistant to Ethylene Oxide ($C.H_2CH_2O$).

d] Resistant to Silanes (SiH4), phosphine (PH3), formaline, oils and hydrogen.

e] Dimensionally stable at elevated temperatures 150°C+

RAPRA - the Technical Research Establishment were contacted and the Plascam 220 material selection programme used. Following is a list of materials that were to be analysed as perspective condidates :-

1] PEEK - Polyetheretherketone - Victrex ICI
2] PES - Polyethersulphone - Victrex ICI
3] ULTEM - Polyetherimides - G E Platics
4] UDEL - Polysulphone - Union Carbide
5] RYTON PPS - Polyphenylensulfide - Phillips Pet.
6] POLYESTER(PTMT) - Polytetramethyleneterephthalate - - Dynamit Nobel
7] NORYL PPO - Polyphenylene Oxide - G.E.Platics

At this point it was decided possibly arbitrarily, to impose a Heat distortion temperature of 200°C, thus ensuring a factor of safety during the steam sterilisation process.

This point alone discounted some of the materials :

MATERIAL (GLASS FILLED GRADES)	HEAT DISTORTION TEMP (°C)	CONTINUOUS USE TEMP (°C)
PEEK	300	250
PES	216	180
ULTEM	210	170
UDEL	181	149
RYTON PPS R4	260	200
RYTON PPS A100	260	200
POLYESTER (PTMT)	200	145
NORYL	150	105

MATERIAL EVALUATION

The remaining materials were analysed and the following assessment made on each grade :-

1] PEEK - Potentially an ideal polymer for the specification but the prevailing factor was price/Kg (too expensive)

2] PES - To be evaluated.

3] ULTEM- Very poor response from the material supplier. In fact, we are still awaiting for their Technical Service Department to respond.

4] RYTON- Both grades to be evaluated.

Therefore, ongoing trials centred around PES and Ryton PPS materials.

DEVELOPMENT

Prototype tools were manufactured to the initial design while samples of the materials were being obtained. At this point detailed discussions took place with the material suppliers with regard to process parameters.

The prototype tools were produced and would at least prove the concept and ensure the dimensional stability required was achieveable.

With the initial design, both materials produced components which proved the design concept and material choice. However, as in all development programmes whilst the initial evaluation was ongoing, two factors changed :-

1. The initial design was considered bland ! (Fig.6). It was therefore considered necessary to produce a styled version capable of being injection moulded.

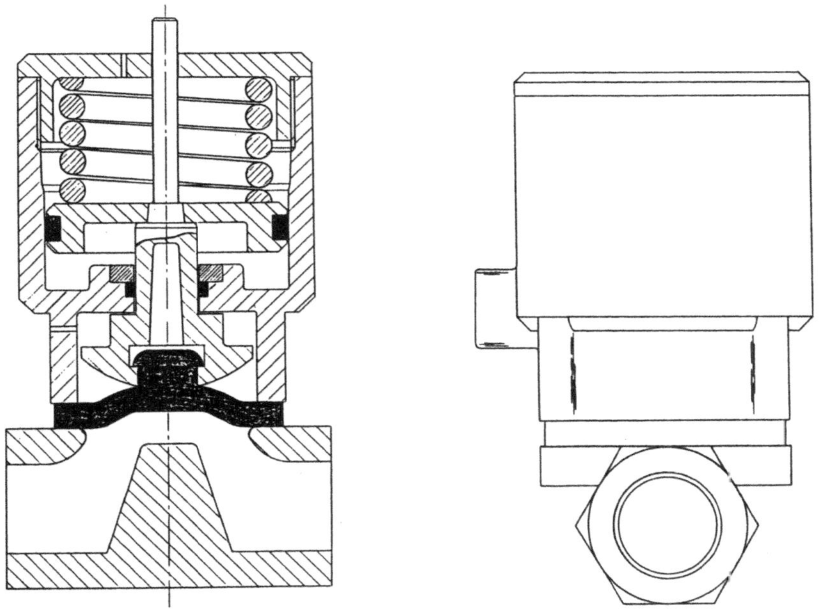

FIG.6

2. The operating parameters for the actuator were increased from single mode operation to triple, i.e., added to spring closing, spring opening and double acting were required.

* FAIL SAFE SPRING OPENING

Fail safe spring opening actuators open the valve to allow the line fluid to flow in the event of failure (or designed withdrawal) of operating pressure. This mode is used when the valve is normally in the open position (to avoid the need for constant application of operating pressure).

DOUBLE ACTING

Air/hydraulic pressure both opens and closes the valve and is used when a failsafe mode is not required.

The prototype tool was modified to the styled version (Fig.7) including the manufacture of additional moulds to incorporate other features.Consultation with both material suppliers at this stage, indicated their uncertainty with the increased section thicknesses brought about by the style change. Both materials were again evaluated with the modified tool to check for porosity and sink in the thick sections.

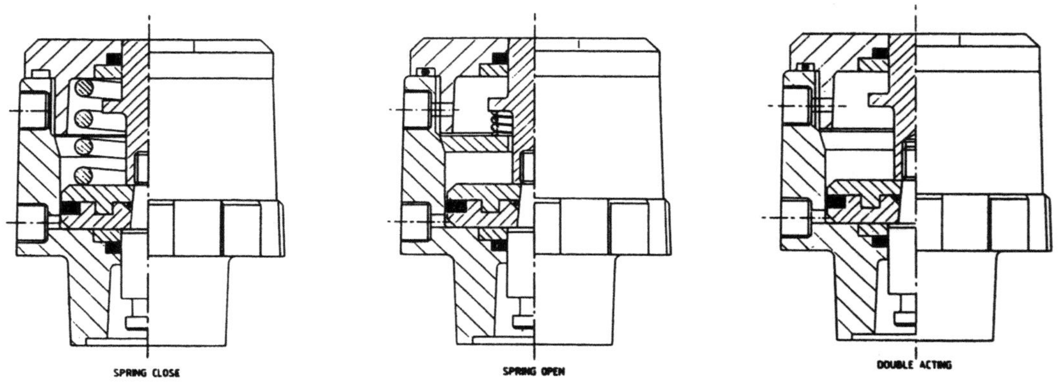

FIG.7

This evaluation led to serious porosity problems with Ryton PPS (Fig.8) (confirming the suppliers concern about the increased sections). PES however, proved to be excellent and even better than ICI believed possible. Microporosity analysis was carried out in the laboratory and found to be non-existant.

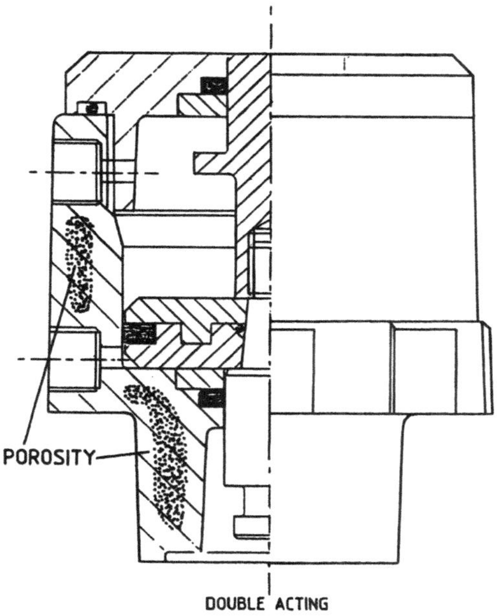

FIG.8

Further re-design was thought necessary to reduce the section(Fig.9) thickness around the bonnet castellations. Although the design chosen was suitable for the smaller sizes in the range. On the larger units the weight would have increased considerably causing potential moulding problems plus an additional cost penalty due to material increase. It was thought that the castellation design was aesthetically pleasing and should be maintained if at all possible. Therefore, the design of the castellations was inverted saving 12% in material usage and making manufacture far easier by reducing section thickness and sharp edges around the bonnet base.

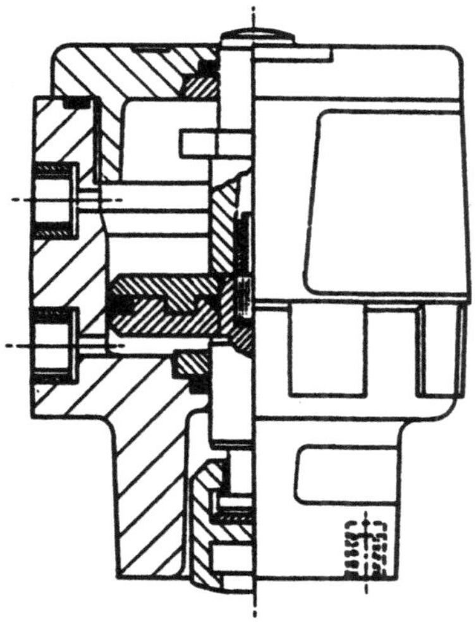

FIG.9

During this trial period it was found that the inserts required in the actuator had to be post moulded and not fitted in-situ during the moulding process, as initially intended.

This led to the introduction of ultrasonic welding which is generally more acceptable on amorphous polymers than a crystalline one.

Phillips Petroleum are currently evaluating the effects of ultrasonic welding on PPS as they do not recommend it at the moment.

However, the ultrasonic welding of PES although creating considerable concern at first, because very little welding experience existed in this country, proved to be no problem at all in practice.

PES has therefore chosen itself to be the material of manufacture.

This material has proved extremely dimensionally stable, and as the project progressed, extensive tests were carried out under simulated service conditions which have proved that the material functions as predicted.

In fact, considerable time and effort was put in to product evaluation. It should be stressed at this stage, that when dealing with advanced materials, it is important that the designers and material suppliers work in unison. The product must be developed with the material supplier and frank and open, yet confidential, discussions about the "do's" and don'ts" of the application are extremely important. Confidence is then built up and provides a base to consider other excursions away from traditional metal parts.

It should be stated at this point however, that although PES was chosen for this particular project on its merits, this in no way reflects that the other materials are inferior for other applications. No doubt many other research and development departments are undergoing similar material analysis which could result in a different material choice being made.

CONCLUSIONS

Following the successful evaluation of PES with the prototype tooling, the Saunair EC Actuator is currently being launched.

It is a unit which after a considerable period of time evaluating its design and moulding characteristics, meets all the parameters laid down and proves the potential for engineering plastics.

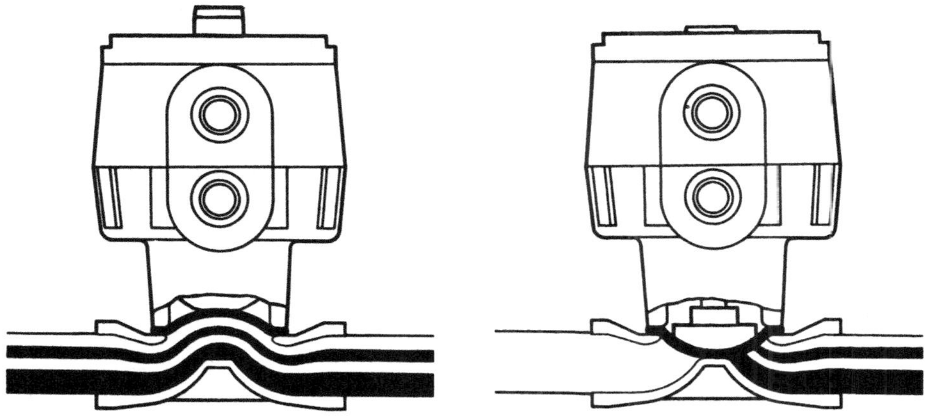

INTRODUCTION TO SAUNAIR EC ACTUATOR RANGE

Saunders Saunair EC range of actuator is designed for use with weir type A and type AFP valve bodies in the size range DN8 - DN50.

The range of actuators is suitable for use with all grades of Saunders rubber diaphragms and gives 100% leak tight closure against the recommended line pressures.

The EC Actuator range directly replaces manual bonnet assemblies and may be ordered as part of a whole valve or separately to convert a manual valve to power operation. Installation may be carried out with the valve body in the pipeline.

Saunair EC Actuators are designed as compact units, with close coupled bonnet assemblies, manufactured by injection moulding in Polyethersulphone (PES). The units are fully suitable for a wide range of sterilization procedures.

THREE MODES OF OPERATION ARE AVAILABLE		
Close on air failure	**Open on Air failure**	**Double Acting**
These actuators close the valve against maximum permissible line pressure in the event of failure (or designed withdrawal) of operating pressure to operating head.	These actuators open a valve to allow fluid flow in the event of failure (or designed withdrawl) of operating pressure.	Operating pressure both opens and closes the valve.
Normal use: when a valve is normally in closed position (to avoid the need for constant application of operating pressure).	Normal use: when a valve is normally in the open position (to avoid the need for constant application of operating pressure).	Normal use: when failsafe mode is undesirable.

ACTUATOR SELECTION

The above operating modes apply to valves DN8, DN10, DN15, DN20, DN25, DN40 & DN50. For each line size actuators are available with NITRILE seals (Standard Model) or FLUOROELASTOMER seals (High Temperature Model)

PERFORMANCE/OPERATING/LINE PRESSURES

Full details of operating pressures required for varying line pressures are given in data sheet DM69. The overall performance figures are :

Operating Pressure - 6 bar
Line Pressure 5.5 bar max
at 0% P.
7.0 bar max
at 100% P.
Temperature ratings:
-10°C to +100°C ambient.

Autoclave maximum 120°C
(nitrile seals)
-150°C
(fluoroelastomer seals)
Line fluid temperature ratings are as for standard weir type line diaphragms (see data sheets DM9 and DM62-64)

APPROXIMATE VOLUMES OF OPERATING FLUID AT SPECIFIED OPERATING PRESSURES (not free air)

Valve Size DN	VOLUME CM^3	
	Upper Chamber (to close)	Lower Chamber (to open)
8	22	8
10	49	16
15	49	16
20	212	62
25	212	62
40	988	244
50	995	336

METHOD OF OPERATION/ COMPONENTS, CLOSE ON AIR FAILURE SAUNAIR EC ACTUATORS

Saunders EC Actuators in close on air failure mode depend upon a piston moving within an airtight chamber to fully open a valve when required. Operating pressure is applied to the lower port (15) (6 bar maximum) causing the piston assembly (16, 17,18) to move upwards, raising the compressor (12) and diaphragm (11) clear of the valve body weir - to allow full fluid flow. Confirmation of valve operations (visually) is given by position indicator (2)

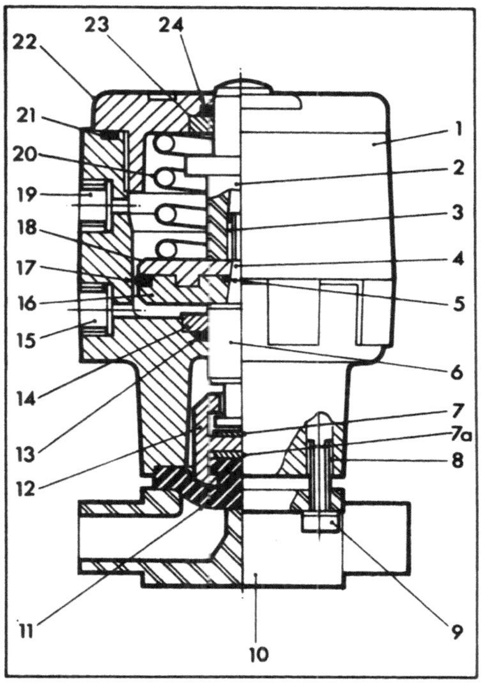

AIR INLET PORTS (15, (19)

It is recommended that the size of fitting used to connect the air supply is 1/8" BSP x 7.5mm max thread length.

ITEM	COMPONENT	MATERIAL OF CONSTRUCTION
1	Bonnet	PES
2	Indicator	PES
3	Indicator Insert	BRASS
4	Spindle Stud Insert	BRASS (NOT USED DN8)
5	Piston Inner Seal	NITRILE
6	Spindle	PES
7	Thrust Disc	NYLATRON (NOT DN8 & DN10)
7a	Thrust Disc	NYLATRON (DN15 SIZE ONLY)
8	Bonnet Base Insert	BRASS
9	Screw	STEEL
10	Body	AS SELECTED (TYPE 'A'ONLY)
11	Diaphragm	AS SELECTED(EXCL 214 COMBIN'N
12	Compressor	SIL ALUM(DN8,DN10-ZINC ALLOY)
13	Spindle Seal	NITRILE OR FLUOROELASTOMER
14	Bonnet Washer	PES
15	1/8" BSP Insert (Lower Port)	BRASS
16	Lower Piston Section	PES
17	Piston Outer Seal	NITRILE OR FLUOROELASTOMER
18	Upper Piston Section	PES
19	1/8"BSP Insert(Upper Port)	BRASS
20	Spring	STEEL
21	Cap Seal	NITRILE
22	Cap	PES
23	Cap Washer	PES
24	Indicator Seal	NITRILE OR FLUOROELASTOMER

METHOD OF OPERATION/COMPONENTS,OPEN ON AIR FAILURE SAUNAIR EC ACTUATORS

Saunders EC Actuators in open on air failure mode depend upon a piston moving within an airtight chamber to fully close a valve when required. Operating pressure is applied to the upper port (19) (6 bar maximum) causing the piston assembly (16, 17,18) to move downwards, lowering the compressor (12) and diaphragm (11) onto the valve body weir - closing the valve leaktight. Confirmation of valve operation (visually) is given by position indicator (2)

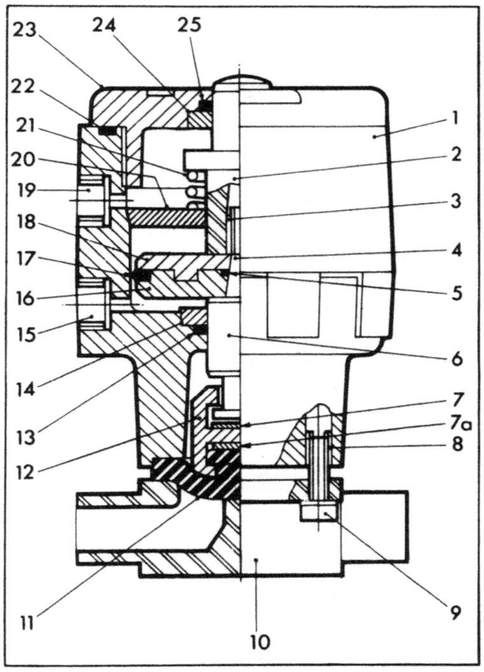

AIR INLET PORTS (15),(19)

It is recommended that the size of fitting used to connect the air supply is 1/8" BSP x 7.5mm max thread length.

ITEM	COMPONENT	MATERIAL OF CONSTRUCTION
1	Bonnet	PES
2	Indicator	PES
3	Indicator Insert	BRASS
4	Spindle Stud Insert	BRASS (NOT USED DN8)
5	Piston Inner Seal	NITRILE
6	Spindle	PES
7	Thrust Disc	NYLATRON (NOT DN8 & DN10)
7a	Thrust Disc	NYLATRON (DN15 SIZE ONLY)
8	Bonnet Base Insert	BRASS
9	Screw	STEEL
10	Body	AS SELECTED (TYPE 'A'ONLY)
11	Diaphragm	AS SELECTED(EXCL 214 COMBIN'N
12	Compressor	SIL ALUM(DN8,DN10-ZINC ALLOY)
13	Spindle Seal	NITRILE OR FLUOROELASTOMER
14	Bonnet Washer	PES
15	1/8" BSP Insert (Lower Port)	BRASS
16	Lower Piston Section	PES
17	Piston Outer Seal	NITRILE OR FLUOROELASTOMER
18	Upper Piston Section	PES
19	1/8"BSP Insert(Upper Port)	BRASS
20	Spring Support Plate	PES
21	Spring	STEEL(DN8-DN15 STAINLESS STL)
22	Cap Seal	NITRILE
23	Cap	PES
24	Cap Washer	PES
25	Indicator Seal	NITRILE OR FLUOROELASTOMER

METHOD OF OPERATION/COMPONENTS, DOUBLE ACTING SAUNAIR EC ACTUATORS

Saunders EC Actuators in double acting mode require operating pressure applied to the upper port (19) to close the valve leak-tight against line pressure or operating pressure applied to the lower port (15) to open the valve to allow full fluid flow. To maintain either open or closed states operating pressure must be consistently maintained (6 bar maximum).

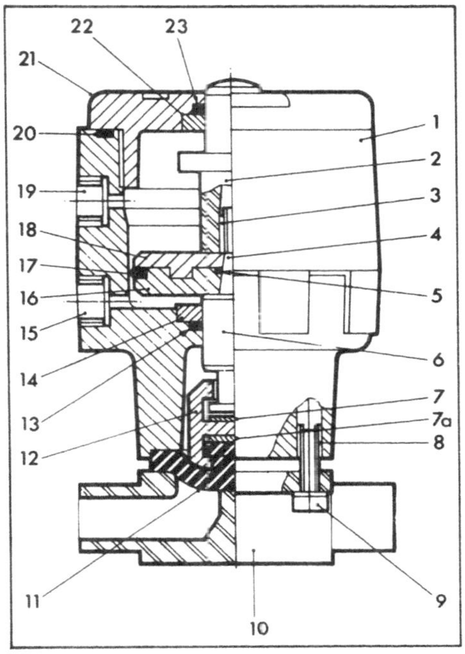

AIR INLET PORTS (15),(19)

It is recommended that the type of fitting used to connect the air supply is 1/8" BSP x 7.5mm max thread length.

ITEM	COMPONENT	MATERIAL OF CONSTRUCTION
1	Bonnet	PES
2	Indicator	PES
3	Indicator Insert	BRASS
4	Spindle Stud Insert	BRASS (NOT USED DN8)
5	Piston Inner Seal	NITRILE
6	Spindle	PES
7	Thrust Disc	NYLATRON (NOT DN8 & DN10)
7a	Thrust Disc	NYLATRON (DN15 SIZE ONLY)
8	Bonnet Base Insert	BRASS
9	Screw	STEEL
10	Valve Body	AS SELECTED (TYPE 'A'ONLY)
11	Diaphragm	AS SELECTED(EXCL 214 COMBIN'N
12	Compressor	SIL ALUM(DN8,DN10-ZINC ALLOY)
13	Spindle Seal	NITRILE OR FLUOROELASTOMER
14	Bonnet Washer	PES
15	1/8" BSP Insert (Lower Port)	BRASS
16	Lower Piston Section	PES
17	Piston Outer Seal	NITRILE OR FLUOROELASTOMER
18	Upper Piston Section	PES
19	1/8"BSP Insert(Upper Port)	BRASS
20	Cap Seal	NITRILE
21	Cap	PES
22	Cap Washer	PES
23	Indicator Seal	NITRILE OR FLUOROELASTOMER

DIMENSIONS, WEIGHTS, SAUNAIR EC ACTUATORS

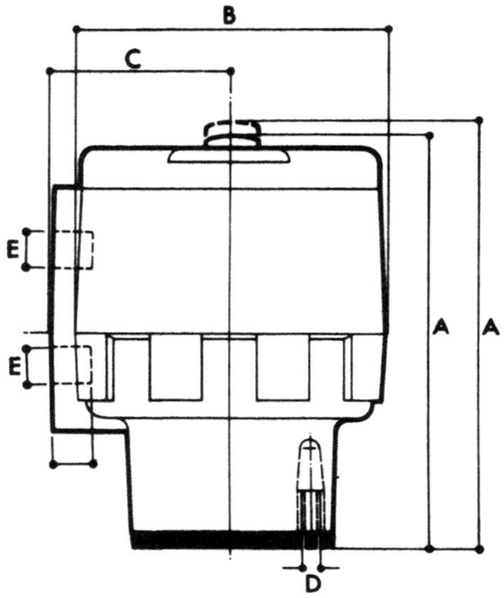

DIMENSIONS (mm)

Valve Size DN	A Open	A Closed	B	C	D (thread)	E Air
8	78	73	58	33	M4 x 2 OFF	1/8" BSP
10	88	82	71	38	M4 x 2 OFF	1/8" BSP
15	101	95	71	38	M5 x 4 OFF	1/8" BSP
20	140	133	103	53	M6 x 4 OFF	1/8" BSP
25	148	138	103	53	M8 x 4 OFF	1/8" BSP
40	204	187	153	78	M10 x 4 OFF	1/8" BSP
50	217	193	153	78	M10 x 4 OFF	1/8" BSP

ACTUATOR WEIGHTS (KG)

Valve Size DN	8			10			15			20			25		
Mode	Closed	Open	D'ble	Closed	Open	D'ble	Close	Open	D'ble	Closed	Open	D'ble	Closed	Open	D'ble
Weight (Kg)	0.25	0.22	0.21	0.42	0.35	0.34	0.45	0.38	0.37	1.20	1.00	0.90	1.29	1.10	1.00
Mode	40			50											
Weight (Kg)	4.0	2.7	2.5	4.3	3.0	2.8									

PRODUCTION, QUALITY TESTING APPLIED TO SAUNAIR EC ACTUATORS

PRODUCTION TEST

ACTUATOR TYPE	TEST TYPE	TEST PURPOSE
Close on air failure Open on air failure	Pressure (6 bar) applied to appropriate chamber to open valve. Durations 6 seconds (8-15mm): 15 secs (20-25mm) 30 seconds (40-50 mm)	Confirm full leak-tight integrity of actuator assembly.
Double acting	Pressure (6 bar) applied consecutively to upper and lower chambers. Duration : as above.	As above
All operation modes	Check travel of indicator spindle.	Confirm full closure/ opening of valve.

Under Saunders Quality Assurance procedures various tests are undertaken to ensure consistent product quality.

These include :

Batch sampling for hydraulic pressure test to destruction (results compared with established standard).

Batch sampling for load test to failure of inserts (results compared with established standards).

Porosity (of moulded actuator body) compared with agreed standard.

PROPERTIES OF 'VICTREX' PES

Property	Test method	Units	PES 3600G/ PES 4100G/ PES 4800G/ PES 5200G	PES 3601GL20/■ PES 4101GL20/ PES 5201GL20	PES 3601GL30/■ PES 4101GL30
GENERAL					
Form	—	—	granules	granules	granules
Appearance	—	—	amber/ transparent	brown/ opaque	brown/ opaque
Relative density	ASTM D792	—	1.37	1.51	1.60
Refractive index	—	—	1.65	—	—
Shrinkage of moulding	—	%	0.6	0.3	0.2
Glass content	—	%	0	20	30
MECHANICAL PROPERTIES					
Tensile strength at 20°C(68°F)	ASTM D638 (DIN 53455)	MPa(psi)	84(12,200)	124(18,000)	140(20,300)
Tensile strength at 150°C(300°F)			55(8,000)	78(11,300)	100(14,500)
Tensile strength at 180°C(355°F)			41(5,950)	60(8,700)	76(11,000)
Elongation at break	ASTM D638	%	40—80	3	3
Flexural strength	ASTM D790	MPa(psi)	129(18,700)	172(24,900)	190(27,500)
Flexural modulus at 20°C(68°F)	ASTM D790	GPa(psi)	2.6(377,000)	5.9(855,600)	8.4(1218,200)
Flexural modulus at 150°C(300°F)			2.5(362,500)	5.8(841,100)	8.2(1189,200)
Flexural modulus at 180°C(355°F)			2.3(333,500)	5.6(812,100)	8.0(1160,200)
Izod impact strength 6.4 mm(0.250 in) specimen notched	ASTM D256	J/m(ft-lb/in)	76(1.42)/84(1.57)/ 110(2.06)/120(2.25)	75(1.40)/80(1.50)/ 84(1.57)	85(1.59)/90(1.69)
6.4 mm (0.250 in) specimen un-notched	ASTM D256	J/m(ft-lb/in)	No break	—/430(8.05)/—	—/540(10.12)
Rockwell hardness	ASTM D785	—	M88	M98	M98
Taber abrasion (1 Kg load CS17 wheel)	ASTM D1044	mg/1000 rev.	6	8	—
THERMAL PROPERTIES					
Heat distortion temperature at 1.82 MPa(264 lbf/in²)	ASTM D648 (ISO R75)	°C(°F)	203(397)	210(410)	216(421)
Vicat softening point					
1 Kg(2.2 lbs)	ASTM D1525	°C(°F)	226(439)	>226(>439)	>226(>439)
5 Kg(11.0 lbs)	ISO R306	°C(°F)	222(432)	>222(>432)	>222(>432)
Coefficient of linear thermal expansion	ASTM D696	per deg C	5.5×10^{-5}	2.6×10^{-5}	2.3×10^{-5}
Underwriters' Laboratories temperature index	UL746	°C(°F)	180(356)	180(356)	180(356)/190(374)

NOTE

■ These materials are anisotropic and properties will vary with direction of measurement.

The property values given were obtained under standard test conditions and are typical values only. They should not be used as a basis for design

Property	Test method	Units	PES 3600G/ PES 4100G/ PES 4800G/ PES.5200G	PES 3601GL20/■ PES 4101GL20/ PES 5201GL20	FES 3601GL30/■ FES 4101GL30
ELECTRICAL PROPERTIES					
Volume resistivity	ASTM D257 IEC 93	Ohm cm	$>10^{17}$	$>10^{17}$	$>10^{17}$
Permittivity 60 Hz	ASTM D150	—	3.7	4.0	4.0
Permittivity 10^6 Hz	IEC 250	—	3.7	4.0	4.0
Loss tangent 60 Hz	ASTM D150	—	0.001	0.002	0.003
Loss tangent 10^4 Hz	IEC 250	—	0.003	0.003	0.004
Dielectric strength at 3.2 mm(0.126 in)	ASTM D149 IEC 243	KV/mm	16	20	20
Corrosion liability factor (CLF)	IEC 426	—	5	—	8
Tracking resistance (CTI)	IEC 112	Volts	150	150	150
Arc resistance (tungsten carbide electrodes)	ASTM D495	—	20—120	—	—
High voltage arc track rate	UL 746	in/min	0.8	5.5	5.0
High amp arc ignition	UL 746	Arcs	34	4	7
FLAMMABILITY AND BURNING BEHAVIOUR					
Flammability rating	ASTM D635	AEB/ATB	4.1/- (immediate extinction)	—	—
Underwriters' Laboratories flammability rating					
0.5 mm(0.020 in)	UL 94	—	V-0	V-0	V-0
0.3 mm(0.012 in)			V-1	—	—
Limiting oxygen index 0.5 mm(0.020 in)	ASTM D286	—	34	—	—
1.6 mm(0.060 in)			38	40	41
Hot wire ignition 3 mm(0.120 in)	UL 746	Seconds	80	140	300
Smoke emission when burning	NBS Smoke Chamber	Specific optical density (DM) (flaming condition 3 mm(0.120 in) sample)	30	—	15

SUMMARY OF PROPERTIES

Property	ASTM Test Method	Units	PEEK 450G	PEEK 450GL20	PEEK 450GL30	PEEK 450CA30
GENERAL PROPERTIES						
Form	—	—	Granular	Granular	Granular	Granular
Density (crystalline)	D792	g/cm³	1.32	1.43	1.49	1.44
(amorphous)			1.265	—	—	—
Colour	—	—	Grey	Brown	Brown	Black
Filler content	—	%	—	20	30	30
Typical level of crystallinity	—	%	35	35	35	35
Processing temperature range	—	°C (°F)	350—380 (660—715)	360—400 (660—715)	370—400 (660—715)	370—400 (660—715)
Water absorption						
24 hours 23°C(73°F)	D570	%	0.5	—	0.11	0.06
Equilibrium 23°C(73°F)	D570	%	0.5	—	—	—
Mould shrinkage	—	%	1.1	0.7*—1.4	0.5	0.1*—1.4
MECHANICAL PROPERTIES						
Flexural modulus 23°C(73°F)	D790	GPa(psi)	3.66(530,800)	6.66(965,900)	10.31(1495,200)	13.0(1885,400)
Flexural modulus 250°C(480°F)	D790	GPa(psi)	0.3(43,500)	1.2(174,000)	2.3(333,500)	3.6(522,100)
Flexural strength 23°C(73°F)	D790	MPa(psi)	170(24,700)**	192(27,800)	233(33,800)	318(46,100) ■
Flexural strength 250°C(480°F)	D790	MPa(psi)	12.5(1,800)**	53.7(7,800)**	70.8(10,300)**	105.2(15,200)**
Tensile strength 23°C(73°F)	D638	MPa(psi)	92(Y)(13,300)	123(B)(17,800)	157(B)(22,800)	208(B)(30,200)
Break or yield 250°C(480°F)	D638 Test speed 5 mm/min. (0.2 in/min)	MPa(psi)	12(Y)(1,700)	24(Y)(3,500)	34(B)(4,900)	43(B)(6,200)
Elongation at break 23°C(73°F)	D638 Test speed 5 mm/min. (0.2 in/min)	%	50	2.5	2.2	1.3
Elongation at yield 23°C(73°F)	D638 Test speed 5 mm/min (0.2 in/min)	%	4.9	—	—	—
Shear strength (ultimate)	D3846	MPa(psi)	95(13,800)	—	97(14,100)	97(14,100) ■
Shear modulus 23°C(73°F)	—	GPa(psi)	1.30(188,500)	—	2.40(348,000)	—
Tensile modulus 1% secant 23°C(73°F)	D638	GPa(psi)	3.6(522,100)	7.0(1015,200)	9.7(1406,800)	13(1885,400)
Compressive strength 23°C(73°F): 0°	D695	MPa(psi)	118(17,100)	162(23,500)	215(31,200)	240(34,800)
Compressive strength 23°C(73°F): 90°	D695	MPa(psi)	119(17,300)	134(19,400)	149(21,600)	153(22,200)
Charpy impact strength 23°C(73°F)	BS 2782 Method 351A					
2 mm(0.080 in) notch radius	—	KJ/m² (ft-lb/in)	34.9 (6.65)	— —	11.3 (0.22)	7.8 (0.15)
0.25 mm(0.010 in) notch radius	—	KJ/m² (ft-lb/in)	8.2 (0.16)	— —	8.9 (0.17)	5.4 (0.1)
Izod impact strength 23°C(73°F)	D256					
Notched (0.25 mm radius, 2.5 mm depth) (0.010 in radius, 0.100 in depth)	—	J/m (ft-lb/in)	83 (1.55)	86 (1.61)	96 (1.8)	85 (1.6)
Unnotched	—	J/m (ft-lb/in)	No break	673 (12.6)	725 (13.6)	749 (14.0)
Poissons ratio 23°C(73°F)	D638	—	0.42	—	0.45	—
Rockwell hardness						
R scale	D785	—	126	125	124	124
M scale	D785	—	99	102	103	107
THERMAL PROPERTIES						
Melting point	—	°C(°F)	334(633)	334(633)	334(633)	334(633)
Glass transition temperature	—	°C(°F)	143(289)	143(289)	143(289)	143(289)
Specific heat	—	cal/g °C	0.32	—	—	—
Coefficient of thermal expansion	D696	m/m/°C	47×10^{-6} 108×10^{-6} at >150°C (>300°F)	24×10^{-6}	22×10^{-6}	15×10^{-6}
Heat distortion temperature 1.82 MPa (264 lbf/in^2)	D648	°C(°F)	160(320)	285(545)	315(600)	315(600)
Thermal conductivity	C177	W/m°C	0.25	0.41	0.43	0.92
Maximum continuous service temperature (Estimated UL-based test)	—	°C(°F)	250(480)	250(480)	250(480)	250(480)

* Mould shrinkage is dependent on fibre orientation and the lower figure refers to shrinkage in the direction of fibre orientation.

** yield value at > 5% strain (B) = Break
(Y) = Yield

Property	Test Method	Units	PEEK 450G	PEEK 450GL20	PEEK 450GL30	PEEK 450CA30
FLAMMABILITY PROPERTIES						
Flammability Rating						
1.45mm(0.057 in) thick sample	UL94	—	V-0	V-0 ■	V-0	V-0
Limiting Oxygen Index						
0.4 mm(0.015 in) thick sample	ASTM D2863	% O_2	24	—	—	—
3.2 mm(0.125 in) thick sample	ASTM D2863	% O_2	35	—	—	—
Temperature Index						
3.2 mm(0.125 in) thick sample	Based on ASTM D2863	°C(°F)	>325(>615)	—	—	—
Smoke Density (DM)c						
3.2 mm(0.125 in) thick sample:	NBS Smoke	—	19	9	—	5
flaming mode non-flaming mode	Chamber	—	2	2	—	1
1.6 mm(0.062 in) thick sample:	NBS Smoke	—	50	—	—	—
flaming mode non-flaming mode	Chamber	—	5	—	—	—
Time to 90% (DM)c						
3.2 mm(0.125 in) flaming mode thick sample:	NBS Smoke	min	18	19	—	19
3.2 mm(0.125 in) non flaming mode thick sample:	Chamber	min	19.6	—	—	—
D(1.5 Min) Value						
Flaming	NBS Smoke	—	0	0	—	0
Non-flaming	Chamber	—	0	0	—	0
D(4 Min) Value						
Flaming	NBS Smoke	—	1	1	—	0
Non-flaming	Chamber	—	0	0	—	0
Toxicity Index 10 g(0.02 lbs) sample						
CO	NES 713	—	0.074	0.065	—	0.051
CO_2	on Cable	—	0.146	0.124	—	0.123
Total for all gases tested		—	0.22	0.19	—	0.17
ELECTRICAL PROPERTIES						
Dielectric strength	ASTM D149 (50µ film)	KV/cm	190	—	—	—
Comparative tracking index 23°C (73°F)	D3638	Volts	150	—	175	—
Loss tangent 23°C(73°F) 1Hz	BS 2782	—	0.003	—	—	—
Volume resistivity 23°C(73°F)	D257	ohm-cm	4.9×10^{16}	—	—	1.4×10^5
Permittivity						
50 Hz-10 KHz, 0-150°C(32-300°F)	BS 2782	—	3.2-3.3	—	—	—
50 Hz, 200°C(390°F)	BS 2782		4.5			

■ Estimated Value

ULTEM® Resin Typical Properties

MECHANICAL	ASTM TEST	UNITS	ULTEM 1000	ULTEM 1010	ULTEM 2100	ULTEM 2200	ULTEM 2300	ULTEM 2400	ULTEM 4000	ULTEM 6000	ULTEM 6100	ULTEM 6200	ULTUM 6202
Tensile strength, yield	D638	N/mm^2	105	105	120	140	160	186	83	103	117	145	97
Tensile modulus	D638	N/mm^2	3.000	3.000	4.500	6.900	9.000	11.700	9000	3000	4500	6900	†
Tensile elongation, yield	D638	%	7-8	7-8	5	-	-	-	-	7-8	-	-	-
Tensile elongation, ultimate	D638	%	60	60	6	3	3	2,5	1,5	30	6	4	6
Flexural strength	D790	N/mm^2	145	145	200	210	230	250	115	145	195	210	145
Flexural modulus	D790	N/mm^2	3.300	3.300	4.500	6.200	9.000	11.700	9.000	3000	4600	6500	3800
Compressive strength	D695	N/mm^2	140	140	160	170	175	200	125	140	-	170	140
Compressive modulus	D695	N/mm^2	2.900	2.900	3.100	3.500	3.800	†	3.300	-	-	-	-
Gardner impact	-	Nm	36	34	-	-	-	-	-	-	-	-	-
Izod impact	D256												
notched (3,2 mm)		J/m	50	30	60	90	100	105	70	40	50	80	40
unnotched (3,2 mm)		J/m	1.300	1.300	480	480	430	430	150	1.300	400	400	350
Shear strength, ultimate	-	N/mm^2	100	100	90	95	100	103	60	100	95	95	95
Rockwell hardness	D785	-	M109	M109	M114	M118	M125	†	M85	-	-	-	-
Taber abrasion (CS 17, 1 kg)	D1044	mg wt. loss./ 1000 cycles	10	10	-	-	-	-	-	-	-	-	-
THERMAL													
Deflection temperature, unannealed	D648												
@ 1.82 N/mm^2 (6,4 mm)		°C	200	197	207	209	210	213	211	215	221	223	215
@ 0.45 N/mm^2 (6,4 mm)		°C	210	207	210	210	212	216	-	221	222	225	†
Vicat softening point, method B	D1525	°C	219	219	223	226	228	234	234	234	240	243	237
Continuous service temperature index (UL Bulletin 746B)	-	°C	170	†	170**	170**	180**	†	-	†	†	†	†
Coefficient of thermal expansion (-18 to 150°C), mould direction	D696	m/m-°C	$5{,}6 \times 10^{-5}$	$5{,}6 \times 10^{-5}$	$3{,}2 \times 10^{-5}$	$2{,}5 \times 10^{-5}$	$2{,}0 \times 10^{-5}$	$1{,}4 \times 10^{-5}$	$1{,}6 \times 10^{-5}$	$5{,}1 \times 10^{-5}$	†	$2{,}5 \times 10^{-5}$	$4{,}5 \times 10^{-5}$
Thermal conductivity	C177	W/m°C	0,22	0,22	-	-	-	-	-				
FLAMMABILITY													
Oxygen index	D2863	%	47	44	47	50	50	54	-	44	44	44	48
Vertical burn (UL Bulletin 94)*	-	-	V-0 @ 0,41 mm 5V @ 1,9 mm	V-0@ 1,6 mm†† †	V-0@ 0,41 mm	V-0@ 0,41 mm	V-0@ 0,25 mm	V-0@ 1,6 mm††	-	V-0@ 1,6 mm††	V-0@ 1,6 mm††	V-0@ 1,6 mm††	V-0@ 1,6 mm††
NBS smoke, flaming mode (1,5 mm)	E662												
D_s @ 4 min		-	0,7	†	-	1,3	-	1,0	-	5	-	4	-
D_{MAX} @ 20 min		-	30	†	-	27	-	20	-	70	-	70	-
ELECTRICAL													
Dielectric strength (1,6 mm)	D149												
in oil		kV/mm	28	28	27,5	26,5	24,8	24,0	-	29	†	23	21
in air		kV/mm	33	33	-	-	30	-	-	-	-	-	-
Dielectric constant @ 1 kHz, 50% RH	D150	-	3,15	3,15	3,5	3,5	3,7	3,7	-	3,0	3,1	3,1	3,1
Dissipation factor	D150												
@ 1 kHz, 50% RH, 23°C		-	0,0013	0,0013	0,0014	0,0015	0,0015	0,0020	-	0,001	0,001	0,001	0,001
@ 2450 MHz, 50% RH, 23°C		-	0,0025	0,0025	0,0046	0,0049	0,0053	†	-	-	-	-	-
Volume resistivity (1,6 mm)	D257	ohm-m	$6{,}7 \times 10^{15}$	$6{,}7 \times 10^{15}$	$1{,}0 \times 10^{15}$	$7{,}0 \times 10^{14}$	$3{,}0 \times 10^{14}$	†	-	$1{,}0 \times 10^{15}$	$1{,}0 \times 10^{15}$	$1{,}0 \times 10^{15}$	$1{,}0 \times 10^{15}$
Arc resistance	D495	seconds	128	†	85	85	85	-	-	127	-	-	140
OTHER													
Specific gravity	D792	-	1,27	1,27	1,34	1,42	1,51	1,61	1,7	1,29	1,35	1,43	1,42
Mould shrinkage	D955	m/m	0,007	0,007	0,005-0,006	0,003-0,005	0,002-0,004	0,001-0,003	0,002-0,003	0,005-0,007	†	†	0,005-0,007
Water absorption	D570												
@ 24 hours, 23°C		%	0,25	0,25	0,28	0,26	0,18	0,12	0,11	0,28	0,24	0,22	0,22
@ equilibrium, 23°C		%	1,25	1,25	1,0	1,0	0,9	0,9	-	-	-	-	-

* This rating is not intended to reflect hazards presented by this or any other material under actual fire conditions.
** Provisional rating. Final rating may be higher.

† Testing in progress.
†† General Electric Company test data.

TYPICAL PROPERTIES OF RYTON PPS INJECTION MOULDING GRADE R-4

PROPERTIES	TESTS*		UNITS	RYTON R-4
	ASTM	DIN		
Density	D1505-68	53479	g/cm^3	1.6
Tensile strength	D 638-72		MPa	135
		53455	MPa	116
Elongation at break	D 638-72		%	1.3
Flexural modulus	D 790-71		MPa	11,730
		53452	MPa	14,740
Flexural strength	D 790-71		MPa	200
		53452	MPa	174
Compressive strength	D 695-69		MPa	145
Compressive strength, axial		53454	MPa	137
transverse		53454	MPa	138
Izod impact strength,				
3.2 mm specimen, unnotched	D256-72a		J/m	430
3.2 mm specimen, notched	D256-72a		J/m	75
Charpy impact strength, unnotched		53453	kJ/m^2	12.8
notched		53453	kJ/m^2	6.6
Rockwell hardness	D 785-65			R-123
Water absorption, 24h at 25°C	D 570-63	53495	%	<0.05
Heat distortion temperature at 1.8 MPa (1)	D 648-72	53461	°C	>260
UL long term temperature index (2)			°C	200
Coeff. of linear thermal expansion (axial)			K^{-1}	$2.2x10^{-5}$
Limiting Oxygen Index	D2863-74		%	46.5
Flammability, UL 94				V-0/5V
Dielectric strength (3.2 mm specimen)	D 149-64		kV/mm	17.7
		53481	kV/mm	15.5
Dielectric constant, 25°C, 1 kHz	D 150-70	53483		3.9
25°C, 1 MHz	D 150-70	53483		3.8
Dissipation factor, 25°C, 1 kHz	D 150-70	53483		0.0010
25°C, 1 MHz	D 150-70	53483		0.0013
Volume resistivity, 2 minutes (1)	D 257-66	53482	ohm.cm	$4.5x10^{16}$
Surface resistivity	D 257-66	53482	ohm	$4.6x10^{16}$
Arc resistance	D 495-73		sec	34
Tracking resistance		53480		KC 180

* The tests were performed on as-moulded injection moulded specimens.
ASTM tests : melt temperature = 315°C, mould temperature = 38°C
DIN tests : melt temperature = 325°C, mould temperature = 140°C

(1) Annealed specimens
(2) Electrical and mechanical

Date of printing : 10/1980.

TYPICAL PROPERTIES OF RYTON PPS INJECTION MOULDING GRADE R-7

PROPERTIES	TESTS* ASTM	TESTS* DIN	UNITS	RYTON R-7
Density	D 1505	53479	g/cm^3	1.9
Tensile strength	D 638		MPa	96
		53455	MPa	111
Elongation at break	D 638		%	0.7
Flexural modulus	D 790		MPa	16,500
		53452	MPa	15,500
Flexural strength	D 790		MPa	158
		53452	MPa	176
Compressive strength	D 695		MPa	158
Compressive strength, axial		53454	MPa	119
transverse		53454	MPa	138
Izod impact strength,				
3.2 mm specimen, unnotched	D 256		J/m	185
3.2 mm specimen, notched	D 256		J/m	53
Charpy impact strength, unnotched		53453	kJ/m2	8.6
notched		53453	kJ/m2	3.9
Heat distortion temperature at 1.81 MPa (1)	D 648	53461	°C	260
UL long term temperature Index			°C	200-220
Flammability, UL 94, 0,71 mm				V-0
Dielectric strength (3.2 mm specimen)	D 149		kV/mm	13.0
Dielectric constant, 25°C, 1 kHz	D 150	53483		5.1
25°C, 1 MHz	D 150	53483		4.6
Dissipation factor, 25°C, 1 kHz	D 150	53483		0.058
25°C, 1 MHz	D 150	53483		0.0088
Volume resistivity	D 257	53482	ohm.cm	5×10^{15}
Arc resistance	D 495		sec	167
Comparative tracking index (CTI)	D 3638		V	225

* The tests were performed on as-moulded injection moulded specimens :
melt temperature = 325°C,
mould temperature = 140°C

(1) Annealed specimens

TYPICAL PROPERTIES OF RYTON A-100

PROPERTIES	TESTS (1)		UNITS	RYTON A-100
	ASTM	DIN		
Density	D 1505	53479	g/cm3	1.66
Tensile strength	D 638		MPa	121
		53455	MPa	127
Elongation at break	D 638		%	1.1
Flexural modulus	D 790		MPa	12000
		53452	MPa	13000
Flexural strength	D 790		MPa	172
		53452	MPa	190
Compressive strength	D 695		MPa	158
Compressive strength, axial		53454	MPa	112
transverse		53454	MPa	114
Izod impact strength,				
3.2 mm specimen, unnotched	D 256		J/m	254
3.2 mm specimen, notched	D 256		J/m	53
Charpy impact strength, unnotched		53453	Kj/m2	11
notched		53453	Kj/m2	3.7
Rockwell hardness	D 785			R-116
Heat distortion temperature at 1.81 MPa (2)	D 648	53461	°C	260
UL long term temperature index			°C	200 - 220
Limiting oxygen index	D 2863		%	47
Flammability, UL 94				V-O/5V

(1) The tests were performed on as-moulded specimens (melt temperature 320-340°C, mould temperature 135°C).

(2) Annealed specimens (2 hrs at 200°C)

TYPICAL PROPERTIES OF UDEL® GF-110, 120, 130 RESINS

Properties	ASTM Test Method	GF-110	GF-120	GF-130
Fabrication				
Melt Flow, dgm/min	D-1238	4	7-8	7-8
Mold Shrinkage, %	D-955	0.4	0.3	0.2
Mechanical				
Tensile Strength, psi (MPa) ...	D-638	11,300 (78.0)	14,000 (96.6)	15,600 (107.6)
Tensile Elongation, %	D-638	4	3	2
Tensile Modulus, psi (MPa) ...	D-638	530,000 (3,655)	750,000 (5,172)	1,070,000 (7,379)
Flexural Strength, psi (MPa) ..	D-790	18,500 (127,6)	21,500 (148.3)	22,4000 (154.5)
Flexural Modulus, psi (MPa) ..	D-790	550,000 (3,793)	800,000 (5,517)	1,100,000 (7,586)
Notched Izod, ft.lbs./in (J/m) ..	D-256	1.2 (64)	1.3 (69)	1.4 (75)
Tensile Impact, ft.lbs./in^2(kJ/m^2)	D-1822	48 (100.8)	54 (113.4)	52 (109.2)
Thermal				
Heat Deflection Temp, 264 psi, °F (°C)...........	D-648	354 (179)	356 (180)	358 (181)
Flammability Rating, UL-94	—	V-0 @ 0.062″ (V-0 @ 1.57mm)	V-0 @ 0.062″ (V-0 @ 1.57mm)	V-0 @ 0.062″ (V-0 @ 1.57mm)
Electrical				
Dielectric Strength, volts/mil (kV/mm)	D-149	475 (19)	475 (19)	475 (19)
Volume Resistivity, ohm-cm ..	D-257	$> 10^{16}$	$> 10^{16}$	$> 10^{16}$
Dielectric Constant	D-150			
@ 60 Hz		3.3	3.4	3.5
@ 1 MHz................		3.4	3.5	3.7
Dissipation Factor	D-150			
@ 60 Hz		0.001	0.001	0.001
@ 1 MHz................		0.005	0.005	0.004
General				
Density, g/cc (Mg/m^3)	D-1505	1.33 (1.33)	1.40 (1.40)	1.49 (1.49)

Mechanical Properties

PTMT and PTMT-G 30 are notable for their good mechanical properties. The low moisture absorption has practically no influence on these properties.

The following table shows the test values according to DIN-standards.

Table 2

Property	DIN-standard	Unit	PTMT	PTMT-G 30
Yield point	53 455	kp/cm^2	600	-
Tensile strength at break	53 455	kp/cm^2	350	1,400
Percentage elongation	53 455	%	200	< 2
Modulus of elasticity (tensile test)	53 457	kp/cm^2	26,000	100,000
Flexural stress at conventional deflection	53 452	kp/cm^2	870	1,900*
Impact strength at	53 453	cmkp/cm^2		
23°C			no fracture	38
0°C			no fracture	
- 20°C			no fracture	
- 40°C			no fracture	35
Notched impact strength at	53 453	cmkp/cm^2		
23°C			approx. 4	11
0°C			approx. 4	
- 20°C			approx. 4	
- 40°C			approx. 4	10
Ball indentation hardness	53 456	kp/cm^2		
10 sec.			1,250	1,725
60 sec.			1,200	1,675
Abrasion resistance	Taber abraser (emery S 33)	mg/100 rev.	17	15

* Flexural strength

Electrical properties

PTMT and PTMT-G 30 have good electrical insulating properties, and their dielectric behaviour is favourable. The electrical properties are not significantly influenced by moisture absorption.

Table 3

Property	DIN-standard	Unit	PTMT	PTMT-G 30
Dielectric constant ε_r at	53 483	-		
10^3 Hz			3.25	3.7
10^6 Hz			3.10	3.4
Dissipation factor $tg\delta$ at	53 483	-		
10^3 Hz			0.002	0.0025
10^6 Hz			0.02	0.0190
Surface resistance	53 482	Ω	$> 10^{13}$	$> 10^{13}$
Resistance between tapered pin electrodes	53 482	Ω	$> 10^{13}$	$> 10^{13}$
Volume resistivity	53 482	$\Omega \cdot cm$	$> 10^{16}$	$> 10^{16}$
Dielectric strength	53 481	kV/mm	50 (0.25 mm sheet) 25 (1 mm sheet)	29 (1 mm sheet)
Resistance to tracking	53 480	KA-method KB-method KC-method	KA 2 >200 >600	KA 1 180 280

2nd International Conference on

Developments in Valves and Actuators for Fluid Control

Manchester, England: 28-30 March 1988

PAPER F1

SIZING OF LOW PRESSURE AIR VALVES FOR SAFE FILLING AND DRAINING OF PIPELINES

Y.Dvir
Consulting engineer, Lehavot Habashan, Israel

Summary

In order to avoid hazardous hydraulic phenomena in water systems, air valves should be included in system design, with carefully calculated orifice sizes.

This paper explains the basic principles of air flow in air valve orifice, and proposes procedures for orifice sizing. It suggests criteria for safe air and water flow velocities, and includes auxiliary tables and examples of sizing calculations.

List of symbols

A - pipe cross-section area
a - orifice cross-section area
a_e - effective area of orifice
d - pipe inside diameter
C - Hazen-Williams coefficient for hydraulic conductivity of pipes
c - acoustic wave celerity in water
E - elasticity modulus of pipe material
g - gravitational constant
H - pressure head
Δh - head rise, head loss
J - hydraulic gradient
K - bulk modulus of water
k - adiabatic constant of air
L - pipe length
P - pipe/atmosphere pressure ratio
p - absolute pressure at orifice
Q - discharge, volume flow ratio
S - slope
T - time
t - pipe wall thickness
V - volume
v - air flow velocity
v_p - velocity of flow in pipe
w - specific weight of air

Indexes: $_1$ - conditions at the inlet of orifice
$_2$ - conditions at the outlet of orifice

1. Air valves and their function

The air valve is an appliance which controls the exchange of air enclosed in a water pipeline with atmospheric air. Two main types of air valves are produced commercially.

- The high pressure, or small orifice valve, which serves to release small amounts of air flowing or accumulated in a pipe, under working pressure.
- The low pressure, or large orifice valve, which serves to introduce atmospheric air into a pipeline, during drainage of the water, or sometimes, in case of interruption of the water column; it also serves to release the air contained in a pipe, while filling the pipeline with water. This valve opens only at nearly atmospheric pressure.

The high pressure valve is sometimes called "automatic-", the low pressure "kinetic air valve", but these terms have no technical meaning. Both types of valves operate entirely automatically, using the principle of a float which closes an orifice, either directly or by means of a lever mechanism.

This paper deals with the low pressure - "kinetic" type only.

2. The air valve in water system design

The design of water systems seldom takes account of dangers which can be caused by inappropriate sizing of air valves. In most cases such valves are chosen at random, without the proper calculations, and, more often than not, without understanding their real function and purpose.

This is due to several causes. The average designer of water systems is not sufficiently acquainted with the principles of air flow. The manufacturers of air valves only seldom provide reliable data for the performance of valves, and even then, the data are given in units of free air, which are unsuitable for design calculations. The valves are manufactured in a few standard sizes and are classified rather by the pipe connection than by orifice size, etc.

The purpose of this paper is to pinpoint the critical factors, which should be used in the sizing of air valves, and to propose the proper procedure for calculating the correct size of the valve orifice for a particular system.

It was found in field observations, that the most serious attention should be paid to the prevention of water hammer, caused by closure of air valve in the final stage of pipe filling. The second most important point is to prevent an excessive vacuum, while the pipe is drained.

In most practical cases, in a properly managed water system, the process of filling and draining a pipeline will be controlled manually (or automatically) by a control valve at the inflow or at the drain end of the line. However, for safety purposes, the design must be such, that even in case of faulty control equipment, or misjudged manual regulation the system will not suffer damage from hydraulic occurences.

3. Sizing of air valves

The difficulty in sizing the orifice of an air valve has several causes. The calculation of the flow of a compressible fluid is more complicated than that of water flow; the critical flow velocity is reached at comparatively low pressure differentials; the difference between the volumetric flow rate of air entering and leaving the orifice baffles the water engineer; the limitations of flow velocity imposed by vacuum or water hammer are not clearly understood and so on.

As mentioned, the low pressure air valve serves mainly for draining and filling pipelines. When a pipe lying on a slope is drained through an open valve, air must be introduced at the upper end of the line, in order to prevent vacuum, which can have a damaging influence on system elements (leakage, cavitation, collapsing etc.) The vacuum will depend on the parameters of the water flow and pipe conditions, and on the size of the air valve orifice. It is customary to ascertain that the vacuum pressure shall not fall below 50 kPa (5 m water column) absolute.

When, again, such a pipeline is filled with water from the source (pump, main etc.), an air valve placed at the upper end will release the air, thus allowing the water to flow along. Velocity of flow then depends on the hydraulic parameters (slope, head at inlet, hydraulic conductivity of the pipe), on the size of the air valve orifice and on air pressure at the orifice. When the water reaches the air valve and all the air has escaped, the water flow stops. The stoppage produces a water hammer, whose magnitude depends on the final flow velocity and the sonic wave celerity (i.e the characteristics of the pipe). In steel and asbestos-cement pipes the approximate pressure rise caused by water hammer can reach about 1000 kPa (100 m) for each 1 m/s of the flow velocity. It is customary to allow for water hammer pressure rise of, at most, 50% the working pressure of the system. The air valve orifice should be sized so, as to control the discharge of escaping air and to prevent excessively high flow velocity of water.

It should be evident, that though the low pressure air valve can serve a dual purpose, i.e. prevention of vacuum at drainage and restraint on the flow velocity at filling, these purposes are incompatible. To reduce the vacuum a larger orifice is required, to reduce the flow velocity the orifice should be smaller. The ideal solution is to use two or more one-way air valves, and to size them accordingly - one for airing, another for air release. However most system designs use two-way valves for both, draining and filling, and this requires serious consideration of the limiting parameters.

In both types of flow, either filling or draining the pipe, consideration concerns the volumetric flow rate and the velocity of water flow. Therefore the flow of air, which evacuates the pipe in filling, or replaces the water in draining must be calculated in terms which are conforming to the water flow parameters, i.e. in terms of compressed or expanded volume, at pipe pressure. Such calculation differs from the standard procedure of computing air flow in terms of atmospheric conditions (free air, N.T.P.). In the following considerations all air flow will be expressed in terms of pipe conditions.

4. Air flow in orifice

4.1. Basic principles

The air flow in water pipe can be considered isothermal, but the flow of air through an orifice of the air valve is assumed to be adiabatic and frictionless and is calculated by a formula derived from Bernoulli's theorem:

$$v = \frac{Q}{a} = \left[2g\,\frac{k}{k-1}\,\frac{p_1}{w_1}\right]^{\frac{1}{2}}\left[\left(\frac{p_2}{p_1}\right)^{\frac{2}{k}} - \left(\frac{p_2}{p_1}\right)^{\frac{k+1}{k}}\right]^{\frac{1}{2}} \tag{1}$$

The critical flow occurs when

$$\frac{p_1}{p_2} = \left(\frac{k+1}{2}\right)^{\frac{k}{k-1}} = 1.89 \tag{2}$$

For all the values of

$$\frac{p_1}{p_2} > 1.89$$

the flow of air depends only on inlet conditions

$$v = \left[2g\,\frac{k}{k+1}\,\frac{p_1}{w_1}\left(\frac{2}{k+1}\right)^{\frac{2}{k-1}}\right]^{\frac{1}{2}} \tag{3}$$

and as in the pipe $\frac{p_1}{w_1}$ is constant, the flow at inlet remains constant for all values of $p_1 > 1.89 p_2$.

These are the basic formulas for air flow through an orifice, ensuing from the flow theory for compressible fluids.

4.2. Experiments

To confirm the theoretical considerations of air flow, a series of experiments was made. The purpose of these was to conclude, if the inflow of air into the orifice remains constant at all inlet pressures above the critical. It was not meant to verify the theoretical results precisely, but merely to receive comparative data. Therefore no scientifically designed and calibrated equipment was used.

The equipment (see Fig.1) consisted of two pressure vessels of approximately 25000 cm³ (25 liters) each. The vessels were connected at the bottom by a short pipe with a shut-off valve. The first vessel was filled with water and pressurized by air through a pressure reducing valve. The second vessel was filled with air and pressurized by the same valve to exactly the same pressure, so that when the connecting valve was opened, no flow ensued in either direction. On top of the air vessel a pressure gauge and a short vertical pipe was installed, with a shut-off valve under a calibrated orifice.

By opening the valve under the orifice the air was permitted to escape, and water filled the second vessel. The period of time from the beginning of flow till the appearance of water was recorded.

Inflow velocity was calculated from the volume of air in the vessel, the discharge time and the orifice area. The results are shown in table 1. Discrepancies in recorded times and calculated velocities are probably due to faulty calibration of the orifices, inexact time measurements and changes in ambient temperature, which varied considerably throughout the day. However the results confirm that the flow rate remains practically constant at all inflow pressures above the critical.

Table 1: Measurements of air velocity in orifices

Air vessel volume: 24650 cm³					
Nominal diameter of orifice mm²		0.6	1.0	1.6	2.5
Effective area of orifice mm² *		0.26	0.70	1.77	4.32
Test pressure kPa gauge	p_1/p_2	Time in seconds			
150	1.5	473.6	178.9	70.5	27.6
250	2.5	475.4	176.9	69.2	28.7
400	4.0	474.6	180.8	71.4	29.3
550	5.5	474.5	177.6	70.4	28.7
Mean velocity calculated m/s		199.8	197.2	197.9	199.7

*Obtained by calibration with water

From this point on, two different instances of air flow will be treated separately.

4.3 Release of air

4.3.1. Theory

The air flows through an orifice from its compressed state at inlet (in the pipe), to atmospheric conditions at outlet.

Here p_1 is the pressure in the pipe, and it can change according to conditions in the system. p_2 is atmospheric, and is assumed to be constant at 100 kPa(10m) absolute. Let

$$\phi = \frac{p_1}{p_2}$$

Assuming conditions in the pipe $\frac{p_1}{w_1}$ constant and substituting the various constant values in (1):

$$v = 756.5 \left(\phi^{-1.42} - \phi^{-1.71}\right)^{\frac{1}{2}} \tag{4}$$

for all values of $\phi < 1.89$. When the absolute pressure in the pipe rises above 189 kPa (89 kPa or 8.9 m manometric), the flow becomes critical. From (3)

$$v = 198 \text{ m/s approximately}$$

which is the constant inlet velocity of air for all pressures greater than 89 kPa (8.9 m) manometric.

In case of a pipeline fed from a pressure source, and not controlled by an inlet valve, the pressure is usually higher than critical. Therefore the velocity of 198 m/s should be generally used for sizing the orifice of an air valve.

However in cases of low pressure sources, gravitational intakes etc. or very low hydraulic conductivity of the pipe, the pressure at the air valve can be lower than 89 kPa (8.9 m) manom. and the air velocity must be calculated from (4) or from auxiliary tables.

The velocity of air at orifice inlet, multiplied by the effective area of the orifice rends the total air discharge, and this equals the rate of volume per unit time of the water flow which fills the pipe. Thus the velocity of water flow can be obtained.

As mentioned before, excessive velocity of flow can cause a considerable water hammer, with very high pressure rise in the pipe. This can be avoided if water velocity is limited by appropriate sizing of the air valve orifice. As the closure of air valve can be considered instantaneous, the permissible filling velocity can be calculated from the Joukovsky formula for maximum surge pressure

$$\Delta h = \frac{c\,v}{g} \tag{5}$$

4.3.2. Sizing procedure

The procedure for sizing the orifice of an air valve shall be the following:
- Determination of the permissible surge pressure (usually not over 150% of static pressure in the system).
- Calculation of wave celerity for the given pipe from the formula

$$c = \left[\frac{Kg}{w\left(1 + \frac{K}{E}\frac{d}{t}\right)}\right]^{\frac{1}{2}}$$

- Insertion of both values into (5) and calculation of the maximum permissible filling velocity.
- Calculation of head loss in pipe and pressure at the air valve in the final filling stages.
- If the pressure is equal or higher than 89 kPa (8.9 m) manometric, the orifice area will be calculated from

$$a_e = \frac{v_p A}{198}$$

- If the pressure is lower than 89 kPa, the velocity of air will be calculated from (4) and the orifice area from

$$a_e = \frac{v_p A}{v_1}$$

In regard to the above procedure it must be taken into account that in pipes with a very moderate slope or in nearly horizontal ones, most of the filling flow can appear in form of enclosed channel flow with varying depth, and therefore it will be almost impossible to calculate pressure at the air valve exactly.

4.4 Intake of air

4.4.1. Theory

The air flows through an orifice from atmospheric conditions at inlet to vacuum conditions at outlet (in the pipe).

Here the pressure p_1 is atmospheric and is assum ι constant, p_2 is less then atmospheric.

Let

$$\phi = \frac{p_2}{p_1}$$

For values of $\phi > 0.528$ the velocity of flow is given by (1), and after substituting the constants

$$v_2 = \left(\phi^{-0.571} - \phi^{-0.286}\right)^{1/2} \qquad (6)$$

When the pressure falls below 52.8 kPa(5.3 m) absolute, the flow becomes critical and the inlet velocity constant. Therefore the mass flow of air entering the orifice from the atmosphere also remains constant for all pressures in the pipe, lower than 52.8 kPa. But the vacuum in the pipe will cause further expansion of air, and thus velocity of flow at outlet of the orifice will rise proportionally to the fall of pressure:

$$v_2 = \frac{198}{\phi} \text{ m/s} \qquad (7)$$

for $p_2 < 52.8$ kPa absolute.

4.4.2. Sizing procedure

The procedure of orifice sizing will be as follows:

- Determination of the maximum permissible vacuum within the pipe (usually at most 50 kPa [5 m] absolute).
- Calculation of the available head differential.
- Calculation of the required flow velocity, from the usual head loss formulas.
- Calculation of air flow from (6) or (7), or from auxiliary tables(See Table 2).
- Calculation of the orifice area from

$$a = \frac{v_p A}{v_2}$$

5. Calculation examples

Some numerical examples show the calculation method of orifice size and the problems that can possibly arise, when the valve is supposed to serve both filling and draining of the pipeline.

5.1.

A 300 mm ID (12") pipe of 750 m length lies on a rising slope of 8/1000. The assumed hydraulic conductivity is C - 130 (Hazen-Williams), and wave celerity 1000 m/s. The water is supplied from a 1750 mm (70") main, at a regulated pressure of 500 kPa (50 m) and a practically unlimited flow rate potential. A 300 mm drain valve is installed near the supply connection, and an air valve should be placed at the high end of the pipe. While the pipe is being filled or drained, all branches and service connections are closed. The air valve should be chosen so that the maximum pressure of surge will not exceed 50% of static pressure and the vacuum at draining will not fall below 50 kPa (5 m) absolute.

See calculation sheet 1. As can be seen, the sizes of orifices required for safe filling and draining are close enough, and there is no problem in installing the right size of air valve, if such can be commercially acquired.

5.2.

A 500 mm (20") seriously corroded steel pipe lies on a slope of 50 m over 3000 m lenght. As in the preceding example the water supply is practically unlimited, at constant 600 kPa (60 m), the permissible pressure rise is 150 kPa (15 m) maximum, the permissible vacuum is 50 kPa (5 m) absolute, hydraulic conductivity of pipe C - 90 and wave celerity 1000 m/s.

See calculation sheet 2. In this case there is a considerable discrepancy between the sizes of the orifices suitable for draining and filling. If the actual valve will be based on drain calculations, the anticipated velocity of flow in filling can reach 0.8 m/s and the ensuing water hammer, together with static pressure, about 1400 kPa (140 m) which will probably have a damaging effect on the corroded pipe. On the other hand, draining the pipe with an air inlet of only 0.00015 m^2(150 mm^2) area will give a relatively low discharge, with a vacuum of 20 kPa (2.m) absolute, again a very dangerous condition.

This problem can be solved in several, though not exactly ideal ways:
- Using two different, one-way air valves, one, smaller, for expelling air, another, larger, for draining.
- using an intermediate orifice size, and deliberately permitting a higher than intended pressure rise, and lower draining pressure, with all the involved risk.
- Installing two or more additional valves along the pipeline, using the smaller orifice size. In this case water velocity will diminish gradually, as the valves close, and the final velocity will be as calculated. However, positioning of the air valves must ensure, that they will all open simultaneously when the drain valve opens, and this can be done only on nearly horiziontal pipelines.

6. Orifice efficiency

In all above calculations orifice area means the functional, effective area of the orifice. This is smaller than the factual cross-sectional area of the orifice opening. The ratio of the effective and the real areas is the coefficient of efficiency and depends on the qualities of the orifice (geometry, smoothness etc.) and on flow characteristics (Re number).

This coefficient can be determined only experimentally; for most round, smoothly machined orifices and for average Reynolds numbers it varies from 0.75 for larger, to 0.9 for smaller orifices, approximately.

Determination of the coefficient for air flow is somewhat difficult, because the involved velocities are high and volumetric mensuration of air is complicated. Therefore the actual calibration of orifices is done by using water, at Reynolds numbers similar to the anticipated characteristics of air flow.

For the critical velocity of air flow, the matching flow velocity of water will be about 13.5 m/s, i.e. a flow under, approximately, 90 kPa (9 m) head differential. Therefore the experimental determination of the orifice coefficient should be done under this head.

The calibration equipment is simple, as shown in Fig.2. It consists of a short pipe, a ball valve or pressure regulator, a pressure gauge and a calibrated tank. The pressure is first regulated with open flow to the required 90 kPa (9 m) at the orifice inlet. The stream is then directed into the tank and time measurement taken. The effective area of the orifice is computed from

$$a_e = \frac{V}{T(2gH)^{1/2}}$$

7. Practical applications

An additional problem is how to apply the calculation of orifice size to an actual air valve. As explained before, commercial low pressure valves are produced with a few standard orifice sizes. In general the orifices are part of the body cover, and are not interchangable.

However over the last years some models of air valves were equiped with an additional threading in the air outlet. This threading was primarly intended for the installation of drain pipe, which removes the occasional splashes of water from the valve. Such threaded outlet can be effectively used for installing an additional, calibrated orifice, of smaller size than the original. Such orifice will not influence the opening or closing of the valve, but it will effectively control the air flow. See Fig.3.

8. Discussion

It should be clearly understood that the calculating method, as outlined above, will not always be valid for all water supply systems. Such conditions as variable inflow pressure (pump source or main with limited flow potential), alternatively rising and falling slope etc. will require modification of the calculating procedure, and adaptation of data to the various stages fo calculation.

However comprehension of the principles governing the air flow in orifices, and the use of air flow terms consistent with terms of water flow, will help to design water systems free from hazardous hydraulic occurences.

The manufacturer of air valves can greatly contribute to the proper system design, by providing hydraulic data for their valves, expressed in terms of compressed air flow, and by supplying effective areas of valve orifices, based on test-bench measurements.

A design of air valves with interchangable orifices, or with a wide range of orifice sizes will also result in safer water system design.

9.Bibliography

1. Gandenberger, W.:"Long Waterlines".(Fernwasserleitungen).Munich, Germany, Oldenburg, 1957.(In German).
2. Dvir, Y.:"Air in pipelines".Tel-Aviv, Israel, ICWA, 1977.(In Hebrew).
3. Dvir, Y.:"Control appliances in water systems", Volume 2. Lehavot Habashan, Israel,1986.(In Hebrew).
4. Volk, W.:"Valves in pipelines".(Absperrorgane in Rohrleitungen). Berlin, Germany, Springer Verlag, 1959.(In German).
5. Webb, T. and Gould, B.W.:"Water hammer".Kensington, Austalia, NSW University Press, 1978.

Calculation sheet 1

Data:
$L = 750$ m, $C = 130$, $d = 0.3$ m (300 mm, 12"), $A = 0.071$ m², $S = 8/1000$
$c = 1000$ m/s, $H = 50$ m, Δh max = 25 m, Δh min = -5 m $\phi = 0.5$ (draining)

Parameter	Formula	Data	Result
1. Filling			
Max. water velocity	$v_p = \frac{\Delta h g}{c}$	$\Delta h = 25$ m $c = 1000$ m/s	$v_p = 0.25$ m/s
Hydraulic gradient	Hazen-Williams	$d = 0.3$ m $C = 130$ $v_p = 0.25$ m/s	$J = 0.26/1000$
Static head at orifice	$H_1 = H - SL$	$H = 50$ m $S = 8/1000$ $L = 750$ m	$H_1 = 44$ m
Residual pressure	$p_1 = 10(H_1 - LJ)$	$H_1 = 44$ m $L = 750$ m $J = 0.26/1000$	$p_1 = 438$ kPa>89 kPa
Orifice area	$a_e = \frac{v_p A}{198}$	$v_p = 0.25$ m/s $A = 0.071$ m²	$a_e = 8.96 \times 10^{-5}$ m² (89.6 mm²)
2. Draining			
Air velocity	$v_2 = \frac{198}{\phi}$	$\phi = 0.5$	$v_2 = 396$ m/s
Hydraulic gradient	$J = S - \frac{\Delta h}{L}$	$\Delta h = 5$ m $L = 750$ m $S = 8/1000$	$J = 1.33/1000$
Water velocity	Hazen-Williams	$d = 0.3$ m $C = 130$ $J = 1.33/1000$	$v_p = 0.6$ m/s
Orifice area	$a_e = \frac{v_p A}{v_2}$	$v_p = 0.6$ m/s $A = 0.071$ m² $v_2 = 396$ m/s	$a_e = 10.76 \times 10^{-5}$ m² (107.6 mm²)

Calculation sheet 2

Data:
$L = 3000$ m, $C = 90$, $d = 0.5$ m(20"), $A = 0.196\ m^2$, $S = 50/3000$
$c = 1000$ m/s, $H = 60-50 = 10$ m, $\Delta h_{max} = 15$ m, $\Delta h = -5$ m, $\phi = 0.5$ (draining).

Parameter	Formula	Data	Result
1. Filling			
Max. water velocity	$v_p = \frac{\Delta h g}{c}$	$\Delta h = 15$ m $c = 1000$ m/s	$v_p = 0.147$ m/s
Hydraulic gradient	Hazen-Williams	$C = 90$ $d = 0.5$ m $v_p = 0.147$ m/s	$J = 0.18/1000$
Residual pressure	$p_1 = 10(H-LJ)$	$H = 10$ m $L = 3000$ m $J = 0.18/1000$	$p_1 = 94.6$ kPa>89kPa
Orifice area	$a_e = \frac{v_p A}{198}$	$v_p = 0.147$ m/s $A = 0.196\ m^2$	$a_e = 14.6 \times 10^{-5} m^2$ ($146\ mm^2$)
2. Draining			
Air velocity	$v_2 = \frac{198}{\phi}$	$\phi = 0.5$	$v_2 = 396$ m/s
Hydraulic gradient	$J = S - \frac{\Delta h}{L}$	$S = 50/3000$ $\Delta h = 5$ m $L = 3000$	$J = 15/1000$
Water velocity	Hazen Williams	$d = 0.5$ m $J = 15/1000$ $C = 90$	$v_p = 2.14$ m/s
Orifice area	$a_e = \frac{v_p A}{v_2}$	$v_p = 2.14$ m/s $v_2 = 396$ m/s $A = 0.196\ m^2$	$a_e = 1.06 \times 10^{-3} m^2$ ($1060\ mm^2$)

Table 2: Air flow velocity in orifice

Pressure at outlet kPa absolute	Velocity* m/s	Pressure at inlet kPa absolute	Velocity** m/s
10	1960	105	87
15	1307	110	117
20	980	115	137
25	791	120	151
30	660	125	162
35	565	130	170
40	494	135	176
45	439	140	182
50	396	145	186
55	355	150	189
60	322	155	191
65	291	160	193
70	260	165	195
75	230	170	196.0
80	200	175	196.8
85	169	180	197.3
90	134	185	197.6
95	92	189	197.7

* Air expanded at pressure as shown.
**Air compressed at pressure as shown.
Atmospheric air at 21°C, sp.w.1.20 kg/m³

Table 3: Influence of ambient temperature on air flow velocity

Temperature °C	Correction factor *	Critical velocity m/s
0	0.964	190.6
1	0.966	191.0
2	0.967	191.2
3	0.969	191.6
4	0.971	191.9
5	0.972	192.2
6	0.974	192.6
7	0.976	192.9
8	0.978	193.3
9	0.979	193.6
10	0.981	194.0
11	0.983	194.3
12	0.985	194.7
13	0.987	195.1
14	0.988	195.3
15	0.990	195.7
16	0.992	196.1
17	0.993	196.3
18	0.995	196.7
19	0.997	197.1
20	0.998	197.3
21	1.0	197.7
22	1.002	198.1
23	1.003	198.3
24	1.005	198.7
25	1.007	199.1
26	1.008	199.3
27	1.010	199.7
28	1.012	200.1
29	1.014	200.5
30	1.015	200.7

* The correction factor is valid for all air velocities as shown in Table 2.

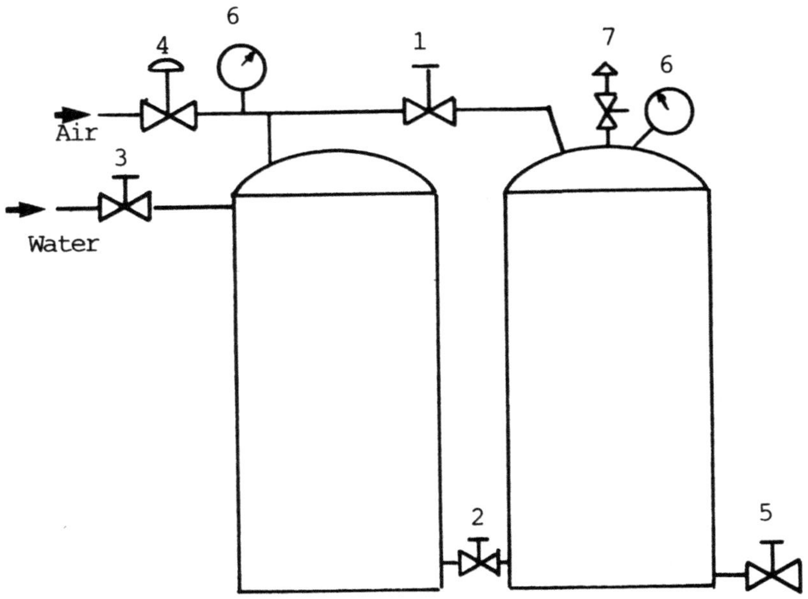

1.Manual valve for primary pressurizing.
2.Connecting valve.
3.Water supply valve.
4.Pressure regulator.
5.Drain valve.
6.Pressure gauges.
7.Orifice & orifice valve.

At test:
Valves 1,3,5 closed
Valves 2,7 open
Valve 4 at test pressure.

Figure 1: Air velocity measurment

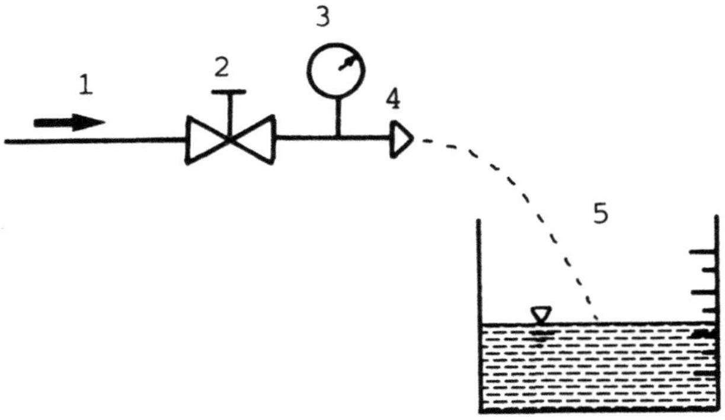

1.Supply piping.
2.Regulating valve.
3.Pressure gauge
4.Orifice (for calibration).
5. Calibrated tank.

Figure 2: Orifice calibration

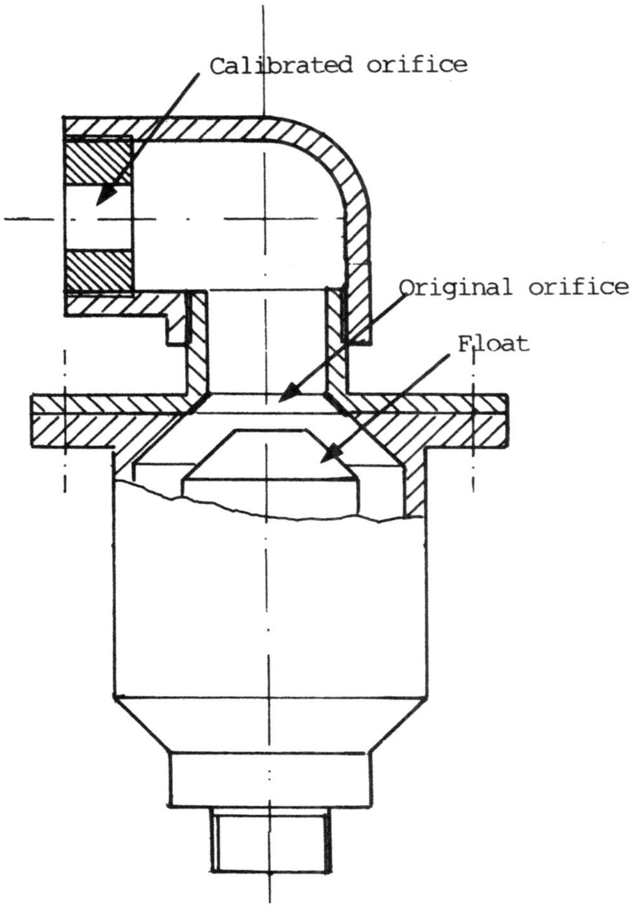

Figure 3: Air valve with an additional orifice

2nd International Conference on

Developments in Valves and Actuators for Fluid Control

Manchester, England: 28-30 March 1988

PAPER F2

A PROPOSED METHOD OF CONTROL VALVE SIZING FOR NON-IDEAL FLUIDS

Allen C. Fagerlund
Fisher Controls International, Inc.
Marshalltown, Iowa USA

Summary

I.

A semi-analytical model has been developed to describe the flow of two-phase fluids through a restriction. Two-component control valve data was used to describe the slip phenomenon between phases. The effects of metastability and the recovery coefficients were also incorporated in the model.

II.

A gas at high pressure can exhibit non-ideal behavior such that its thermodynamic properties cannot be evaluated by assuming the ideal gas law without a significant error resulting. The effect of this on valve sizing is determined and a correction factor for current sizing procedures is defined.

Held in Manchester, England. Organised and sponsored by BHRA, The Fluid Engineering Centre and co-sponsored by the British Valve Manufacturers Association and the Institution of Mechanical Engineers Process Industries Division

Nomenclature

A —Area

C_{vf} —Valve flow coefficient

C_{VR} —Required valve flow coefficient

C_p^0 —Ideal constant pressure specific heat

C_p —Constant pressure specific heat of real gas

C_v^0 —Ideal constant volume specific heat

C_{va} —Constant volume specific heat of real gas

C_g —Fisher's gas sizing coefficient

C_1 —C_g/C_{vf}

C_2 —Fisher's specific heat ratio coefficient

G —Specific gravity

K_m —Recovery coefficient (F_L^2)

k_0 —Ideal gas specific heat ratio ($C^0{}_p/C^0{}_v$)

n —Isentropic volume exponent

P —Absolute pressure

P_c —Absolute critical pressure of gas

P_r —Reduced pressure (P/P_c)

P_v —Vapor pressure

ΔP —Valve pressure drop

R —Ideal gas constant

T —Absolute temperature

T_c —Absolute critical temperature of gas

T_r —Reduced temperature (T/T_c)

u —Velocity

v —Specific volume

W —Mass flow rate

x —Quality $W_g/(W_g+W_f)$

y —Slip ratio

Z —Compressibilty factor

η —Pressure ratio

Subscripts

I.

0 —Inlet condition

1,2 —Model designation

g —Gas or vapor phase

f —Liquid phase

II.

1 —Inlet condition

I. Two-Phase Flow

Introduction

The flow of two-phase mixtures through restrictions is a complex phenomenon that to date has not been fully described analytically. It is an area that received a geat deal of attention because of its application to nuclear reactor technology. The majority of the work done in this area considered ideal geometries such as nozzles, orifices and straight pipes. In the area of control valves very little work has been done. Brockett & King [1] studied subcooled water. Stiles [2] looked at subcooled freon. Martinec [4] compared subcooled freon in valves with ideal geometries. Sheldon & Schuder [3] looked experimentally at air/water mixtures through valves that resulted in a sizing procedure. Fagerlund [10] presented an analytical model that required use of the Sheldon & Schuder data to establish the behavior of valves as opposed to more ideal geometries. However, the data used was limited to a single valve travel.

Fagerlund & Storer [11] have expanded this to include several valve travels that further generalizes the technique.

It is the intent of this paper to summarize a practical approach to sizing valves for two-phase service that may be reduced to either a graphical or calculator procedure.

Discussion of Analysis

A fundamental assumption in this method is that the quality remains constant between the inlet and the vena contracta. For gas-liquid flows it is obvious providing vaporization does not occur. However, for vapor-liquid flows it requires an assumption that a metastable condition can exist, that is, that the fluid process lags equilibrium by a time sufficient to preclude increasing quality between the inlet and the vena contracta.

If the phases are treated separately using the standard gas and liquid sizing equations, [5,6] a condition of slip is automatically specified at the throat. The gas sizing equation allows sonic velocity to exist at the throat while the liquid phase behaves as a Bernoulli flow. A limited two-phase factor, presented by Sheldon & Schuder [3] is currently used to correct for this inherent slip condition in valves.

It appears that some actual slip does exist, however, not to the extent shown inherently with the sizing equations.

Fagerlund [10] derived a correction factor for valve sizing which corrects for the inherent and actual slip. Only the results are reprinted here as the following relationship.

$$\frac{C_{VR}}{C_{vg} + C_{vf}} = \frac{\left[1 + \left(\frac{1 - x}{x}\right) \frac{v_f}{v_{g1}} \frac{1}{Y_1}\right] \sqrt{1 + \left(\frac{1 - x}{x}\right) Y_1^2}}{\left[1 + \left(\frac{1 - x}{x}\right) \frac{v_f}{v_{g2}} \frac{1}{Y_2}\right] \sqrt{1 + \left(\frac{1 - x}{x}\right) Y_2^2}} \qquad (1)$$

where subscript 1 refers to the actual conditions and subscript 2 refers to the inherent or ideal conditions.

It remains to define the terms in Equation 1. All are to be evaluated at the throat. Quality and specific volumes for both phases at the inlet are assumed known. The specific volume of the liquid is taken to be constant while the gas phase will follow a polytropic process from the inlet to the vena contracta.

Tangren [7] offers the following comments relative to this polytropic process:

> "At the low qualities the mass and specific heat of the liquid are so large relative to the gas that the gas may remain in thermal equilibrium with the liquid during expansion without appreciably lowering the liquid temperature. Therefore, the mixture behaves as though the gas process were isothermal."

At high qualities the gas phase dominates and an isentropic process is followed.

The ratio of the inlet and vena contracta specific volumes of the vapor is inversely proportional to the pressure ratio from the inlet to the vena contracta.

$$v_{g1} = v_{g0} \frac{P_0}{P_1} = v_{g0} \frac{1}{\eta_1} \quad (2a)$$

$$v_{g2} = v_{g0} \frac{P_0}{P_2} = v_{g0} \frac{1}{\eta_2} \quad (2b)$$

For the present discussion it will be assumed that critical flow exists since critical flow data is used throughout. Since the gas valve sizing equation is based on ideal gas considerations, the critical pressure ratio (η_2) may be calculated from the standard form.

$$\eta_2 = \left[\frac{2}{\gamma + 1}\right]^{\frac{\gamma}{\gamma-1}} \quad (3)$$

With the isothermal assumption this reduces to a limiting value of $\eta_2 = 0.6065$ as $\gamma \rightarrow 1$.

Using the continuity equation the inherent slip may be expressed as:

$$y_2 = \frac{u_{f2}}{u_{g2}} = \left(\frac{1 - x}{x}\right) \frac{v_{f2}}{v_{g2}} \frac{C_{vg}}{C_{vf}} \quad (4)$$

where the sizing coefficients are assumed proportional to the flow areas of the two phases. Attempts to incorporate the recovery coefficients have had mixed success.

The actual slip and pressure ratios in equation (1) must be developed. The actual slip ratio was determined from the Sheldon & Schuder [3] air-water data. Equation 1 may be solved for y_1, which results in finding the roots of a fourth order polynomial. The ratio of C_{vR} to ($C_{vf} + C_{vg}$) is known from the test data. Slip ratio models have been the subject of both theoretical and experimental research for many years and still remain unresolved in a strict sense. One central debate is whether the slip ratio is proportional to the square root, cube root or in fact another power of the phase density ratio as suggested by Simpson [13]. The data taken on valves in Figure 1 shows a suitable collapse with the cube root. Density ratios were varied using different inlet pressures. Also included on this figure is the air-water data of Fauske [8] which was taken on nozzles discharging to atmosphere. The agreement with this and valve data offers encouragement for having a generalized method. A fit of this data yields the following expression for the slip ratio,

$$y_1 = 3 \cdot x^{-0.06} \left(\frac{v_f}{v_{g0}} \eta_1\right)^{1/3} \quad (5)$$

In all cases the slip ratio will have an upper limit of unity which is homogeneous flow. For gas-liquid mixtures Sheldon & Schuder [3] saw a lower limit close to $y_1 = 0.4$.

One-component mixtures would have a higher value for a given quality and density ratio because of the increased pressure ratio η_1, as shown in equation (5). Calculated values outside of these limits will be assumed to have the limiting value.

The final term which requires definition is the actual critical pressure ratio. Like slip this cannot be exactly formulated without empiricism. Currently the critical pressure ratio is considered known for saturated liquids and is specified by r_c and is defined as the pressure ratio which provides the maximum mass flow rate through a restriction. This value is based on equilibrium assumptions which do not allow a constant quality for vapor-liquid flows. Therefore, a critical pressure ratio shown in Figure 2 as a function of quality and pressure was generated by using the value of r_c up to the quality representing the throat quality had an equilibrium process been followed. The high quality (greater than 0.3) region was an extension to the value proper for the 100% vapor condition. Figure 2 is based on steam while Figure 3 is a general form of the same curve as a function of reduced pressure.

This completes the presentation of all factors required to evaluate equation (1).

Discussion of Results

Due to the close correlation of nozzle and valve slip ratios it was decided to check the overall approach against the nozzle data in the literature.

Data is taken from Henry & Fauske [9].

A. Steam - water at 3448 kPa (500 psia) - Figure 4.

Various models are included on the graph. Information from Figure 3 is plotted on Figure 4a to show that the assumed form of the critical pressure compares well with data. Figure 4b shows a comparison of theory with data. The errors at very low qualities could be due to a breakdown in continuity of the vapor phase as a bubble flow begins to exist. At this point normal liquid sizing procedures could be followed. Future work will have to determine a guideline as to where this behavior begins. For this case the flow is essentially homogeneous.

B. Steam - water at 121 kPa (17.6 psia) - Figure 5.

Figure 5a is again a check on the proposed critical pressure ratio where correlation is quite good. Mass velocity calculations also compare well, with an actual slip ratio at the lower allowable limit.

C. Two-phase carbon dioxide at 1379, 3448 kPa (200, 500 psi) - Figure 6.

Critical pressure ratios were found from Figure 3 and applied as with the steam data. Mass velocity calculations compare very well to show that scaling with the critical pressure does seem appropriate.

Conclusions and Recommendations

A semi-analytical method has been developed to describe one and two component, two-phase flow through a restriction. Two-component control valve data was used to describe the slip phenomenon between phases. This information was then used to predict the behavior of one-component flows through more ideal geometries such as orifices and nozzles. Calculations were successfully made for steam-water and two-phase carbon dioxide mixtures. Various assumptions were made to simplify the analysis such as;

a) Constant quality from the inlet to the vena-contracta.

b) Insothermal expansion of the gas phase.

c) Slip ratios have limits, $1 \geq Y \geq .4$.

d) The phase flow area ratio is equal to the ratio of the gas C_v and liquid C_v.

All steps in this method may be reduced to either a graphical or calculator procedure, thus facilitating day to day use.

Valve geometry differences are taken into account in the initial C_v calculations for each phase.

The procedure developed in this report represents a practical approach to sizing valves for two-phase service which provides a smooth transition from low qualities into saturated and subcooled liquid service.

Further work must be undertaken to determine the effects of valve lift on the slip terms of Equation (1). Also a better definition of the actual critical pressure ratio (η) is required as well as further clarification as to the limits of metastability.

II. Non-Ideal Gas

Introduction

Applications have been on the increase in recent years for high pressure gas service. This is due to higher pressure oil and gas fields, higher operating pressures in power plants and other severe service applications.

Currently valve sizing procedures for compressible flow are based on the assumption of an ideal gas which undergoes an isentropic process. Though it has been long estalished that high pressures give rise to non-ideal characteristics in real gases there has been little impetus for determining its total impact on valve sizing other than including the compressibility factor "Z". This factor corrects the ideal density for real gas behavior as a function of temperature and pressure. However, it does not correct the isentropic process itself which is affected by change a real gas creates in the isentropic exponent.

Before going further, clear definitions of what is meant by the terms ideal gas and real gas are in order.

Ideal Gas -

A) Intermolecular spacing is great enough that intermolecular forces may be considered negligible.

B) All gas internal properties (entropy, enthalpy, specific heat, etc.) are functions of temperature only.

C) Obeys ideal gas law.

$$PV = RT \qquad (6)$$

Real Gas -

A) Intermolecular forces are no longer negligible.

B) Gas internal properties are functions of both temperature and pressure.

C) Deviation from ideal gas expressed with compressibility factor.

$$PV = ZRT \qquad (7)$$

In the following sections a method will be developed to account for real gas effects on the isentropic exponent and then applied to current valve sizing techniques.

The basic analysis was taken from Scheonthaler [14]. All parameters are defined in the nomenclature.

Analysis

When describing the flow through nozzle orifices, valves or other restrictions, it is usually assumed that the thermodynamic process from the inlet to the vena contracta is isentropic and can be represented by:

$$PV^n = \text{constant} \tag{8}$$

In general the isentropic exponent may be written as (2):

$$n = -\frac{V}{P}\left(\frac{\partial P}{\partial V}\right)_T \left(\frac{C_p}{C_v}\right) \tag{9}$$

For an ideal gas it is readily shown that the isentropic exponent "n" is equal to the ratio of specific heats "k".

However, under those conditions where the ideal gas law does not hold, the expression for the exponent becomes a complex function of the temperature and pressure. It may be determined for those fluids where extensive property data exists such as steam [16].

However, to develop a more general approach requires that an equation of state which reflects the real gas properties of fluids be utilized.

Various equations of state have been developed relating pressure, temperature and density though some are accurate over only limited temperature and pressure ranges and for only a few substances.

The Redlich-Kwong equation [17] is written as:

$$P = RT/(V - b) - a/VT^{0.5}\ (V + b)$$

$$a = 0.4278 \frac{T_c^{\,2.5} R^2}{P_c},\ b = 0.867 \frac{T_c R}{P_c} \tag{10}$$

and represents an improvement to Van der Waal's equation and the coefficients can be evaluated from the critical properties. Since, this equation is commonly used in the hydrocarbon industry where it is anticipated that valve sizing corrections will be needed, it was selected for use in this study.

Therefore, Eqn. (10) will be used to evaluate the parameters in Eqn. (9) to allow a determination of the isentropic exponent.

It follows directly that

$$\frac{V}{P}\left(\frac{\partial P}{\partial V}\right)_T = -\frac{Z}{(Z-B)^2} + \frac{A}{Z(Z+B)} + \frac{A}{(Z+B)^2} \tag{11}$$

where A and B are functions of the reduced temperature and pressure as well as the Redlich-Kwong coefficients and are written as:

$$A = .4278 \frac{P_r}{T_r^{2.5}}, \qquad B = .0867 \frac{P_r}{T_r} \tag{12}$$

It now remains to express the ratio of specific heats as a variable function of both temperature and pressure.

ΔC_p is commonly defined in terms of excess or residual enthalpy.

$$\Delta C_p = C_p - C_p^0 = \left[\frac{\partial(\Delta H)}{\partial T}\right]_P \tag{13}$$

where ΔH is in turn defined by the activity coefficient $\emptyset$.

$$\Delta H = H - H_o = -RT^2 \left(\frac{\partial \emptyset}{\partial T}\right)_P \tag{14}$$

and

$$\emptyset = \int_o^P \left(\frac{Z-1}{P}\right) dP \tag{15}$$

This integral was given by Redlich-Kwong as

$$\emptyset = Z-1 - \ln(Z-B) - A/B \ln(1+B/Z) \tag{16}$$

when solved with their equation of state as given by Eqn. (10). Thus working through this sequence numerically yields C_p. C_v is then defined from the standard thermodynamic relationship;

$$C_v = C_p - T\left(\frac{\partial P}{\partial T}\right)_v \left(\frac{\partial V}{\partial T}\right)_P \tag{17}$$

The partials in the above equation are obtained through differentiation of the Redlich-Kwong equation of state.

Therefore, once C_p is found it can be used with Eqns. (17) and (11) to solve for the isentropic exponent in Eqn. (9).

Discussion

The results developed in the previous section can now be applied to the valve sizing problem.

A calculation format was chosen which requires the reduced temperature and pressure as well as the low pressure ratio of specific heats to be known. These quantities are generally available with the given sizing information.

Figure 7 shows the typical results found from the analysis with the isentropic exponent plotted as a function of reduced temperature and pressure for Methane (k_o = 1.3).

Since the low pressure ratio of specific heats is an input parameter it is interesting to note how varying this can affect the end results. This is shown in Figure 8 which indicates that the change is not large, however, still enough to require being taken into account. Therefore, several charts will be required to minimize extrapolation.

Schuder & Buresh [5] developed a correction "C_2" for isentropic exponents which differed from the values for ideal air (i.e., n=k=1.4). It was written with the Universal Gas Sizing equation as,

$$Q = \sqrt{\frac{520}{GT}}\, P_1 C_2 C_g \sin\left[\frac{3417}{C_1 C_2}\sqrt{\frac{\Delta P}{P_1}}\right]_{Deg.} \tag{18}$$

$$C_2 = \frac{\sqrt{\frac{n}{n+1}\left[\frac{2}{n+1}\right]^{\frac{2}{n-1}}}}{.4839} \tag{19}$$

Though originally it was intended that this would be a correction for various ratios of specific heat it can now be used as the correction factor for non-ideal isentropic exponents.

Figure 9 shows C_2 as a function of reduced temperature and pressure for a gas with a low pressure specific heat ratio of 1.3.

In a similar vein if the ISA sizing equations [6] are used the correction for varying ratios of specific heat is,

$$F_k = \frac{k}{1.4} \tag{20}$$

This can now be written as,

$$F_n = \frac{n}{1.4} \tag{21}$$

to allow for the non-ideal isentropic exponent.

Figure 10 is a comparison between theory and data taken at the Fisher Controls high pressure flow facility. Though this data is limited it does indicate the proper trend in predicting flowrate. It is anticipated that more extensive data will be collected in the near future.

Conclusions

It has been shown that current gas sizing equations which are based on ideal gas assumptions may be corrected for the effects of real gas behavior.

Correcting the density with the compressibility factor is proper, however, it does not address the effects of a real gas on the isentropic process which is assumed between the inlet and the throat.

Therefore, to eliminate this sizing deficiency curves are presented for the non-ideal isentropic exponent under a wide variety of conditions which can be used in any of the commonly used valve sizing equations.

References

1. Brockett, G.F. and King, C.F., "Sizing Control Valves Handling Flashing Liquids", (1953), Fisher Governor Co. Technical Bulletin.

2. Stiles, G.F., "Sizing Control Valves for Choked Conditions Due to Cavitation or Flashing", Flow - Its Measurement and Control in Science and Industry, (1974), Vol. 1, Part 2, pp 1097.

3. Sheldon, C.W. and Schuder, C.B., "Sizing Control Valve for Liquid-Gas Mixtures", Instruments and Control Systems, (1965), Vol. 38, January 1965.

4. Martinec, E.J., Jr., "Two-Phase Critical Flow Saturated and Subcooled Liquids Through Valves", (1979), Reactor Analysis and Safety Division, Argonne National Laboratory.

5. ISA Standard, "Control Valve Sizing Equations for Compressible Fluids, ISA-S39.3, 1973.

6. ISA Standard, "Control Valve Sizing Equations for Incompressible Fluids", ISA-S39.1, 1972.

7. Tangren, R.F., et. al., "Compressibility Effects in Two-Phase Flow" J. of Applied Physics, (1949), Vol. 20, p. 736.

8. Fauske, H.K., "Two-Phase, Two- and One-Component Critical Flow", Proc. of Symp. on Two-Phase Flow, (1965), University of Exeter, Devon, England, Vol. 3 SG101.

9. Henry, R.E. and Fauske, H.K., "The Two-Phase Critical Flow of One-Component Mixtures in Nozzles, Orifices and Short Tubes", J. of Heat Transfer Division, (1971), Vol. 93 No. 2, May p 179.

10. Fagerlund, A.C., "Sizing Valves for Two-Phase and Subcooled Liquids Flows", Presented at the A.S.M.E. Winter Annual Meeting, Heat Transfer Division, (1980), Paper 80WA/HT-64, November 16-22, 1980.

11. Fagerlund, A.C. and Storer, W.J., "Progress in Valve Sizing for Two-Phase Flows", Proc. of ISA Spring Symposium, (1983), Houston, TX.

12. Wallis, G.B., "One-Dimensional Two-Phase Flow", (1969), New York, New York

13. Simpson, H.C., Rooney, D.H., Callander, T.M.S., "Pressure Loss Through Gate Valves with Liquid - Vapour Flows" 2nd International Conference on Multi-Phase Flow, London, England, June, 1985. Sponsored by BHRA Fluid Engineering.

14. Schoenthaler, J.L., "Adiabatic Exponents of Natural Gases at High Pressure and Their Relation to Compressor Performance", Standard Oil of California, September, 1967.

15. Edmister, W.C., Applied Hydrocarbon Thermodynamics, Vol I., 1961.

16. ASME STEAM TABLES, American Society of Mechanical Engineers, 5th Ed., 1983.

17. Redlich, O. and Kwong, J.N.S., "On the Thermodynamics of Solutions, Chemical Reviews, Vol. 44, 1949.

18. Buresh, J.F. and Schuder, C.B., "Development of a Universal Gas Sizing Equation for Control Valves", ISA Transactions, Vol. 3, No. 4, October, 1964.

19. Instrument Society of America, ANSI/ISA-S75.01, 1985, "Control Valve Sizing Equations".

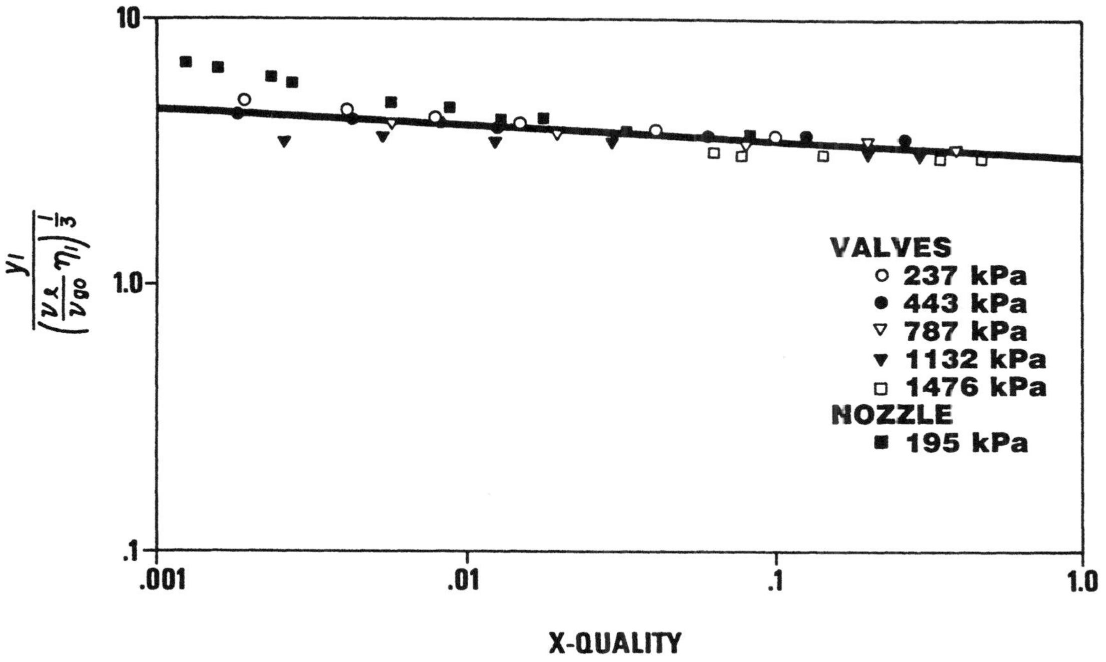

Figure 1. Slip Ratio vs. Quality

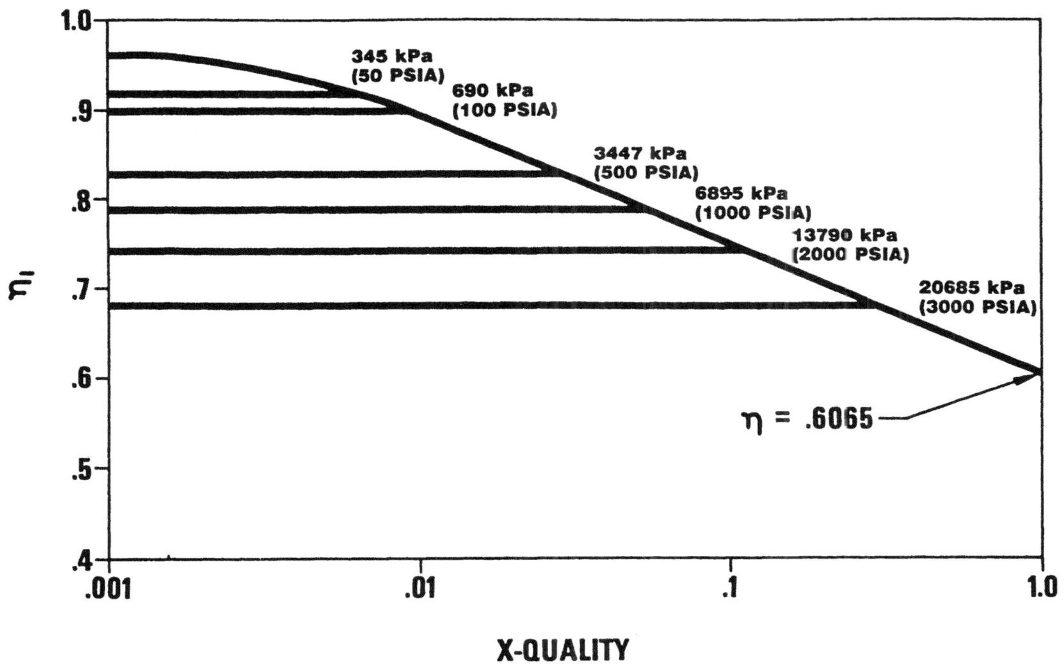

Figure 2. Critical Pressure Ratio for Water vs. Quality

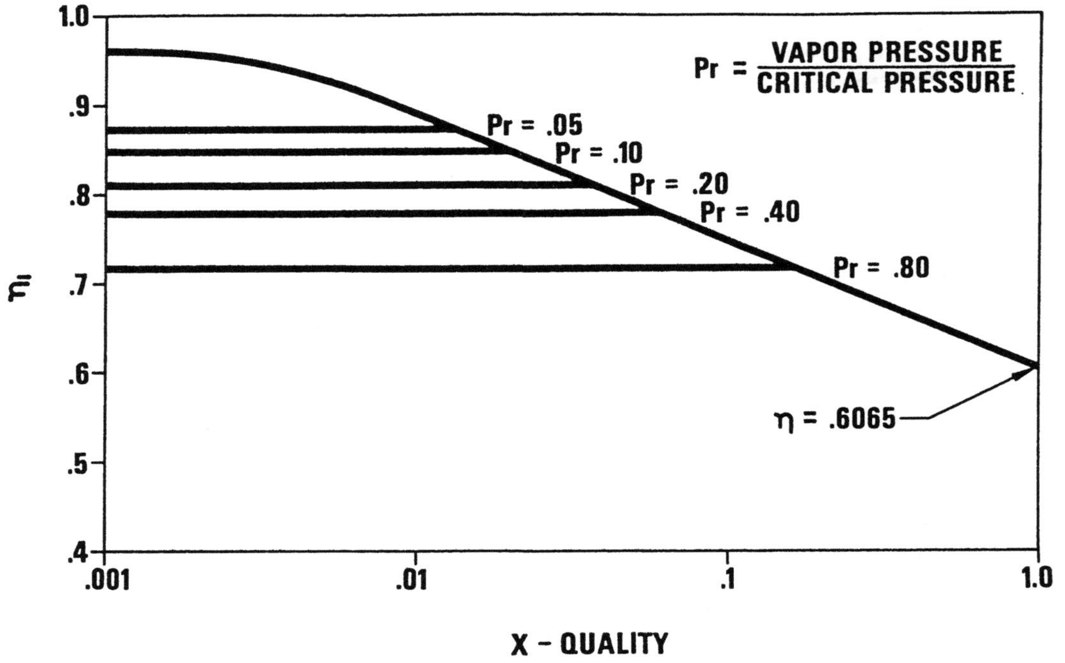

Figure 3. General Critical Pressure Ratio vs. Quality

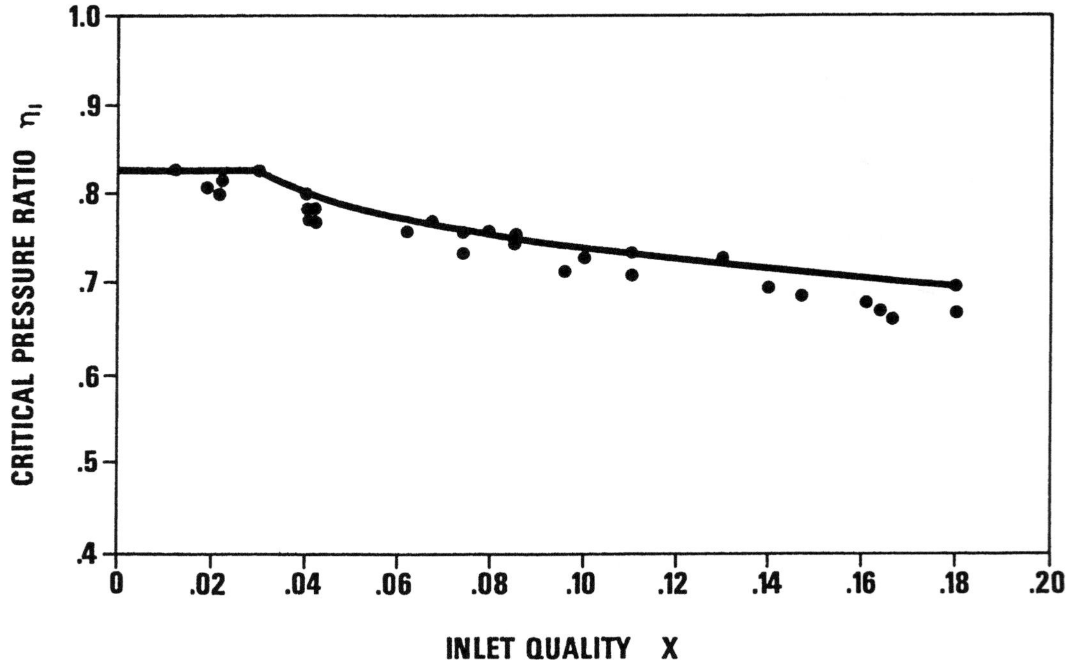

Figure 4a. Steam 3448 kPa (500 PSIA)
Critical Pressure Ratio vs. Quality

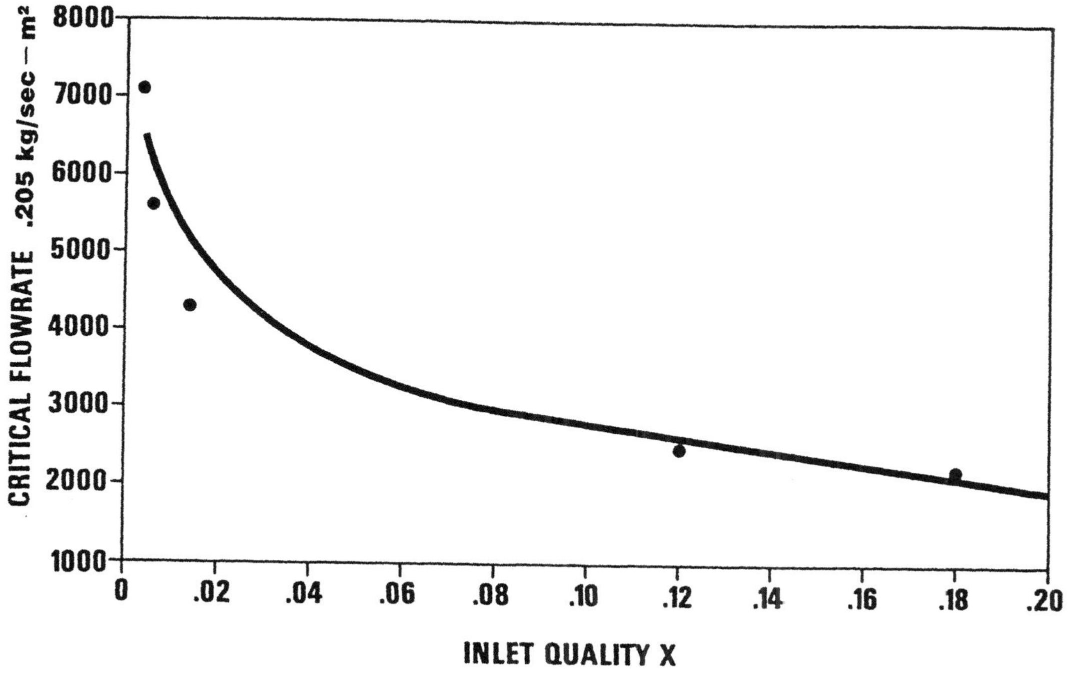

Figure 4b. Steam 3448 kPa (500 PSIA)
Critical Flowrate vs. Quality

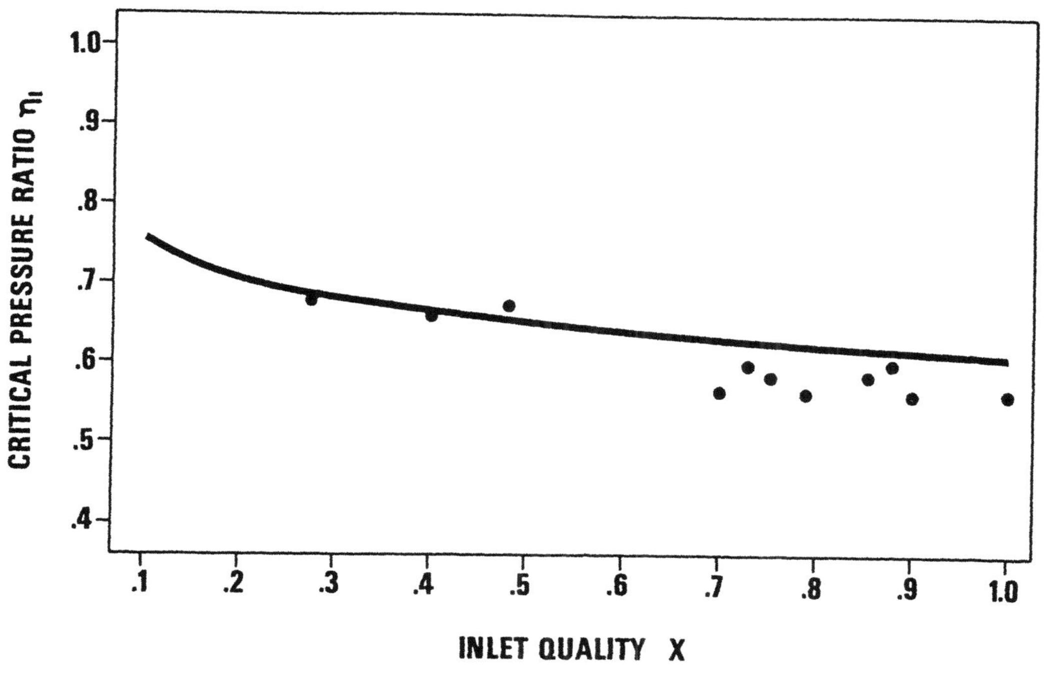

Figure 5a. Steam 121 kPa (17.6 PSIA)
Critical Pressure Ratio vs. Quality

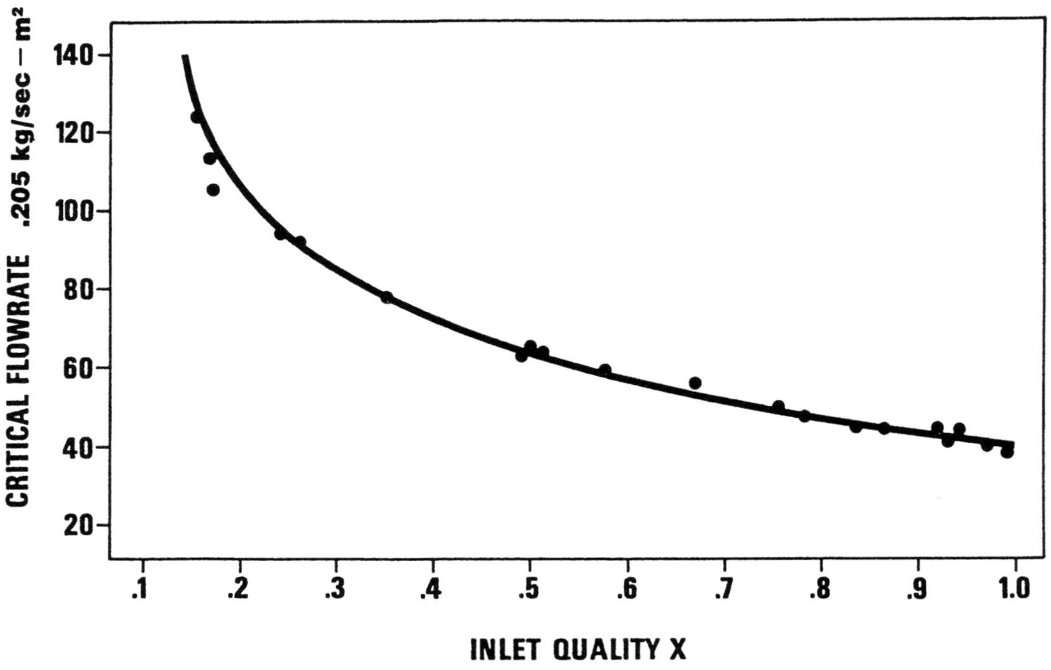

Figure 5b. Steam 121 kPa (17.6 PSIA)
Critical Flowrate vs. Quality

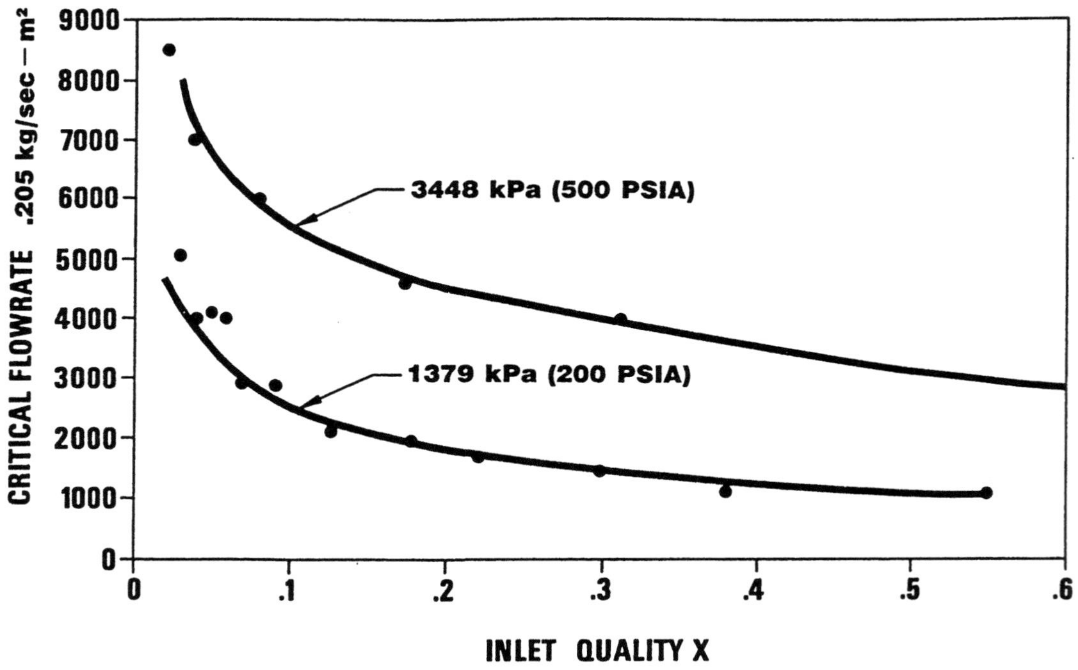

Figure 6. Carbon Dioxide
Critical Flowrate vs. Quality

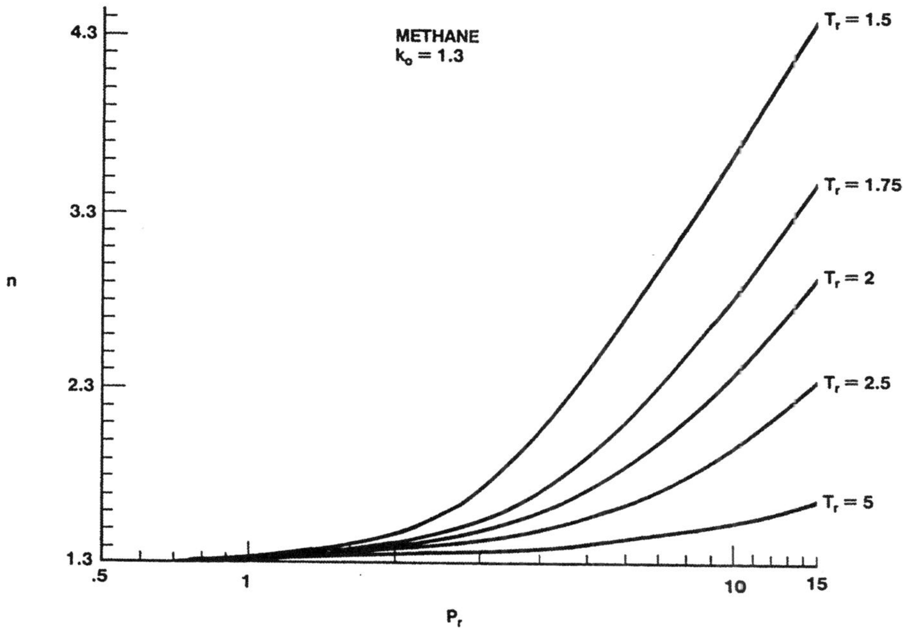

Figure 7. Isentropic Exponent vs. Reduced Temperature and Pressure (k_0 = 1.3)

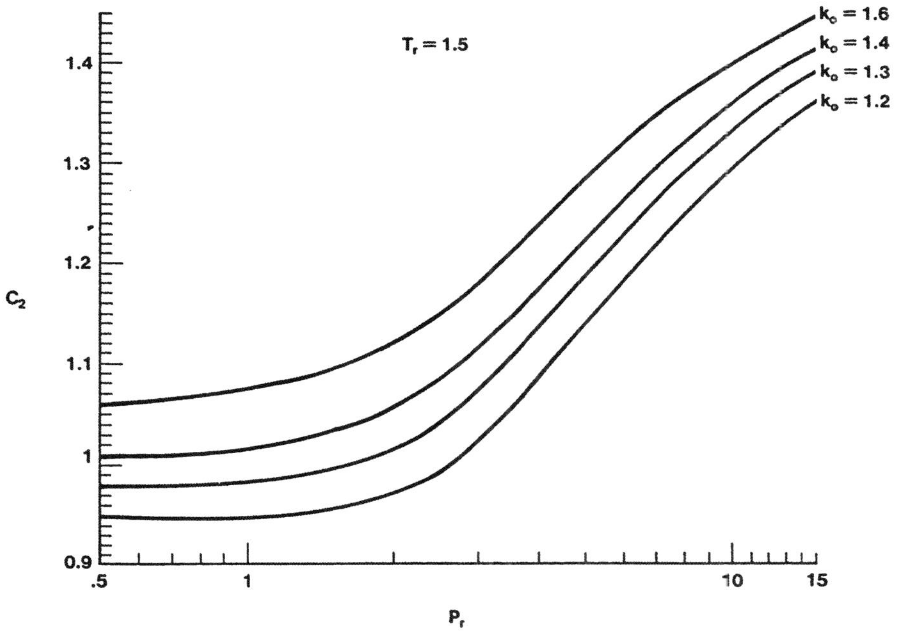

Figure 8. Variation in Correction to Universal Gas Sizing Equation vs. Reduced Pressure and Ratio of Specific heats (T_r = 1.5)

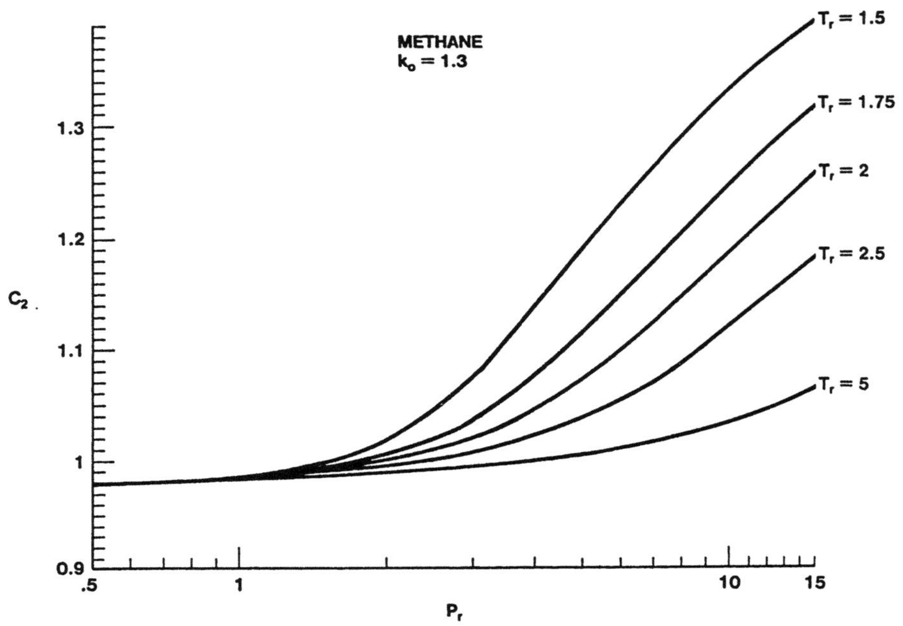

Figure 9. Correction to Universal Gas Sizing Equation vs. Reduced Temperature and Pressure (k_0 = 1.3)

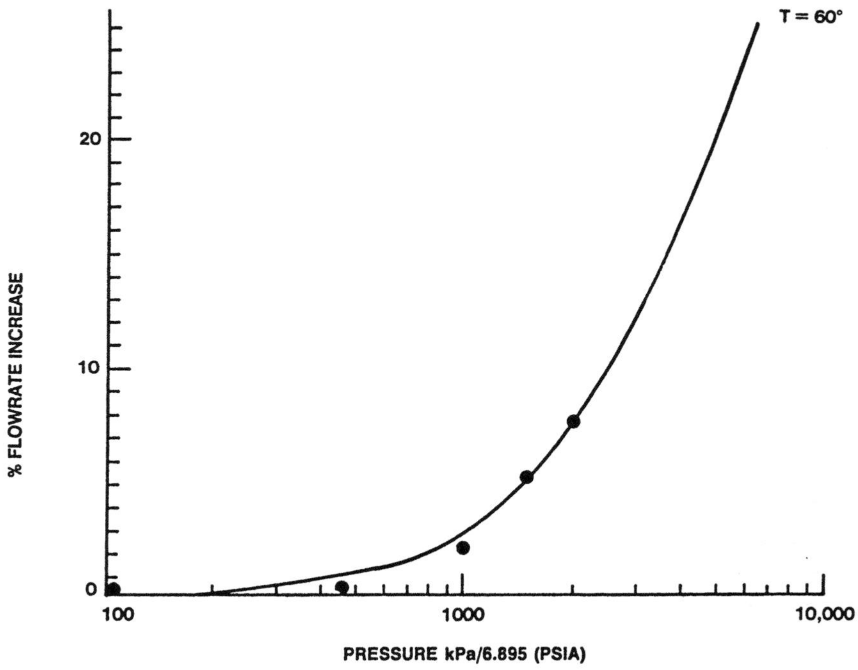

Figure 10. Percent Flowrate Increase vs. Pressure (Air k_0 = 1.4)

2nd International Conference on

Developments in Valves and Actuators for Fluid Control

Manchester, England: 28-30 March 1988

PAPER G1

PREDETERMINATION OF THE NOISE EMISSION OF VALVES IN FLUID SERVICE

T. Dobben

SAMSON AG
Mess- und Regelungstechnik
6000 Frankfurt am Main
(Federal Republic of Germany)

1. Summary

The paper presents the current status of the standardization efforts in the Federal Republic of Germany with regard to the predetermination of the noise emission of valves in fluid service and provides a basis for experimental research on the internal acoustic power and its spectral distribution.

2. Symbols and Units

A_i	pipe cross sectional area	m^2
A_0	reference area	m^2
A_M	barrel jacket area	m^2
c	sound velocity	m/s
c_i	sound velocity in the fluid	m/s
c_R	dilatational wave velocity	m/s
c_0	sound velocity in the air	m/s
d_a	pipe outer diameter	m
d_i	pipe inner diameter	m
E	modulus of elasticity	N/m^2
F_F	factor for critical pressure ratio	-

Held in Manchester, England. Organised and sponsored by BHRA, The Fluid Engineering Centre and co-sponsored by the British Valve Manufacturers Association and the Institution of Mechanical Engineers Process Industries Division

f_G	cut-off frequency	Hz
F_L	factor for pressure recovery	-
f_m	octave band mean frequency	Hz
f_r	ring frequency	Hz
K_d	diffusivity correction factor	dB
K_v	valve flow coefficient	-
K_z	impedance correction factor	-
l	pipe length	m
L_{pa}	external sound pressure level	dB
L_{pi}	internal sound pressure level	dB
L_{Wa}	external acoustic power level	dB
L_{Wi}	internal acoustic power level	dB
m	mass flow	kg/s
n	number of increments	-
p	pressure	Pa
p_i	internal sound pressure	N/m^2
p_v	Vapour pressure	Pa
p_0	reference sound pressure	N/m^2
p_1	pressure upstream from the valve	Pa
p_2	pressure downstream from the valve	Pa
R	transmission loss	dB
s	wall thickness	m
t	temperature	°C
w_D	velocity at the restriction	m/s
x_F	pressure ratio $(p_1 - p_2) / (p_1 - p_v)$	-
y	valve load	%
z	pressure ratio at start of cavitation	-
η	acoustic efficiency with y = 75 %	-
ρ_0	density of air	kg/m^3
ρ_i	density of the fluid	kg/m^3
ρ_R	density of pipe wall material	kg/m^3

3. Introduction

In connection with the revision of the VDMA-Einheitsblatt 24422 (Guidelines for the Calculation of Noise Generated by Control and Shutoff Valves) a method of calculation is proposed which starts out from the internal acoustic power of the valve, proceeds to the calculation of acoustic insulation of the pipe and finally leads to the external airborne sound level.

In this process, only the octave band range of 500 to 8000 Hz is considered, because the impact of the other octave bands is low due to the so-called A-weighting and the reduced pipe radiation at low frequencies and the increased airborne sound absorption at high frquencies.

4. Calculation of the internal acoustic power based on measured data

Assuming a plane wave field, the acoustic power W_i radiated from a valve to the acoustically inert pipe connected can be calculated from the mean sound pressure across the pipe cross section by the equation below:

$$W_i = \frac{p_i^2 \cdot A_i}{\rho_i \cdot c_i} \qquad [W] \qquad (1)$$

Using the equation

$$W_0 = \frac{p_0^2 \cdot A_0}{\rho_0 \cdot c_0} \qquad [W] \qquad (2)$$

the internal acoustic power level related to airborne sound can be derived:

$$L_{Wi} = 10 \lg \frac{Wi}{W_0}$$

$$= 20 \lg \frac{p}{p_0} + 10 \lg \frac{\rho_0 \cdot c_0}{\rho_i \cdot c_i} + 10 \lg \frac{A_i}{A_0} \qquad [dB] \qquad (3)$$

In the case of sound at frequencies above a cut-off frequency [1] of

$$f_G = \frac{0{,}585 \cdot c_i}{d_i} \qquad [Hz] \qquad (4)$$

determined by the inner pipe diameter d_i and the sound velocity in the fluid c_i, also higher modes are formed besides the plane sound waves (00-mode) owing to radial resonances so that the sound field assumes a diffuse character. When determining the acoustic power level, the level increase caused by the radial sound components is taken into account by subtracting an empirically derived correction factor K_d [2] according to Fig. 1.

Since according to [2, 3] there is no difference between the mean sound pressure level calculated across the pipe cross section and the sound pressure level measured along the inner wall, this sound level measured at a distance of ten times the pipe diameter downstream from the valve can be substituted for 20 lg (p/Po) in equation (3). As regards individual octave bands, the following equation applies:

$$L_{Wi}(f_m) = L_{pi}(f_m) + 10 \lg \frac{\rho_0 \cdot c_0}{\rho_i \cdot c_i} + 10 \lg \frac{A_i}{A_0} \qquad [dB] \qquad (5)$$

where $\rho_0 \cdot c_0$ = 408 Ns/m³ is the reference impedance, and A_0 = 1 m² is the reference area. The characteristic acoustical impedance of the fluid $\rho_i \cdot c_i$ is difficult to determine in the case of cavitation and evaporation because both the density ρ_i and also the sound velocity c_i vary as the temperature (Figs. 2 and 3) and the gas content change.

The dependency of the sound velocity on the contents of undissolved gases is calculated according to [4] , with the pipe wall transmission being negligibly small

$$c_i = \sqrt{\frac{\rho_F}{K} + \frac{\varepsilon \cdot \rho}{\varkappa \cdot p_i \, [\rho/\rho_W + \varepsilon \, (1 - \rho/\rho_W)]^2}} \qquad [\frac{m}{s}] \qquad (6)$$

and is shown in Fig. 4 for a water/air mix as a function of the mass concentration of the gas.

Moreover, owing to the small difference of the impedances of the fluid and the pipe wall in acoustically narrow pipes, the sound velocity is substantially lower than in the unbounded medium [5] .

The equation

$$\frac{c}{c_i} = \frac{1}{\sqrt{1 + \frac{2\rho_i \cdot c_i}{E} \cdot \frac{1 + (1 - 2s \,/\, d_a)^2}{1 - (1 - 2s \,/\, d_a)^2}}} \qquad [\frac{m}{s}] \qquad (7)$$

yields the velocity conditions shown for water in Fig. 5, dependent on the ratio of pipe wall thickness s to the pipe inner diameter d_i.

Because of the complex relationships, the characteristic impedance of the fluid in equation (5) is assumed to be 1.485 . 10^6 Ns/m³ (water 10 °C), and the influences of temperature, gas content and pipe (equations 6 and 7) are taken into account by a factor K_z

$$\rho_i \cdot c_i = 1{,}485 \cdot 10^6 \cdot K_z \qquad [Ns/m^3] \qquad (8)$$

whose calculation is left out here.

The acoustic power level obtained from measured data (see equation 5) then contains an error of +10 lg K_z which, as will be shown later, is inapplicable to the calculation of the external sound level.

5. Empirical equations for the predetermination of the internal acoustic power spectrum

For single-stage valves with V-port or parabolic plug, the equations presented below are derived by means of equation (5) and from measurement results. They are based on the valve-related characteristics z and F_L.

F_L is the factor for the pressure recovery which is measured according to IEC 534, Parts 2 and 3, dependent on the valve load. The z-value, which is also determined as dependent on the load according to [6] represents the pressure condition where cavitation begins.

5.1 Flow free of cavitation

In flow free of cavitation ($X_F < z$), the acoustic power is proportional to the radiated power:

$$W_i = \eta \cdot \overset{\circ}{m} \cdot \frac{w_D^2}{2} \qquad [W] \qquad (9)$$

The velocity at the restriction w_D is obtained from the energy equation with the factor F_L for the pressure recovery:

$$w_D = \sqrt{\frac{2}{\rho_i} \cdot \frac{p_1 - p_2}{F_L^2}} \qquad [\frac{m}{s}] \qquad (10)$$

When the units $[\mathring{m}]$ = kg/h, [p] = bar and $[\rho_i]$ = kg/m³ are used in equation (9), and applying equation (10) the following equation is obtained for the acoustic power

$$W_i = 27{,}78 \; \frac{\eta \cdot \mathring{m} \cdot \Delta p}{F_L^2 \cdot \rho_i} \qquad [W] \qquad (11)$$

or, expressed as acoustic power level with the reference power $W_0 = 10^{-12}$:

$$L_{Wi} = 10 \lg \eta + 10 \lg \mathring{m} + 10 \lg \Delta p - 10 \lg \rho_i$$

$$- 20 \lg F_L + 134{,}4 \qquad [dB] \qquad (12)$$

The acoustic efficiency η is dependent on the style and load of the valve. For standard valves with parabolic or V-port valves, η amounts to $\sim 10^{-8}$ at a load of 75 %. For standard valves at 75 % load, the octave levels can be approximated by the equation

$$L_{Wi} (f_m) = L_{Wi} - \frac{f_m}{200} - 7 \left(\frac{X_F}{z} \right)^4 \qquad [dB] \qquad (13)$$

(see Fig. 6).

5.2 Flow with cavitation

With the pressure conditions $X_F = z$ the liquid starts to cavitate, and the flow noise caused by turbulences is superimposed by the cavitation noise resulting from the bubble collaps spread over time.

The radiated acoustic power first increases as the pressure rises (increasing number of cavitation bubbles), but when a maximum is exceeded at full cavitation, it decreases again. At $X_F = 1$ the cavitation noise disappears completely, because the pressure in the bubble ($\sim P_v$) is equal to the ambient pressure (p_2), and the bubbles do no longer collaps.

For standard valves the radiated acoustic power can be approximated by the equation

$$L_{Wi} = 10 \lg \eta + 10 \lg \mathring{m} + 10 \lg \Delta p + 10 \lg \rho_i$$

$$- 20 \lg F_L - \frac{z_y^{0{,}0625} \; (1-X_F)^{0{,}8}}{X_F^{\;z_y}} \cdot 180$$

$$\lg \frac{1{,}001 - X_F}{1 - z_y} + 134{,}4 \qquad [dB] \qquad (14)$$

where

$$\Delta p = F_L^2 \; (p_1 - F_F \cdot p_V) \qquad [bar] \qquad (15)$$

The spectral distribution for standard valves at 75 % load is obtained approximatively by the equation

$$L_{Wi}\,(f_m) = L_{Wi} - \frac{f_m}{400}\left[\frac{X_F - z_y}{1 - z_y}\right]^2 + 4{,}6\left[\frac{X_F - z_y}{1 - z_y}\right] \quad [\mathrm{dB}] \qquad (16)$$

The difference $L_{Wi}\,(f_m) - L_{Wi}$ is shown in Fig. 6.

5.3 Flow with evaporation

With the pressure conditions $X_F > 1$, i.e. $p_v > p_2$ partial evaporation of the liquid occurs. Thus the average density ρ_m of the diphasic fluid decreases while the velocity of flow increases.

The internal acoustic power is obtained by the method similar to the one described in sub-section 5.1, taking into account the flow limitation according to equation (15)

$$L_{Wi} = 10\ \lg \eta + 10\ \lg \mathring{m} + 10\ \lg\,(p_1 - F_F \cdot p_v)$$

$$10\ \lg \rho_i + 134{,}4 \quad [\mathrm{dB}] \qquad (17)$$

The spectral distribution of the internal acoustic power, which depends on the valve style and load, can be calculated for standard valves approximatively by the equation:

$$L_{Wi}\,(f_m) = L_{Wi} - \frac{f_m}{400} - 2{,}4 \quad [\mathrm{dB}] \qquad (18)$$

6. Acoustic insulation of the pipe

The acoustic insulation R of the pipe is calculated according to the equation given for airborne sound in VDI Guideline 3733, which is based on the work of Crämer [7] and Heckel [8] and which has been modified by practical measurements [2, 9] on natural gas pipelines:

$$R\,(f_m) = 10 + 10\ \lg \frac{c_R\ \ \rho_R \cdot s}{c_i \cdot \rho_i \cdot d_i}$$

$$+ 20\ \lg\ \left[\frac{f_r}{f_m} + \left(\frac{f_m}{f_r}\right)^{1,5}\right] \quad [\mathrm{dB}] \qquad (19)$$

where

$$f_r = \frac{c_R}{\pi \cdot d_i} \qquad (20)$$

the ring frequency at which acoustic insulation reaches its minimum.

More recent measurements [10] show that equation (19) can also be used for calculating the insulation of liquid-filled pipes.

Substituting the characteristic impedance $c_i \cdot \rho_i$ of the medium in equation (19) by the value for water at 10 °C (1.485 . 10^6 Ns/m²) and taking into account the influence of temperature, gas content and pipe wall by the factor K_z as described in Section 4, whose calculation is left out here, a value for R is obtained which is too high by +10 lg K_z.

7. The external acoustic power level

For a short pipe section of unit length directly downstream from the valve and neglecting losses along the pipeline the frequency-dependent external acoustic power level is obtained by the equation

$$L_{Wa}\,(f_m) = L_{WM}\,(f_m) - R\,(f_m) \qquad [dB] \qquad (21)$$

where $L_{Wa}\,(f_m)$ is the power impinging on the inner pipe surface ($A_M = d_i \cdot \pi \cdot 1$),

which, because of $p_i^2 = p_M^2$ (see Section 4) and taking into count the measurement area, is obtained from the internal acoustic power level:

$$L_{WM}\,(f_m) = L_{Wi}\,(f_m) + 10\,\lg \frac{A_M}{A_i} \qquad [dB] \qquad (22)$$

From the power levels calculated for the five octave bands (range 500 Hz to 8000 Hz) the total external acoustic power for the pipe section considered is obtained by the equation

$$L_{Wa} = 10\,\lg \sum_{n=1}^{5} 10^{(L_{Wa_n}/10\ dB)} \qquad [dB] \qquad (23)$$

When the A-weighted power level is required, the individual octave bands shall be weighted according to the table below prior to the addition according to equation (23)

t_m [Hz]	500	1000	2000	4000	8000
L_{Wa} [dB]	-3,2	0	+1,2	+1,0	-1,1

8. The external sound pressure level

Under free-field conditions and with cylindrical radiation, the sound pressure level 1 m away from the pipe section of unit length is obtained approximatively by the equation

$$L_{p_a} = L_{Wa} - 10\,\lg \frac{\pi \cdot 1\,(d_i + 2)}{1_0} \qquad [dB] \qquad (24)$$

As regards other conditions, reference is made to [6] .

9. Test results

The tests were performed on an experimental plant outlined in Fig. 7, which also permits tests under flashing conditions. In this arrangement the water (max. 6000 kg/h) is heated up by injecting steam (max. 5000 kg/h, 270 °C) and is expanded across the test specimen. The steam thus generated is exhausted to the atmosphere, while the condensate remains in the cycle. The quantity of circulating water is kept constant by a level control facility in the steam separator.

Of course, diphasic flows can be realized or different steam conditions can be produced upstream from the valve.

The tests were performed on a valve DN 80, PN 40 (Fig. 8) with reduced seat bore. During the tests the outlet direction, the plug style (parabolic and V-port plugs) and the load (25 % and 75 %) were varied. This resulted in eight z-values in the region between 0.25 to 0.6.

With a constant upstream pressure (8 bar) and a constant upstream temperature (25 °C) the downstream pressure of 6.5 bar was reduced to 1.5 bar, and thus the pressure ratio X_F varied between 0.19 and 0.82. With a constant pressure differential (6.5 bar) the temperature was subsequently raised to 125 °C, and thus the pressure ratio X_F increased to 1.14.

The internal acoustic power levels of the above-mentioned combinations predetermined according to equations 12, 14, and 17 with an acoustic efficiency = 10^{-8} are shown in Fig. 9. The internal acoustic power calculated according to equation 5 from the sound pressure levels measured along the inner wall deviates by not more than ± 6 dB from the predetermined values.

The deviations of the measured octave band levels from those predetermined according to equations 13, 16, and 18 are slightly higher (± 8 dB).

On the other hand, the measured external acoustic power levels of a pipe section 1 m long deviated by a maximum amount of +4 dB/-7 dB from those predetermined according to equations 21 and 23, taking into account the transmission loss (Fig. 10), and for the A-weighted sound pressure level 1 m away from the pipe section predetermined according to equation 24, the maximum deviation amounts to +2 dB/-5 dB.

10. References

[1] Skudrzyk, E. : Die Grundlagen der Akustik Kapitel VII, Springer Verlag Wien, 1954

[2] Sinambari, G.R. : Ausströmgeräusche von Düsen und Ringdüsen in angeschlossene Rohrleitungen. Ihre Entstehung, Fortpflanzung und Abstrahlung Dissertation Universität Kaiserslautern, 1981

[3] Petermann, I. : Akustische Untersuchung der Einflüsse von Durchmesser und Strömungsgeschwindigkeit auf die Schalldämmung zylindrischer, geflanschter Stahlrohre endlicher Länge Dissertation Universität Kaiserslautern, 1985

[4] Allievi, L. : Allgemeine Theorie über die veränderliche Bewegung des Wassers in Leitungen Springer Verlag Wien, 1909

[5] Kahl, W. : Die Eigenschaften wassergefüllter Rohre für Widerstands- und Schallgeschwindigkeitsmessungen Acustica 3 (1953), S. 111, 123

[6] VDMA 24423 : Geräuschmessung an Ventilen
Juni 1981

[7] Cremer, L. : Theorie der Schalldämmung zylindrischer Schalen
Acustica 5 (1955), S. 245/256

[8] Heckel, M. : Experimentelle Untersuchungen zur Schalldämmung von Zylindern
Acustica 8 (1958), S. 259/65

[9] Stüber, B. : Schalldämmende Rohrleitungsummantelungen hoher Pegelsenkung
Forschungsbericht Nr. 5500 der Müller-BBM GmbH
i.A. des Ministeriums für Arbeit, Gesundheit und Soziales des Landes Nordrhein-Westfalen, 1978

[10] Dobben, T.; Schmitt, A.; Kretzschmar, H.; Siemers, H. : Schallmessungen an wassergefüllten Stahlrohrleitungen

[11] VDI 3733 : Geräusche bei Rohrleitungen
Entwurf 1983

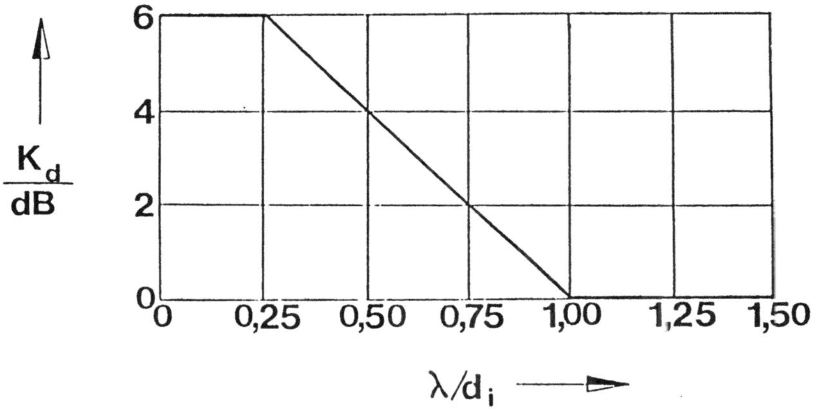

Figure 1: Correction factor K_d

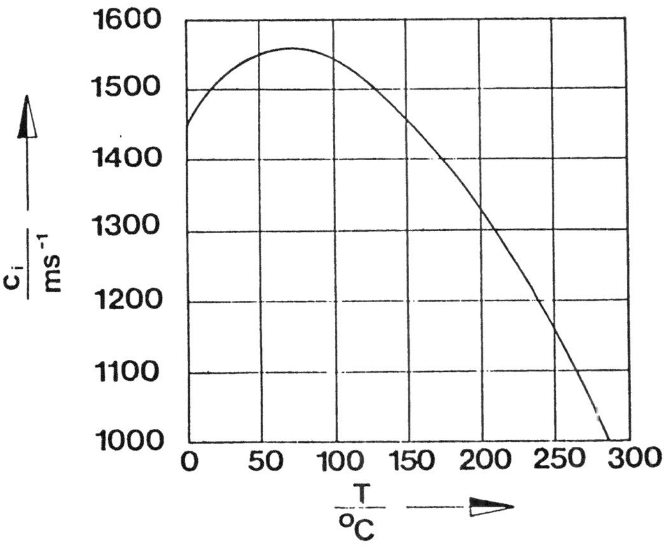

Figure 2: Sound velocity in water dependent on the temperature

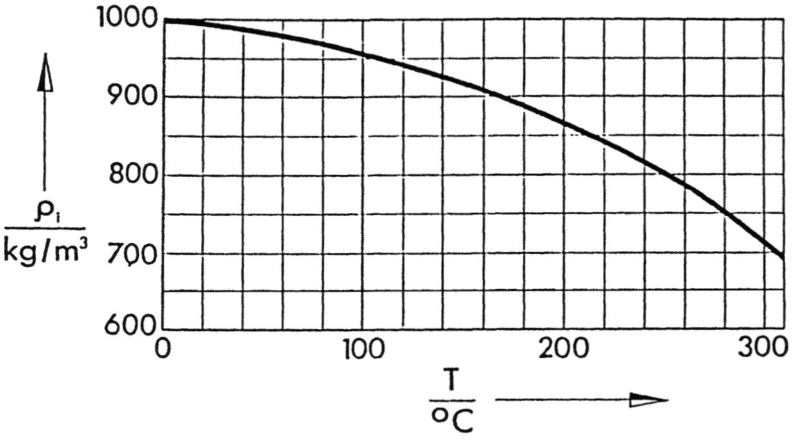

Figure 3: Density of water dependent on the temperature

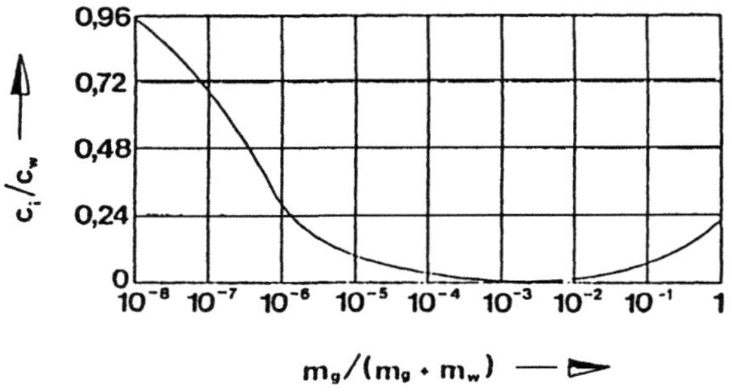

Figure 4: Sound velocity in a water-air mix dependent on the mass concentration in the fluid

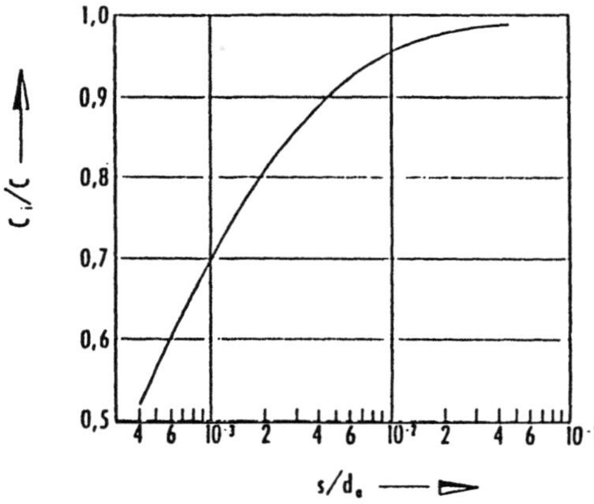

Figure 5: Ratio of the sound velocity in water in steel pipes versus the sound velocity in the unbounded fluid

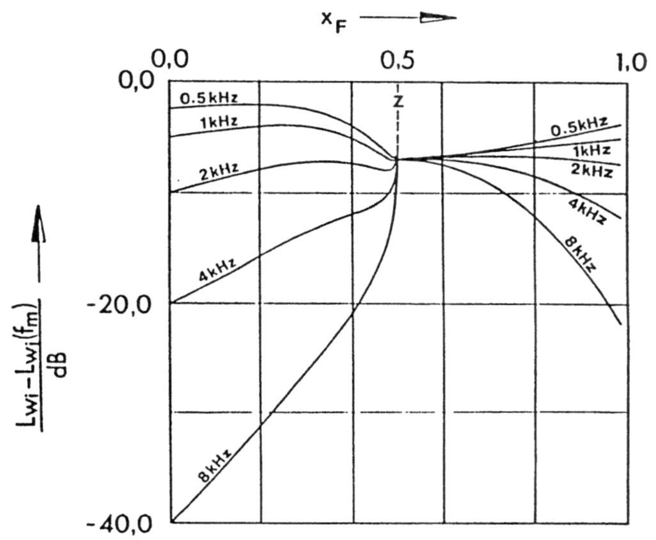

Figure 6: Spectral distribution of the internal acoustic power

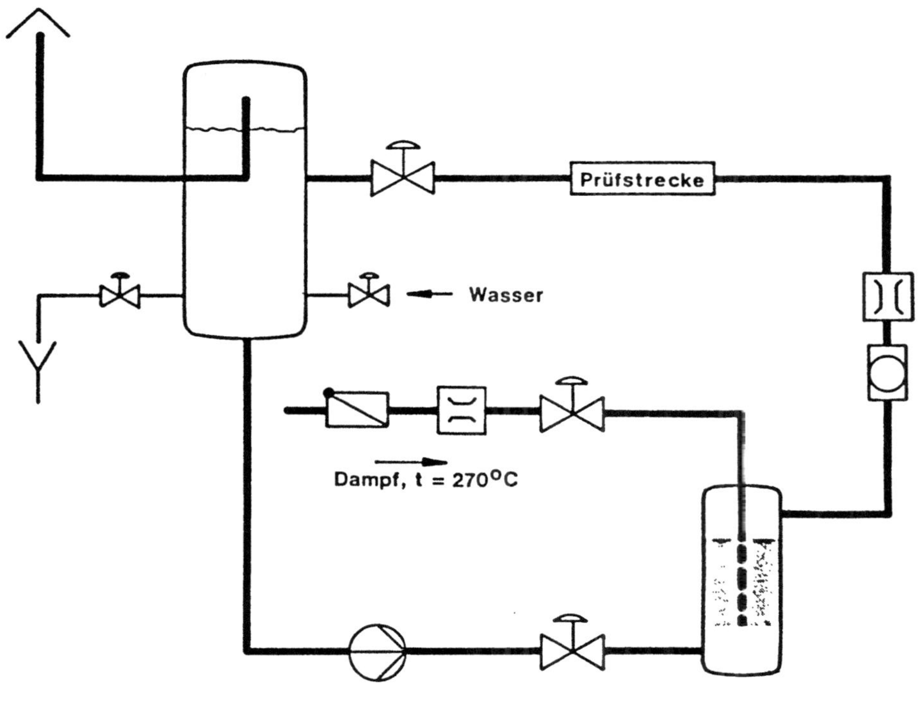

Figure 7: Experimental plant

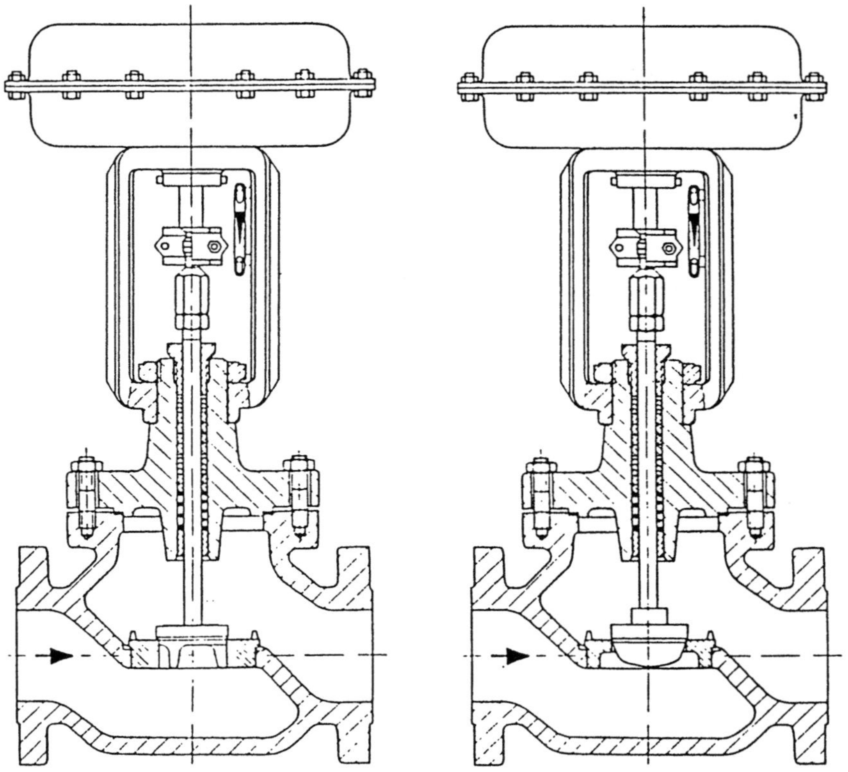

Figure 8: Control valve with parabolic plug and V-port plug

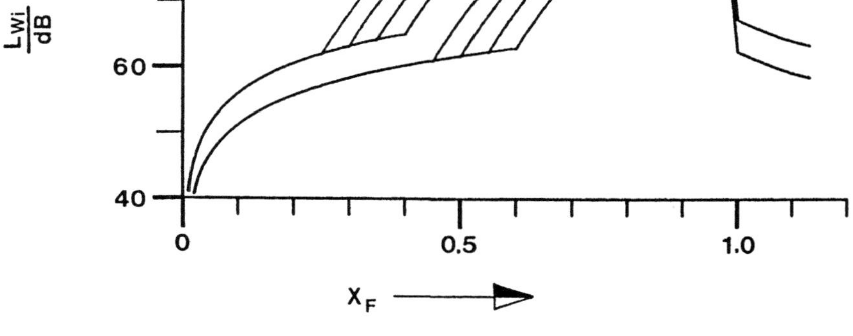

Figure 9: Predetermined internal acoustic level of the combination tested

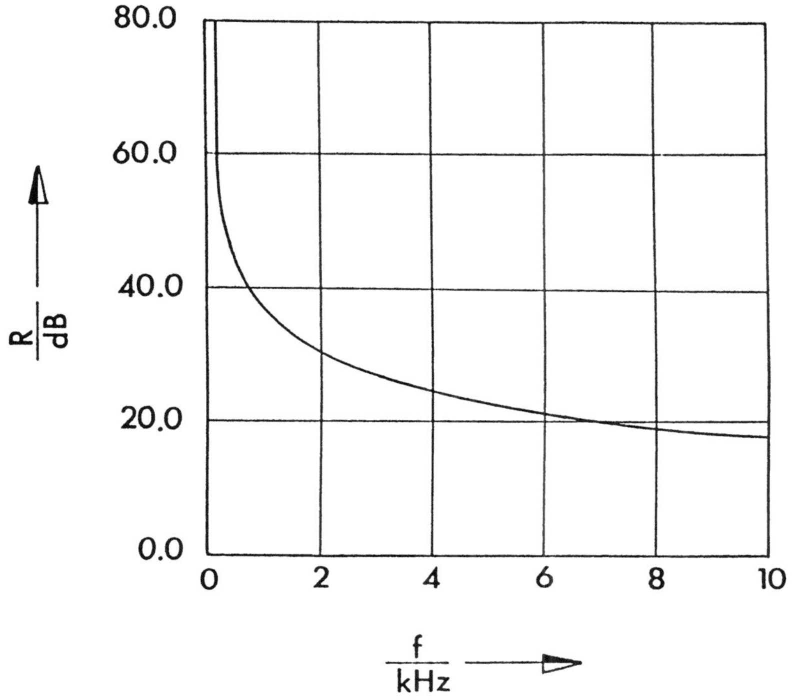

Figure 10: Transmission loss for a steel pipe DN 80, s = 2.9 mm

2nd International Conference on

Developments in Valves and Actuators for Fluid Control

Manchester, England: 28-30 March 1988

PAPER G2

PROPOSED UNIVERSAL METHOD FOR QUANTIFYING THE SEVERITY OF A CAVITATING CONTROL VALVE SERVICE.

Roger W. Barnes
Vice President of Engineering
Valtek Incorporated, USA

Fred M. Cain
Product Engineering Manager
Valtek Incorporated, USA

SUMMARY

Although the widely used K_c, cavitation index, (1) gives an "on/off" indication of whether or not a control valve will be experiencing cavitation, no universally accepted method exists to measure the damage potential of a cavitating condition in a control valve. This paper proposes such a method which allows users and vendors to quantify the service much the same as a K_v (C_v) calculation quantifies the capacity required. Once a numerical measure of the intensity of the service is calculated, then valve style and trim can be selected.

Guidelines are also presented on what types of valve trim would be suitable for different levels of cavitation.

Held in Manchester, England. Organised and sponsored by BHRA, The Fluid Engineering Centre and co-sponsored by the British Valve Manufacturers Association and the Institution of Mechanical Engineers Process Industries Division

NOMENCLATURE

ΔP = pressure drop P_1-P_2 or P_u-P_d, bar (psi)
ΔP_{choked} = The pressure drop, for a given P_1 and T_1, above which no increase in volumetric flow rate occurs increase ΔP.
ρ = fluid density
σ = cavitation sigma
σ_1 = $(P_u-P_v)/(P_u-P_d)$, upstream pressure based sigma
σ_2 = $(P_d-P_v)/(P_u-P_d)$, downstream pressure based sigma
σ_{rc}= reference critical σ
σ_{rd}= incipient damage reference sigma for style of valve
σ_s = service sigma $K_pK_s(P_d-P_v)/(P_u-P_d)$
ν_f = kinematic viscosity of fluid, cm^2/sec (ft^2/sec)
ν_{water} = kinematic viscosity of water, cm^2/sec (ft^2/sec)
C_d = discharge coefficient
d = small diameter valve, mm (inches)
D = large diameter valve, or valve size, mm (inches)
F_L = $[(\Delta P_{choked})/(P_1-F_FP_v)]^{0.5}$
G_f = specific gravity of fluid
I = cavitation intensity factor = $K_vK_FK_d(\sigma_{rd}/\sigma_s)$
K = $(P_u-P_d)/(P_u-P_v)$, cavitation index
K_c = $(P_1-P_2)/(P_1-P_v)$
K_d = duty scale factor for intensity
K_f = overall fluid scale factor for Intensity = $K_{fv}K_{fT}K_{fo}$
K_{fo} = other fluid effects scale factor for Intensity
K_{fT} = fluid temperature scale factor for Intensity
K_{fv} = fluid viscosity scale factor for Intensity
K_p = Pressure Scale Factor
K_s = Size Scale Factor
N = exponent for size scale factor K_s
P_1 = upstream static pressure 2 pipe diameters upstream of the valve
P_2 = downstream static pressure 6 diameters downstream of the valve
P_d = downstream static pressure at valve outlet, bar (psi)
P_u = upstream static pressure at valve inlet, bar (psi)
P_v = vapor pressure of the fluid at the upstream temperature
Q = volumetric flow rate, m^3/hr(gallons/minute)
T_{avg} = $(T_B + F_F)/2$, ^0C (^0F)
T_B = boiling temperature, ^0C (^0F)
T_F = freezing temperature, ^0C (^0F)
T_s = fluid service temperature, ^0C (^0F)
V = average fluid velocity, m/sec (ft/sec)
V_c = Velocity Factor for Intensity
Vo = threshold velocity for damage, m/sec (ft/sec)
X = factor accounting for the style of valve in K_p.

1. Introduction

Whenever a control valve is required to operate continuously in a high pressure drop liquid application, the potential for cavitation damage exists. Most control valves are sized using the ISA's (Instrument Society of America's) sizing equations. These equations are fairly accurate for predicting choked flow and moderately accurate for predicting the onset of cavitation within the valve. However, the severity of that cavitation and the subsequent steps which the control valve user must take to control cavitation damage are not indicated by the F_L and K_c factors.[1]

2. Objective

Our objective is to move the sizing of control valves for cavitating service away from the "art" environment toward the science environment. Some of what is presented is based on test data, while some is based on the authors' judgement and experience. The purist may object to the use of the soft factors that are not substantiated by hard test data. However, the application engineers who are sizing control valves in cavitating service must make "judgement calls" based on their experience. They make a judgement on the use of hardened trim, single-stage or multi-stage based on duty cycles, criticality of the service, fluid properties, etc. in addition to pressure drop and vapor pressure factors. The authors' objective is to quantify some of this experience, moving cavitation sizing from a learned art toward a empirical methodology based on scientific fundamentals. It is hoped that this will stimulate a dialogue with other experts in order to refine the process. This quasi-expert system might be programmed on a hand held calculator or microcomputer to be used by control valve specifiers. Our paper presents a challenge to researchers to apply their expertise to develop a cavitation severity evaluation system with practitioner appeal. The specialist may be uncomfortable with this soft approach. The non-specialist practitioner, however, will welcome some attempt to quantify the "art."

---

1 F_L, liquid pressure recovery factor, is a measure of the ability of the valve to convert the kinetic energy of the flow stream at the vena contracta back into pressure and is $F_L = [\Delta P_{choked}/(P_1-F_FP_v)]^{1/2}$. The factor F_F is determined experimentally as specified in ISA standard S39.02. By rearranging the equation as $\Delta P_{choked} = F_L^2(P_1-F_FP_V)$, the pressure drop at which the valve would be fully choked can be estimated. In liquid service, if the valve is fully choked, then cavitation is present (2). K_c is a dimensionless ratio experimentally determined from a plot of Q versus $(\Delta P)^{1/2}$, It is used to describe the point of initial departure from a linearly proportional relationship of Q versus square root of pressure drop. At the inflection point, the pressures are determined and are used to define the index $K_c = (P_1-P_2)/(P_1-P_v)$ where P_1 = inlet pressure; P_2 = valve outlet pressure; P_v = vapor pressure of liquid at the valve inlet temperature. The pressure drop defined by the rearranged equation $P_1-P_2 = K_c(P_1-P_v)$ is generally accepted as the onset of cavitation.

3. Intensity

Robertson (3) suggested the concept of an intensity measure I. We will define our intensity in the form

$$I = V_C K_F K_D (\sigma_{rd}/\sigma_S) \qquad (1)$$

where

$$\sigma_s = K_p K_s (P_d - P_v)/(P_u - P_d) \qquad (2)$$

σ_{rd} = incipient damage reference sigma for style of valve
V_C, K_F, K_D = modifiers (defined below)

The Intensity Factor "I" can be considered an intensity or as a life reduction factor. The magnitude represents an approximation of how many times faster material will be removed over the threshold damage rate for the given operating conditions and service environment. Once a numerical measure of severity of the service is calculated, valve styles and trim can be selected. This separates the vendor-specific data from the service factors.

3.1 Cavitation Sigma (σ)

A factor widely used in the civil and mechanical engineering communities, sigma (with no subscript) has been used for many years to quantify a cavitating condition. A form of it was first introduced by Thoma in 1920 (3, 4). Although cavitation is an extremely complex phenomenon, related to many variables such as pressures, temperature, thermodynamics, fluid properties (surface tension, viscosity, etc), size, velocity, etc., much of this data can be correlated by using one of several sigma factors.
A pressure based sigma is convenient for valve sizing, since the upstream and downstream pressures are generally known. Several forms which are commonly used are summarized by Rahmeyer (5):

$$\sigma_1 = (P_u - P_v)/(P_u - P_d) \qquad (3)$$

$$\sigma_2 = (P_d - P_v)/(P_u - P_d) \qquad (4)$$

$$K = (P_u - P_d)/(P_u - P_v) \qquad (5)$$

$$\sigma_1 = \sigma_2 + 1 = 1/K \qquad (6)$$

where (all pressures are absolute)

P_u = upstream static pressure at the valve inlet
P_d = downstream pressure at the valve outlet[2]
P_v = vapor pressure of the fluid at the upstream temperature

---

2 The locations of the upstream and downstream measurements differ with th location of the pressure taps referred to in ISA S75.02-1981 Control Valv Capacity Test Procedures (11) which specifies 2 nominal pipe diameter upstream and 6 nominal pipe diameters downstream for P_1 and P_2 measurements Whether or not the pressures should be corrected for manifold losses betwee the taps or not is a subject upon which it is difficult to reach agreemen even within the ISA committee itself. (The authors are active in IS Committee SP75.06). From a practical standpoint, it makes little differenc for globe and other low recovery valves. For high recovery valves (ball an disk) operating near 100% open, an attempt should be made to correct th downstream pressure for pipe losses. This is especially important fo downstream pressures near the vapor pressures where small changes in greatly affect the magnitude of σ_2. The practitioner, however, will generall not have the luxury of locating pressure taps anywhere he likes and mus estimate piping losses anyway to arrive at valve pressure drops.

Another sigma form is based on velocity. It is useful when pressures are not readily measurable. It is most often used for evaluating cavitation caused by surface roughness, submerged bodies, and offsets in boundaries (6).

$$\sigma = [2(P_u-P_v)]/\rho V^2 \quad (7)$$

where

ρ = fluid density
V = the mean or other relevant velocity

Care must be taken not to use scale effects (described in sections below) derived using data based on the velocity form of sigma with sigmas based on pressures.

Since the major function of a control valve is to control fluid by taking a pressure drop, the pressure based sigmas (equations 3 and 4) are more appropriate for control valves. The downstream pressure based sigma (equation 4) is preferred by the authors, because typical values may span several orders of magnitude making numerical evaluations of cavitation intensity more sensitive. This form of sigma (equation 4) will be used for the rest of the paper.

Sigma may be considered as a dimensionless ratio of forces tending to suppress cavitation (the system pressure) to the forces tending to cause cavitation (the pressure drop) (8). When either the difference between the downstream pressure and the vapor pressure gets small or the overall pressure drop gets very large, sigma becomes smaller and the cavitation becomes more severe. For control valves operating at typical pressures of near 10 bar (145 psi) and typical sizes of up to 50 mm (2-inch), the guidelines for trim styles shown in Table 1 have proven successful through years of field experience.

The major benefits of the multi-hole, parallel path design retainer or cage shown in Figure 1 and referred to in Table 1, is to direct the vena contracta and pressure recovery away from the metal boundary, thus directing the collapsing bubble away from metal boundaries. Since the bubble must be generally within one bubble diameter of the metal boundary to cause damage (4,9), these cavitation bubbles are less likely to cause damage. Also, by breaking the flow up into smaller streams, the physical size of the cavitation bubbles is smaller. The implosions of these smaller bubbles are less violent, thus adding a second measure against cavitation noise and damage. If this method does not provide adequate protection, multi-stage devices such as those shown in Figure 2 are necessary.

Table 1 shows typical ranges of values for the service sigmas (σ_s defined in equation 2) of various styles of trim for globe-style control valves. Although the sigma (σ_2) in equation 4 provides a much better indication of the severity of the cavitation than F_L or K_C, it also falls short of taking into account various factors such as velocity, pressure, fluid effects, size scale effects, etc. Scale effects factors are needed to adjust sigma (σ_2) to account for differences between actual operation conditions and a tested, reference sigma. The service sigma (σ_s) and Intensity Factor (I) implement factors for scale effects.

3.1.1 Pressure Scale Factor, K_p

It is possible to have the same sigma (σ_2) for two entirely different intensities of cavitation. For example: P_u = 11.0 bar (160 psi), P_d = 1.0 bar (14.5 psi), P_v = 0.1 (1.45 psi) yields a σ_2 = (1.0-.1)/10.0 = 0.09; P_u = 110.0 bar (1600 psi), P_d = 10.0 bar (14.5 psi), P_v = 1.0 bar (14.5 psi), also yields a sigma of σ_2 = (10.0-1.0)/100.0 = 0.09. However, one's intuition and experience would lead one to believe that the cavitation in the 100 bar (1450 psi) pressure drop service would be much more severe than that in the 10 bar (145 psi) differential. This, in fact, is supported by research of Tullis (6) and Rahmeyer (5,8). The following equation adjusts sigma (σ_2) for pressure scale effects. This relationship is supported by a significant body of research too extensive to repeat here (5,6,8), but which was performed on valves and orifices at pressures up to 17 bar (250 psi). There is only limited data to indicate whether or not the equation is valid at higher pressures.

$$K_p = [(P_u-P_v)/(P_u-P_v)_r]^X \quad (8)$$

where (all pressures are absolute):

$(P_u - P_v)r$ = reference pressures for the model testing
X = factor accounting for the style of valve (Table 2)
$(P_u - P_v)$ = actual operating pressures

Equation 8 can be used to scale the results of a test at one pressure to apply at another set of pressures. Since the premise of this method is based on a "typical" control valve test at about 7 bar (102 psi), substituting a 7 bar (102 psi) reference pressure drop into equation 8 yields:

$$K_p = [(P_u - P_v)/7]^X \quad \text{(pressures in bar)} \tag{8a}$$

$$K_p = [(P_u - P_v)/102]^X \quad \text{(pressures in psi)} \tag{8b}$$

Exponents, X, for critical cavitation and incipient cavitation damage are shown in Table 2 (6).

3.1.2 Size Scale Factor, K_s

The size of the cavitation bubble is proportional to the size of the flow passage where the cavitation bubble was created (4,10). The larger the bubble, the more violent and damaging is the collapse. Cavitation for a given unadjusted sigma (σ_2) causes less damage in small passages than in large passages (10,11). This is another reason why the multi-hole (parallel path) devices are effective in controlling cavitation. For example, a 150 mm (6-inch) control valve would have a 125 mm to 150 mm (5 to 6-inch) flow path. A multi-hole trim would break the flow path up into hundreds of 5 mm (1/4-inch) flow streams. This effectively reduces the damage and vibration potential due to the size effects from a 150 mm (6-inch) valve to a 5 mm (1/4-inch) valve. The following equation adjusts the service sigma based on a nominal reference valve size. This equation assumes that conventional trim (either single or double port) is used. If any type of device is incorporated to break the flow into multiple passages, this equation is no longer based on the valve size. Instead, the diameter of the individual path must be used.

$$K_s = \{1 - [\log(D/d)/10^{(.40 - .52 \log Cd)}]\}^N \tag{9}$$

where D = diameter of large valve mm (inches)
d = diameter of small valve mm (inches)
Cd = discharge coefficient (equations 10)[3]
N = -2 when scaling from a large reference valve to smaller one
N = +2 when scaling from a small reference valve to a larger one
log = base 10 logarithm

A valve size of 50 mm (2-inch) is normally used in tests as a small reference valve by the authors. If the 50 mm (2-inch) valve data is used for predicting cavitation in larger valves, equation, 9 becomes:

$$K_s = \{1 - [\log(D/50)/10^{(.40 - .52 \log Cd)}]\}^2 \tag{9a}$$

$$K_s = \{1 - [\log(D/2)/10^{(.40 - .52 \log Cd)}]\}^2 \quad \text{(U.S. Customary)} \tag{9b}$$

$$C_d = [(D^4/625K_v^2) + 1]^{-1/2} \tag{10a}$$

$$K_v = Q(G_f/\Delta P)^{1/2}$$

$$C_d = [(890D^4/C_v^2) + 1]^{-1/2} \quad \text{(U.S. Customary)} \tag{10b}$$

$$C_v = Q(G_f/\Delta P)^{1/2} = 1.156\ K_v \tag{11}$$

---

3 NOTE: There are several forms of discharge coefficients. Only Cd as defin[ed] by equation 10a or 10b may be used in equations 9, 9a and 9b.

where Q = volumetric flow rate, m^3/hr(gallons/minute)
G_f = specific gravity of fluid
D = valve size in mm (inches)
ΔP = pressure drop bar (psi)

Cd will range from 0 to 1 depending on valve style and valve opening. The Cd for all valve styles converge to zero at zero percent open. It should be noted that K_v is the metric equivalent of the I.S.A. value flow coefficient C_v which uses U.S. customary units; the relationship between K_v and C_v is shown in equation 11.

3.1.3. Application of Scaled Sigma

Once all the scale effect modifiers are calculated, they should be used to modify the basic sigma ratio $(P_d - P_v)/\Delta P$ to yield the service sigma (σ_s) using equation 2. These sigmas can then be compared to the guidelines in Tables 1 and 4. A condition called "critical" cavitation exists when incipient cavitation becomes light and steady and beyond which the noise and vibration sharply increase. The point of inflection will be designated by the value of sigma at that point ($\sigma_{r\ critical}$). Service sigmas greater than $\sigma_{rcritical}$ are normally considered a good application limit (12). In the case of multi-staged trim, each manufacturer has the responsibility to indicate how many stages are required for sigmas less than 0.2, since the efficiency of each stage varies based on the individual design. As an example, ChannelStream valves manufactured by Valtek require stages indicated in Table 3 for sigmas below 0.2.

3.2 Intensity Scale Factors

If the recommended application guidelines for any style of valve are exceeded, then damage, vibration, etc. can be expected roughly proportional to the Intensity Factor defined by equation 1 and repeated here.

$$I = V_c K_F K_d \ \sigma_{rd}/\sigma_s \qquad (1)$$

The basis for interpreting values of I is that a value of one (1) indicates "normal" wear, noise, etc for the valve in the particular service. Values between zero and one mean proportionally less wear, noise, etc. Values greater than one indicate greater than normal wear, damage, vibration and noise in proportion to the value of I. The Intensity Factor is not defined for values less than or equal to zero.

The Intensity Factor is also a function of scale effects. These scale effects listed below are less empirical than K_p and K_s listed in equation 2, only because past testing programs have not produced sufficient data to confirm an empirical relationship. The incipient damage reference sigma (σ_{rd}) can be determined from model tests or tests on valves in service with means provided to inspect damage coupons. Reference values for σ_{rd} and $\sigma_{r\ critical}$ for different valve styles are listed in Table 4.

3.2.1 Velocity Factor, V_c

Nominal flow velocities in the valve are very important in assessing potential damage. Indeed, this may be the most important factor, because it is a function of the difference between the pitting threshold velocity (V_o) and the characteristic velocity for the valve (V) (i.e. the average flow velocity in the valve port). Knapp et. al (4) and Tullis (6) have shown that once damage or pitting has commenced at the incipient damage sigma or velocity, the rate of pitting is a strong function of the difference $V-V_o$. The authors suggest the following velocity correction factor for intensity which appears to be in general agreement with published data (4,6) on pitting rates vs. velocity.

$$V_c = 1 + 10^{2(V-V_o)} \quad (V,\ V_o\ \text{m/sec}) \qquad (12a)$$
$$V_c = 1 + 10^{0.61(V-V_o)} \quad (V,\ V_o\ \text{ft/sec}) \qquad (12b)$$

where V = average fluid velocity through valve outlet area or other characteristic area

V_o = threshold velocity (average) for damage through the valve's characteristic area

From equation 12, it can be observed that values of V_c are essentially unity until valve velocity approaches V_o. When V equals V_o, the value of V_c becomes 2.0. We believe this conservatism is justified in the uncertainties of typical service and test data. As V exceeds V_o, values of K_v increase exponentially, which reflects the generally observed trends of pitting rate and material removal rate in the cavitation zone with sigma values less than incipient damage sigma. Due to the sensitivity of K_v to the velocity difference ($V-V_o$), the value of V_o should be verified by testing. Care should be taken in such tests to allow sufficient incubation time for pitting to be observed in the test materials (6). The importance of testing each valve style for its cavitation parameters (i.e. $\sigma_{r\ critical}$, σ_{rd}, V_o) cannot be overemphasized, because minor differences in design geometry can result in significant differences in valves' cavitation parameters. Lacking actual test values for V_o, 10 m/sec (33 ft/sec) can be used as a conservative estimate.

3.2.2 Fluid Factors, K_f

Most of the cavitation research is done using water as the medium. This is done for experimental convenience and because of the frequent need to take cavitating pressure drops in water process equipment. As a result, the data relating fluid properties of other fluids to the cavitation research done on water is limited. However, some research and observations can be incorporated into some "rules of thumb." The total fluid factor, K_f, is the product of several sub-factors described below.

$$K_f = (K_{fv})(K_{fT})(K_{fo}) \qquad (13)$$

K_{fv} is the fluid viscosity factor. Very viscous fluids tend to show less cavitation damage than do non-viscous fluids. The high viscosity tends to limit bubble growth and slow bubble collapse, thereby reducing cavitation damage. Hydraulic oils for example show less damage at a given sigma than does water. An approximate modifier for this effect is

$$K_{fv} = [\nu_{water}/\nu_f]^{1/2} \qquad (14)$$

where ν_{water} = kinematic viscosity of water at 20C (68F)

ν_f = kinematic viscosity of fluid

The authors could find no experimental data that will support the equation for this factor, nor could an independent review of the literature by Hammitt (10). This factor is based on the authors' judgement and experience from visual observations of cavitation in Freon

11 and applications of valves in hydrocarbon services. We encourage research to support or revise this factor.

K_{fT} is the fluid temperature factor. Research data indicates (10) that, at fluid temperatures roughly half-way between the freezing point and boiling point (based on upstream pressure), cavitation damage is approximately three times more severe, i.e. the rate of material removal is three times faster. At low temperatures, this has been well documented, however, the reasons for it are not well understood. At higher temperatures, theory suggests that the thermodynamic effects retard bubble growth and slow its collapse. At high temperatures, the fluid is near its saturation temperature; the temperature difference is small between the vapor in the cavitation bubble and the surrounding fluid. The rate at which heat can be transferred is therefore less. This low heat transfer rate retards both bubble growth and collapse. Equation 15 can be used as an approximate factor to take this into account.

$$K_{fT} = 3[1 - |T_s - T_{ave}|/(T_B - T_{ave})] \qquad (15)$$

where
T_B = boiling temperature at upstream pressure
T_F = freezing temperature at upstream pressure
T_s = fluid service temperature
$T_{ave} = (T_B + F_F)/2$
$|T_s - T_{avg}|$ = absolute value of $(T_s - T_{ave})$

K_{fo} accounts for any other factor based on experience or test data. The overall fluid factor can then be added together to yield an over all fluid factor.

$$K_f = (K_{fv})(K_{fT})(K_{fo}) \qquad (16)$$

3.2.3 Duty Factor, K_d

The duty cycle of the control valve within the cavitating condition (i.e. continuous, intermittent, rare) can also be taken into account with the factor K_d. If the valve will experience the cavitating condition only during a rare upset, and since the damage effects are time dependent, short operation in this condition would not jeopardize the overall performance of the valve. Flow capacity limitations due to choked flow, of course, must be taken into account. If the valve cavitates only during certain service conditions, such as start-up, a more intense cavitating condition could be withstood, than in a continuously throttling valve. If the valve is continuously throttling and/or in critical service, a more conservative factor must be taken. Table 3 represents a proposal of some possible values based on the authors' experience; they are not substantiated by extensive tests or field research. This is an area where additional research and experience are needed.

4. Examples

The following examples are taken from actual control valve service conditions. They serve to illustrate the application of this cavitation evaluation method.

Boiler Feed Water

Line size = 200 mm (8-inch)
P_u = 110 bar (1600 psia)
P_d = 103.5 bar (1500 psia)
Q = 272,000 kg/hr (600,000 #/hr)
T = 234°C (454°F)
Fluid = water
P_v = 5.8 bar (84 psia)

First, standard valve sizing equations (1) would be used to obtain the following values:

K_v = 118 (Cv = 136), assuming $P_1 = P_u$ and $P_2 = P_d$
Valve size = 150 mm (6-inch)
Velocity at valve outlet = 6.6 m/sec (22 ft/sec)
Reduced trim = 75 mm (3-inch)

Next calculate the basic σ_2 using equation 4.

$$\sigma_2 = (P_d - P_v)/(P_u - P_d) = (103.5-5.8)/(110-103.5) = 15$$

This σ_2 must now be modified by K_p (Pressure Scale Factor) and by K_s (Size Scale Factor) to obtain σ_s (service sigmas for comparison with recommended values in Table 1.

$$K_p = [(P_u - P_v)/7]^x = [(110-5.8)/7]^{-.30} = 0.44$$

NOTE: Boiler feed water valves are normally globe style, so a value of -0.30 for X is chosen from Table 2 correspondingly to "Globe Style Incipient Damage" since those valves in Table 1 are "incipient damage" guidelines.

Next, the size scale effects factor (K_s) must be calculated from equation 9. In order to calculate K_s, the discharge coefficient must be calculated according to equation 10.

$$C_d = [(D^4/625K_v^2)+1]^{-1/2} = [(150^4/(625)(118)^2)+1]^{-1/2} = 0.13$$

$$K_s = \{1-[\log(D/50)/10^{(.40-.52 \log Cd)}]\}^N$$
$$= \{1-[\log(150/50)/10^{(.40-.52 \log 0.13)}]\}^2 = 0.873$$

The service sigma can now be calculated using equation 2 and 4.

$$\sigma_s = K_S K_P \sigma_2 = (.44)(.887)(15) = 5.8$$

The lower value of $K_P = .44$ indicates a strong pressure effect at these service conditions. The higher value (i.e. close to 1) of K_S means that size scale effects have a smaller effect in this case. This is the result of the relatively small K_v. Had the valve been operating at higher flow rates closer to the maximum capacity of a 6-inch valve with a K_v of 350, the K_s would have been as follows

$$C_d = [(150^4/(625)(350)^2)+1]^{-1/2} = 0.36$$

$$K_s = \{1-[\log(150/50)/10^{(.40-.52 \log .36)}]\}^2 = 0.79$$

Since smaller values of σ_s indicate more severe cavitation, a K_s of 0.79 versus 0.875 would yield a lower σ_s which indicates a more severe condition.

Returning to our original example, a service sigma (σ_s) of 5.8 is well above the value listed in Table 1 for "no cavitation damage expected." Therefore, one would expect a standard trim to function well. However, boiler feed valves are subjected to start up conditions with much higher pressure drops while the boiler is being brought up to temperature. Suppose the valve controls the following condition during start up.

P_1 = 110 bar (1600 psia)
P_2 = 20.7 bar (300 psia)
Q = 90,900 kg/hr (200,000 #/hr)
T = 121°C (250°F)
P_v = 2.05 bar (29.8 psia)
K_v = 13.4 (Cv = 15.5), assuming ΔP_{choked} = 48.7 bar (706 psi)
Velocity = 1.917 (6.29 ft/sec)

Calculate σ_2 from equation 4.

$$\sigma_2 = (20.7-2.05)/(110-20.7) = 0.21$$

Modify for K_s and K_p from equations 8 and 9.

$$K_p = [(110-2.05)/(7)]^{-0.30} = 0.44$$

Note that K_p did not change. One would expect this, because P_u did not change and P_v only slightly changed.

$$C_d = [(150^4/(625)(13.4)^2)+1]^{-1/2} = 0.014$$

$$K_s \{1-[\log(150/50)/10^{(.40-.52 \log .014)}]\}^2 = 0.96$$

Note that the K_s factor is larger indicating a smaller effect due to size in this condition. This is due to the reduced flow rate and smaller valve opening.

The service sigma σ_s can now be calculated by equation 2.

$$\sigma_s = K_p K_s \sigma_2 = (.44)(.96)(.21) = .089$$

Referring to Table 1, a $\sigma_s = .089$ requires a multi-staged trim to prevent cavitation damage. However, this is only a start up condition, and the user may want to chose hardened trim ($1>\sigma_s>.5$) or a multi-parallel path ($.5>\sigma_s>.2$) instead of the expensive multi-staged device. If the user decides to allow cavitation during start up conditions, he should calculate the Intensity Factor, I, to evaluate the relative severity of this choice.

Calculate V_c, the velocity factor, from equation 12.

$$V_c = 1 + 10^{2(V-Vo)} = 1 + 10^{2(2.86-10)} = 1$$

Note that for any value of velocity below V_o (10 m/s), V_c is about 1 and does not need to be calculated. Only for those values equal to or greater than V_o is V_c relevant.

Calculate the fluid factor, K_f. The fluid is water, so no fluid viscosity factor K_{fv} needs to be calculated, i.e. $K_{fv} = 1$. K_{fT}, the fluid temperature factor from equation 15 is

$$K_{fT} = 3[1- |121-563| /(1126-563)] = 0.64$$

where
$T_s = 121^0C$
$T_B = 1126^0C$
$T_F = 0^0C$
$T_{ave} = 563^0C$

Since K_{fT} is less than 1, the damage would occur at a rate of 64 percent of the reference rate. The overall fluid factor from equation 16 is

$$K_f = (K_{fv})(K_{fT})(K_{fo}) = (1)(.64)(1) = 0.64$$

Calculate K_d, the Duty Factor. For startup conditions from Table 5, $K_d = 0.5$ to 0.8. We will assume 0.5 for relatively infrequent startup cycles.

We can now calculate the Intensity Factor. We will use r_{rd} (reference damage σ for globe valves) from Table 4 = 0.43. From equation 1:

$$I = V_c K_F K_d \sigma_{rd} / \sigma_s = (1)(0.64)(0.5)(0.43)/(0.089) = 1.53$$

The 1.55 means that the valve is likely to wear out 1.55 times faster than a valve operating at the damage threshold.

Often in this application a hardened trim such as Stellite would be chosen. Since the shape of trim has not changed, the $\sigma_{rcritical}$ (see Table 4) has also not changed but r_{rd} would change. The new Intensity Factor would be calculated using equation 1 with $r_{rd} = 0.35$ from Table 4, thus

$$I = (1)(0.64)(0.5)(0.35)/(0.089) = 1.26$$

This lower Intensity value indicates that the trim is roughly 20 percent better than standard trim (non-hardened).

Suppose a multi-parallel path trim were chosen with $\sigma_{rd} = 0.23$. Then equation 1 yields

$$I = (1)(0.64)(0.5)(0.23)/(0.089) = 0.83$$

An $I = 0.83$ indicates that the parallel path trim undergoing infrequent startups could last longer than the standard trim at the threshold of damage.

5. Conclusions

The authors propose the foregoing as a simple, universal method for quantifying the severity of cavitation in control valves. Service sigma, σ_s is a sigma modified for both pressure and size scale effects and provides a much better measure for evaluating cavitation than does K_c. The Intensity Factor, I, gives the user some guidelines for quantifying his service when he must operate the valve in a cavitating flow regime. It is admittedly simplified and is not as accurate as other, more sophisticated methods could be. However, for users specifying hundreds of control valves, a simple method of evaluating the service conditions is extremely useful. This method, although simple, provides a more accurate reflection of the service conditions than does the more commonly used F_i and F_L. Further testing and research is encouraged to build a better data base for more accurate scaling of the Intensity Factor and sigma.

REFERENCES

1."ISA Handbook of Control Valves", 2nd edition. J.W. Hutchinson (Ed-in-chief).

2.Wislicenus, G.F.: "Remarks on the. History of Cavitation as an Engineering Problem". In: Proc ASME Fluids Engineering and Applied Mechanics Conference, Cavitation State of Knowledge (Northwestern University, Evanston, ILL, June 16-18, 1969).

3.Robertson, J.M.: "Cavitation Today - An Introduction". In: Proc ASME Fluids Engineering and Applied Mechanics Conference, Cavitation State of Knowledge, (Northwestern University, Evanston, ILL, June 16-18, 1969).

4.Knapp, R.T., Daily, J.W., and Hammitt, F.G.: "Cavitation". Institute of Hydraulic Research, University of Iowa, Iowa City, Iowa.

5.Rahmeyer, W.J.: "Cavitation Testing of Control Valves". ISA '83 International Conference Proceedings, Instrument Society of America, pp. 1253-1261.

6.Tullis, J.P.: "HYDRAULICS OF PIPELINES - Pumps, Valves, Cavitation, Transients". C June 1984, Third Draft.

7."Control Valve Capacity Test Procedure," Standard ISA S75.02 Instrument Society of America 1981.

8.Rahmeyer, W.J.: "Test Procedures for Determining Cavitation Limits in Control Valves". Research and Technology, Journal AWWA November 1986, pp. 55-58.

9.Morgan, Wm.B. and Parkin, B.R.: International Symposium on Cavitation Inception. In: Proc ASME Winter Annual Meeting, (New York, New York, December 2-7, 1979).

10.Hammitt, F.G.: "Cavitation and Multiphase Flow Phenomena". McGraw-Hill Book.

11.Tullis, J.P.: Testing Valves for Cavitation: Presented September 1974 International Conference on Cavitation, Herist-Walt University Edinburgh Scotland.

12.Tullis, J.P.: Cavitation Data for Valves and its Application: Presented September 1974 International Conference on Cavitation, Herist-Walt University Edinburgh Scotland.

Table 1. Service Sigma Application Guidelines for Globe-Style Valves*

σ_s	Recommended Trim Style
$\sigma_s \geq 1$	- no damaging cavitation expected
$1 > \sigma_s \geq .5$	- hardened trim e.g. Stellite No. 6
$.5 > \sigma_s \geq .2$	- multi-parallel path (small hole) retainer or cage
$\sigma_s < .2$	- multi-staged (serial path) letdown device

* 50 mm (2-inch) globe style, P_u = 7 to 15 bar (100 to 200 psi)

Table 2: Pressure Scale Effects Exponents for Control Valves.

Valve	Cavitation Limit	X
Full Ball	Critical[4]	-.27
	Incipient Damage[5]	-.20*
Segmented Ball	Critical	-.32*
	Incipient Damage	-.28*
Butterfly	Critical	-.28
	Incipient Damage	-.18
Globe (S type)	Critical	-.14
	Incipient Damage	-.30

* Estimated

Table 3. Typical Minimum Values of Service Sigma for a Multi-Staged Control Valve Trim. (ChannelStream)

Number of Stages	Minimum σ_s
2	0.200
3	0.080
4	0.025
5	0.007
6	0.002

---

4 Critical cavitation is the threshold of cavitation roughly defined by the K_c factor used by the Instrument Society of America (1). It would be normally considered as "cavitation-free" operation (12). At this point, noise is steady but light. The intensity is similar to that of "bacon on a hot grill." Normal conversation is not impaired even when standing adjacent to the valve. This limit is generally accepted for applications where a limited amount of noise can be tolerated, but loud noise, erosion and vibration must be avoided.

5 Incipient damage is the service condition at which the surface of metal boundaries will begin to be damaged by cavitation. Experimentally, this is determined by inserting soft (annealed) aluminum plates at suspected cavitation damage areas. A pitting rate of 0.155 pits/centimeter^2/minute (1 pit/inch^2/minute) in soft aluminum is generally accepted as a limit below which a device may be operated continuously if constructed out of normal engineering materials such as steels, stainless steels and hardened alloys (11,4).

Table 4. Reference Values of Sigma-damage and Sigma-critical for Various Styles of Control Valves

Valve Style	σ_{rd}(damage)[6]	$\sigma_{r\ critical}$[7]
Globe		
standard	0.43	0.54
hard	0.35	0.54
multi-path	0.23	0.29
multi-stage*	<0.23	<0.29
Angle	0.71	0.87
Butterfly-60° open	1.86	2.33
90° open	2.50	3.20
Ball - full	2.4	3.0
Ball - segmented	2.8	3.5

* Depends on design

Table 5. Range of Values of Duty Factor, K_d

Frequency of Occurrence	K_d[8]
Rare Upset	0.1 to 0.3
Start up	0.5 to 0.8
Continuously Throttling	1.0
Critical/Continuous	2.0 to 3.0

6 These values are approximate. They depend greatly on subtle changes in valve geometry and changes in material. These values are based on 316 SS valve trim. The damage rate for a given material is also related to corrosion effects which are accelerated by cavitation and velocity. The cavitation intensity factor I should be reduced by a factor of 20 if Stellite is used and increased by 8 if mild steel is used (6).

7 Both σ_{rd} and $\sigma_{rcritical}$ vary with valve opening. If K or K_c is known as a function of stroke for a particular valve or trim, σ_{rd} and $\sigma_{rcritical}$ can be estimated using the following equation $\sigma_{rcritical} = (1/K)-1$ and $\sigma_{rd} = .8\ \sigma_{rcritical}$

8 Since the damage rate is approximately linearly time dependent after an initial incubation period (4) the user can establish his own factors based on a knowledge of the service.

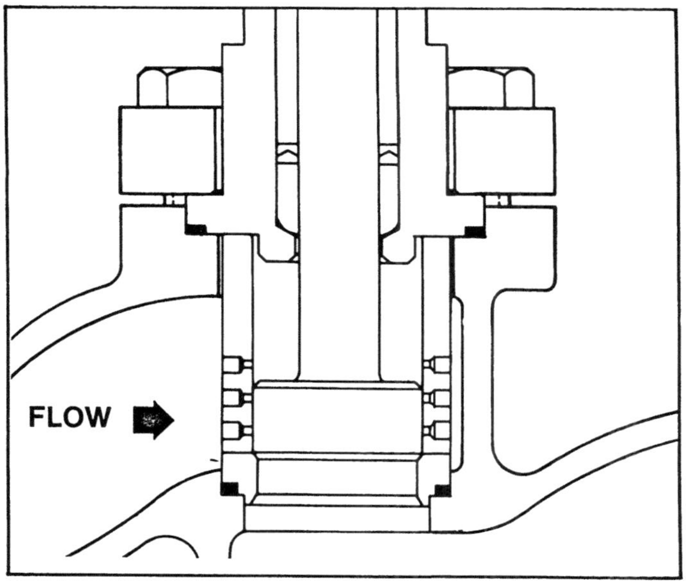

Figure 1.

A typical multi-hole (parallel path) valve trim for minimizing the effects of cavitation. The flow streams are directed inward where pressure recovery and subsequent implosions of vapor bubbles are removed from metal surfaces, reducing the risk of cavitation damage. The small flow passages also reduce noise and vibration.

Figure 2a

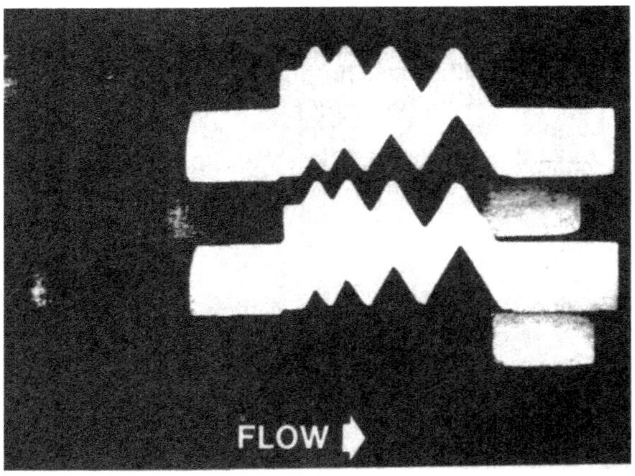

Figure 2b

These multi-stage devices are used to reduce pressure in small stages. Figure 2a uses a series of restrictive channels and interconnecting plenum-holes.
Figure 2b uses a series of machined disks to create expansion and contraction areas. Both devices maintain the pressures at each stage above that which is likely to cause cavitation damage.

2nd International Conference on

Developments in Valves and Actuators for Fluid Control

Manchester, England: 28-30 March 1988

PAPER H1

TESTS OF CHECK VALVES AT EDF : DEVELOPMENT OF A DAMPED CHECK VALVE

M. PANET
EDF Research and Development Division
77250 Moret-sur-Loing - FRANCE

R. MARTIN
VANATOME - 7 rue Vidal - 07100 Annonay - FRANCE

Summary

the paper presents the results of tests using a new type of check valve designed to damp fluid hammering which may occur in fluids subjected to high pressure during some transient flows.

the tests, consisting of sudden closure by depressurization, indicate the efficiency of the prototype valve, the pressure drop of which is measured in steady flow.

the results are compared with those obtained using classical lift check valves in identical conditions, and validate the operation concepts of the prototype.

1.0 - Introduction.

The damping of hammering phenomena by the closure of check valves [Ref. 1] raises both theoretical and technological difficulties, in particular in the case of circuits under high pressure which are suddenly depressurized by the unexpected rupture of a pipe.

This situation led to an agreement between EDF and VANATOME to develop a component which would markedly level out the peak of the overpressure created by the valves operation.

Held in Manchester, England. Organised and sponsored by BHRA, The Fluid Engineering Centre and co-sponsored by the British Valve Manufacturers Association and the Institution of Mechanical Engineers Process Industries Division

2.0 - Hydrodynamic operation hypothesis.

The interaction between the check valve and the fluid is studied during sudden closures induced under the special test conditions of the ECLAIR testing facility (See Figure 1). A water tank under high pressure feeds a pipe which is sealed by two membranes at its other extremity. Bursting of these membranes causes a flow which must be interrupted by the check valve, minimising hammering phenomena. A single punched membrane (See Figure 2) is also used, getting a clean break and a faster depressurization, which allows a more accurate estimate of the closure time of the valve.

The approximate laws of tank drainage indicate that an incompressible fluid set in motion follows a hyperbolic velocity versus time law (See Figure 3).

During closure of the check valve, placed to prevent drainage of the tank, the velocity of the fluid and its mass flow rate follow on OAB trajectory. This causes overpressure which is the higher the more closely the conditions below apply :

- if the flow velocity is high at point A,
- if the motion time of the check valve is short during the AB fraction of closure,
- if the rate of flow through the check valve remains high with small opening positions.

The operating principle of the prototype described below consists of damping the three phenomena above mentioned by developing :

- a valve with a small inertia, shortening the time OA of closure,
- damped motion of the valve during time AB,
- a regularly decreasing rate of flow during closure time AB.

3.0 - The prototype.

3.1 - Design criteria

The prototype is designed to meet the requirements of nuclear installations :

a) Ability to function in the normal operating conditions of a pressurized water reactor :

pressure : 18 MPa, temperature : 300°C.

b) Marked attenuation of overpressure and satisfactory mechanical behaviour during sudden closures under high pressure.

c) As limited a pressure drop as possible for normal operating conditions.

d) Stainless steel leaktight seat without cobalt coating to reduce activated debris in the reactor system.

3.2 - Construction of the prototype.

The characteristics of the prototype are as follows (See Figure 4) :

a) Material used : austenitic steel Z 6 CND 17-12.

b) Cylindrical shape providing higher resistance against internal pressure and easy mounting on the pipes with flanges.

c) Profiled valve offering low resistance to flow.

d) Possible installation on a vertical or horizontal pipe by balancing the weight of the valve with a spring.

e) Free motion of the valve during its movement on OA (See Figure 3) followed by slower motion, by hydraulic damping, during movement AB.

f) Attenuation of the reverse flow by forced flow in a diffuser acting during the movement AB.

4.0 - Sudden closure test methodology.

4.1 - Prototype configurations.

The experimental study of the prototype operation leads to investigate several component configurations. This is obtained with several diffusers and pistons successively used for tests (See Figure 5).

The diffusers D1 and D2 offer the same radial flow hole area, equal about to a third of the pipe section area. The holes are differently distributed : upward or evenly located on each diffuser D1 and D2. The item D3 gives no significant head loss effect and shows a flow area of the same order of the adjacent pipe one.

The effects of the internal clearance of the hydraulic damper and of the piston thickness are studied with two pistons of different dimensions P1 and P2. The damping effect is light or significant whether the piston P1 or P2 is used.

The prototype configurations are referred to test numbering, according to the Figure 5, Tables 2 and 3.

4.2 - Prototype instrumentation.

The damping effect of the DN 68.9 ANSI class 300 prototype is investigated with two high natural frequency transducers which measure the pressure applied on each side of the piston inside the casing valve (See Figure 4 points 3 and 4), allowing to calculate the stress acting on the stem during closure.

These data are completed with the electrical signal transmitted by a copper wire coil wound on the DN 50 ANSI class 1500 prototype, which is induced by the magnetic flux rate crossed by a piston check worked in a ferromagnetic steel (See Figure 6).

Preliminary fast closure tests are realized out of water. The casing valve is quickly moved with a spring compressed with a force ranging from 5 to 10 daN. The coil signal is compared with the displacement versus time curve of the valve, provided by a L.V.D.T. sensor. The coil signal increases with the speed of the piston check, and its sens of variation depends of the acceleration which is graphically described by the curvature of the previously mentioned curve. This proposal is well verified by the prototype fitted with the piston P2 ; the greater damping effect is obtained showing an up and down motion causes by the valve kinetic energy absorption by the damper pressurized air volume located around the point 3 (See Figures 4 and 6).

4.3 - Test methodology.

At the initial moment, pressure is uniform in the installation. The check-valve is completely open, held by a thin wire. Opening the tap which depressurizes the volume between the two bursting membranes (tests n° 1 to 6, 16 and 17) or using the single punched disk alternative device (tests n° 7 to 15, 18 and 19) connects suddenly the exhaust of the pipe up to the air pressure. The fluid is set in motion and attains a high flow velocity.

Pressure fluctuations are measured on either side of the check valve (points 1 and 2) to describe the transient flow fully.

5.0 - Sudden closure test results.

This above-mentioned operating mode was applied to the check valves fitted in EDF reactors for their ability to resist accident conditions caused by rupture of pipes, as well as to the prototypes DN 50 ANSI class 1500 and DN 68,9 ANSI class 300 which afford direct comparison with lift check valves having the same characteristics.

5.1 - Classical technology check-valves.

The nuclear power plan equipment requires the qualification of the classical technology components subjected to sudden depressurization conditions. Giving a satisfactory behaviour with a high mechanical strength, these components induce the well known heavy water hammers which could damage the hydraulic circuits. Tests n° 5 and 16 (See Figures 7 and 12) illustrate these phenomena.

The results (See table 1) show that for a given piping system, the maximum pressure depends on the diameter of the check valve and the initial pressure. For conventional valves the ratio of maximum to initial pressure is between 2 and 5.

5.2 - VANATOME DN 50 class ANSI 1500 prototype.

The research development action began with test n° 6 (See Figure 7) using the two disk bursting device and the less damped configuration of the prototype fitted with piston P1 and diffuser D2. The maximum pressure only exceeds the initial value by 33 %, whereas the test n° 5 of a conventional valve having the same characteristics indicated a 240 % increase.

However, the delay between the bursting of the two membranes is deducted from the variation of the pressure at point 1, which increases slightly before equalling definitively the air pressure value. It makes doubtful wether the valve is open or not before the end of the depressurization.

The tests n° 7 to 10 (See Figures 8 and 9) show the prototype operation with various initial pressure values from 0,5 MPa to 12,3 MPa. The check valve is fitted with the diffuser D1 and the more efficient damper provided by the piston P2. The single punched disk device is used which gives reliable test data.

The pressure at point 1 drops instantaneously, then follows roughly the homologous flow trajectory assumption OAB (See Figure 3) with a maximum value probably depending of the flowrate, and a regular decrease while the piston check moves towards the seat, inside the diffuser, reducing smoothly the rate of flow.

The pressure at point 2 follows the initial pressure drop and rises quickly to the initial pressure test value. Two main peaks of pressure are generally observed : the first one must correlate with the piston check position slightly driven inside the diffuser (point A on Figure 4), which divide by three the flow area ; the second peak must be caused by the total closure of the check valve (point B on Figure 4), and marks the end of closure of the prototype.

The maximum to initial pressure ratio ranges from 1,26 to 1,39 with a short closure time between 250 ms and 105 ms due to the initial pressure increasing from 2,7 MPa to 12,3 MPa (See Table 2).

The coil signal rises to a peak value during the free motion OA (See Figure 4) of the piston-check, and vanishes as soon as the damper is acting, after the piston P2 seals the slots located inside the casing-valve during the motion AB.

These are very favourable results showing the efficiency of the new prototype operation concepts with regard to conventional check-valve technology.

The tests n° 10-11-12 (See Figures 9 and 10) are performed with the same initial pressure equal to 12,3 MPa. The three diffuser are successively used with the piston P2. The different radial hole area repartition of the diffusers D1 and D2 explains that the pressure at point 1 vanishes faster during the test n° 10 : closer time shorter with the item D1 (105 ms) than the D2 one (167 ms). Howewer the pressure peak value is not altered, rising about to 17 MPa in both of the cases.

Using the item D3 modifies the pressure transient (See Figure 10). The pressure at point 1 remains constant during the most part of the valve closure, even when the damper is acting (no coil signal). This results of the membrane head loss unbalanced by the one of the check-valve subjected to a high reverse flowrate. The effective part of the closure happens in a time equalling the sixth (40 ms) of the total closure time (250 ms), which causes an important peak of pressure (28 MPa).

The coil signal indicates an acceleration of the piston-check motion with a second peak value correlated with a high differential pressure applied on the valve at the end of the closure.

The test n° 12 with a significant damping effect and a reduced head loss illustrate the efficiency of the concept which associates a damper with a significant valve pressure drop subjected to reverse flow conditions.

The tests n° 13-14-15 (See Figure 11) confirm the hereabove results, with various values of the initial pressure from 0,38 MPa to 6,8 MPa. The association of the diffuser D1 with the piston P1 is chosen, giving a significant pressure drop but a very slight damping effect under reverse flow conditions.

A time of closure between 150 ms to 24 ms is revealed (See Table 2) which induces hammering with pressure peaks equalling about three times the initial pressure. The coil signal correlates well with these closer times, showing by its gradient the lower effect of damping induced by the increasing initial pressure.

5.3 - VANATOME DN 68,9 class ANSI 300 Prototype.

The second VANATOME DN 68.9 class ANSI 300 Prototype is tested with a view of estimating the compressive stress applied to the stem of the damper when the valve is closing.

The test n° 17 (See Figure 12), performed with two bursting disks, diffuser D1 and piston P1 giving low damping effect, indicate a 200 % increase of initial pressure, which is much lower than that produced by test n° 16 (400 %) of the conventional type lift check valve used for comparison (See Table 3).

The tests n° 18 and 19 (See Figure 13 and 14), performed with the single punched disk device, the diffuser D1 and the piston P2 providing a significant damping effect, show similar to test n° 7 to 10 pressure transient results. This confirm the reliability of this check-valve based on new operating concepts.

It is interesting to observe the satisfactory prototype operation (See Figure 13) at low initial pressure value (0,66 MPa) wich could protect pumps from reverse flow. The ratio of maximum to initial pressure at point 2 is close 1,40.

The closure time reduces from 435 ms to 175 ms when the initial pressure rises from 0,66 MPa to 3,1 MPa.

The pressure at point 3 versus time curve confirms the initial free motion of the valve with a pressure drop inside the damper, and the damped motion provided by a marked overpressure resulting of the water compression around the point 3 inside the hydraulic damper.

The overpressure is located between the two main, pressure peaks measured at point 2, involving a good timing of the diffuser and damper functions. The resulting stress acting on the stem ranges from 514 daN to 2137 daN according to whether the initial pressure value is 0,66 MPa or 3,1 MPa.

6.0 - Hydraulic characterisation tests.

The study of the characteristic curves of flow shows the respective hydraulic behaviour of the VANATOME prototypes and the lift check valves used for comparison (See tests N° 5 and 16).

The variables used are non-dimensional. The rate of flow - Cv - is compared with the maximum value - CV max - measured when the check valves are completely open.

The definition of the rate of flow is :

$$Cv = 0.367\ Q\sqrt{\frac{d}{\Delta P}}$$

Q : m^3/h : flow
d : kg/m^3 : water density
ΔP : MPa : pressure drop created by the check valve.

This coefficient is determined from measuremeents in steady flow. The flow is opposite to the DN50 valve opening (See Figure 15 upper graphic) and is in the sens of the opening of the DN68.9 valve (See Figure 15 lower graphic). This provides no significant differences of the valve flow behaviour. The valve is held in its position by a fastening screw.

The check valves with which the comparison is made reduce latterly the flow rate, since the rate of flow coefficient remains close to the maximum value over a large portion of the path (See Figure 15).

By contrast, the prototypes develop substantial attenuation immediately they begin to close, thus reducing the fluid acceleration whose mass flow rate is therefore reduced.

It is interesting to observe the high value of the rate of flow coefficient of the prototype DN50 ANSI class 1500 : Cvmax = 65. This stems from the highly profiled shape of the valve, and is to be compared with the valve Cvmax = 39 of the conventional check valve.

7.0 - Conclusion.

The purpose of the study was to examine whether the transient pressure induced by the closure of a check valve in a pressurized fluid could be reduced.

To solve this problem, a new technological approach has been developed, which takes basic fluid mechanics in account. This led to deduct a new operating principle : using a damping effect coupled with a high pressure drop, during the check-valve closure.

The sudden closure and hydraulic characterisation tests performed provide a range of positive results, whose coherence is an incentive to develop special check valve techniques to deal with the problem. The action undertaken by EDF to this end is supported by effective collaboration by the manufacturers of cocks, taps and fittings.

Previously worked to operate with depressurisation of high magnitude, the prototypes are able to function even with low pressure conditions, and can protect pumps from reverse flows as well as nuclear installation from pipe failures.

From a technological point of view, the enhancement of the already promising performance of this type of component requires further investigations to verify its operation subjected to high temperature conditions. Complementary instrumentation should be provided to evaluate the flow transients and the motion of the valve in view of a complete knowledge of its interaction with the fluid.

8.0 - References.

[1] BS. Valibouse - PH Verry of Neyrtec Division - Modelling check valve slamming - March 1984.

[2] SL. Collier - CC. Hoerner - A facility and approach to performance test of check valves - Transaction of the ASME.

[3] Provoost, G.A. - "The dynamic behaviour of non-return valves". Paper J1. Procs. 3rd International Conference on Pressure Surges, BHRA 1980.

[4] Provoost, G.A. - "The dynamic characteristics of non-return valves". 11th IAHR Symposium of the Section on Hydraulic Machinery, Equipment and Cavitation; Operating Problems of Pump Stations and Power Plants, Amsterdam, Sept. 1982.

9.0 - Acknowledgements.

The authors would like to express their gratitude to the Mechanics and Component Technology Department of the E.D.F. Research Development Division which provides the necessary facilities to develop this experimental study.

TABLE 1 – CLASSICAL TECHNOLOGY VALVE SUDDEN CLOSURE TEST RESULTS.

Test type	Valve type	Test n°	Diameter (mm)	ANSI Class (Lb)	Pressure (MPa)		Closure time (ms)
					Initial	Maximum	
Two fast burst disks	Swing check valves	1	250	1500	12	60	45
		2	100	600	9	41	100
		3	80	1500	12	29	22
		4	80	1500	3	13	/
	Lift check valves	5	50	1500	12	29	40
		16	68,9	300	3	12	37

TABLE 2 – VANATOME DN 50 PROTOTYPE SUDDEN CLOSURE TEST RESULTS.

Test type	Piston	Diffuser	Test n°	Prototype Diameter (mm)	Prototype ANSI Class (Lb)	Pressure (MPa)		Closure time (ms)
						Initial	Maximum	
Two fast burst disks	P1	D2	6	50	1500	7	9,3	48
Single fast burst disk	P2	D1	7			0,5	~0,5	~300
			8			2,7	3,4	250
			9			6,6	8,9	187
			10			12,3	17,2	105
		D2	11			12,3	17,4	167
		D3	12			12,3	28	250
	P1	D1	13			0,38	0,83	150
			14			2,7	7,3	39
			15			6,8	20,6	24

TABLE 3 – VANATOME DN 68,9 PROTOTYPE SUDDEN CLOSURE TEST RESULTS.

Test type	Piston	Diffuser	Test n°	Prototype Diameter (mm)	Prototype ANSI Class (Lb)	Pressure (MPa)		Closure time (ms)
						Initia	Maximum	
Two fast burst disks	P1	D1	17	68,9	300	3	6	78
Single fast burst disk	P2	D1	18			0,66	0,95	435
			19			3,1	4,3	175

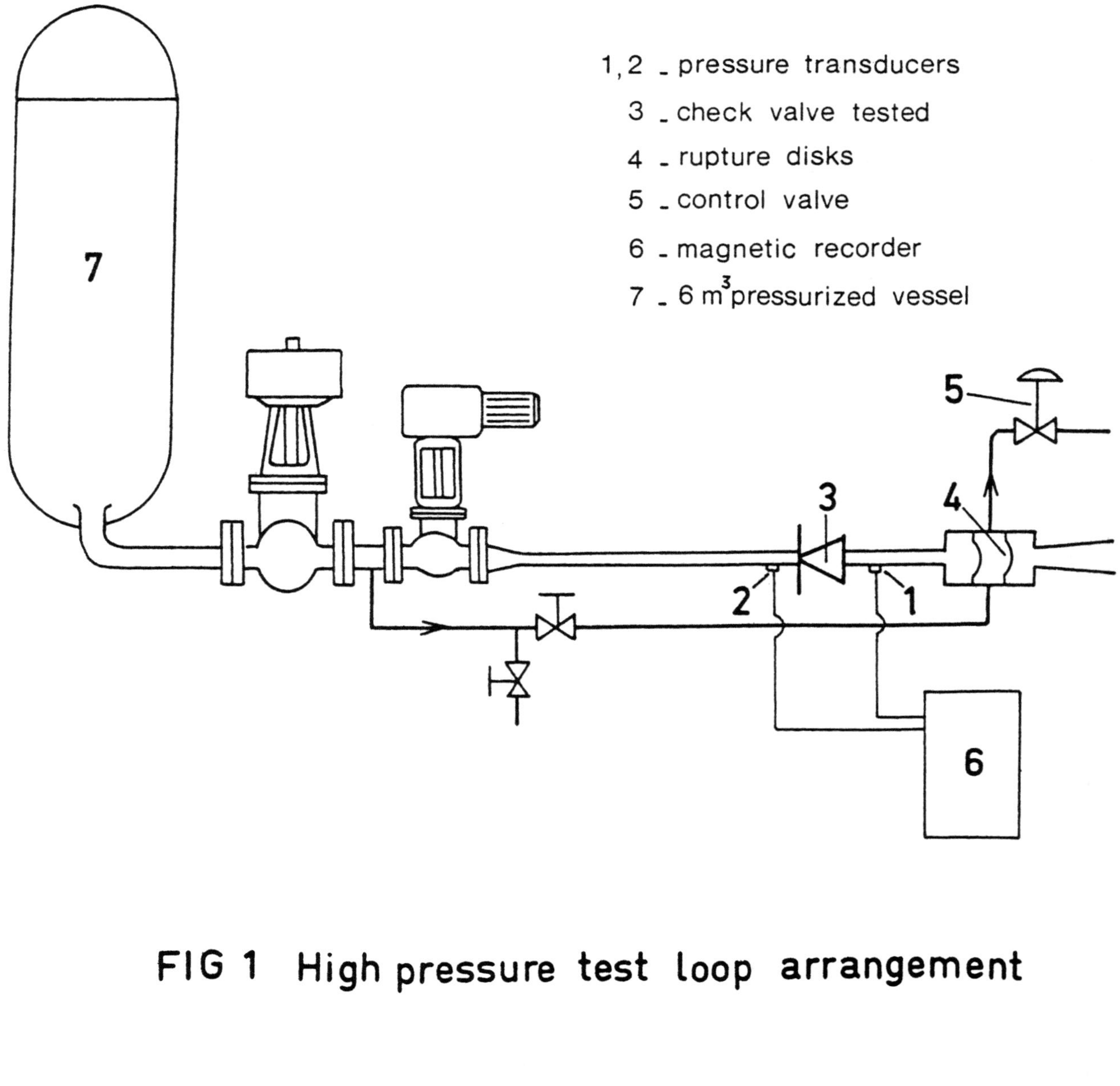

FIG 1 High pressure test loop arrangement

tests n° 7 to 15 - 18-19

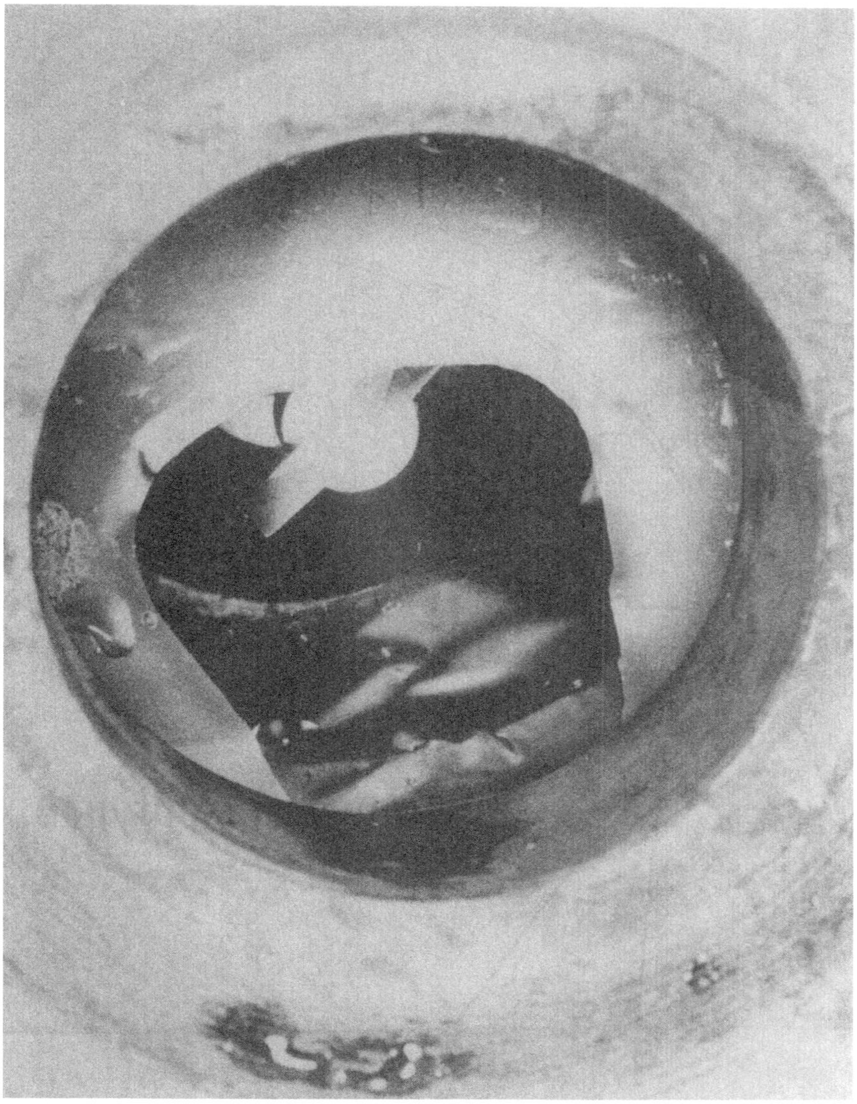

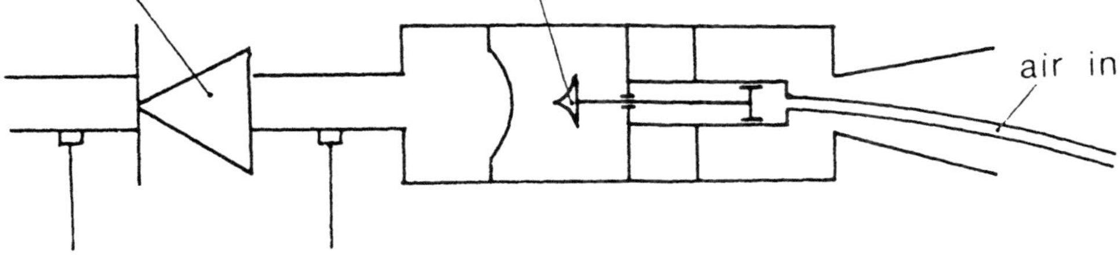

FIG 2 Single punched disk device

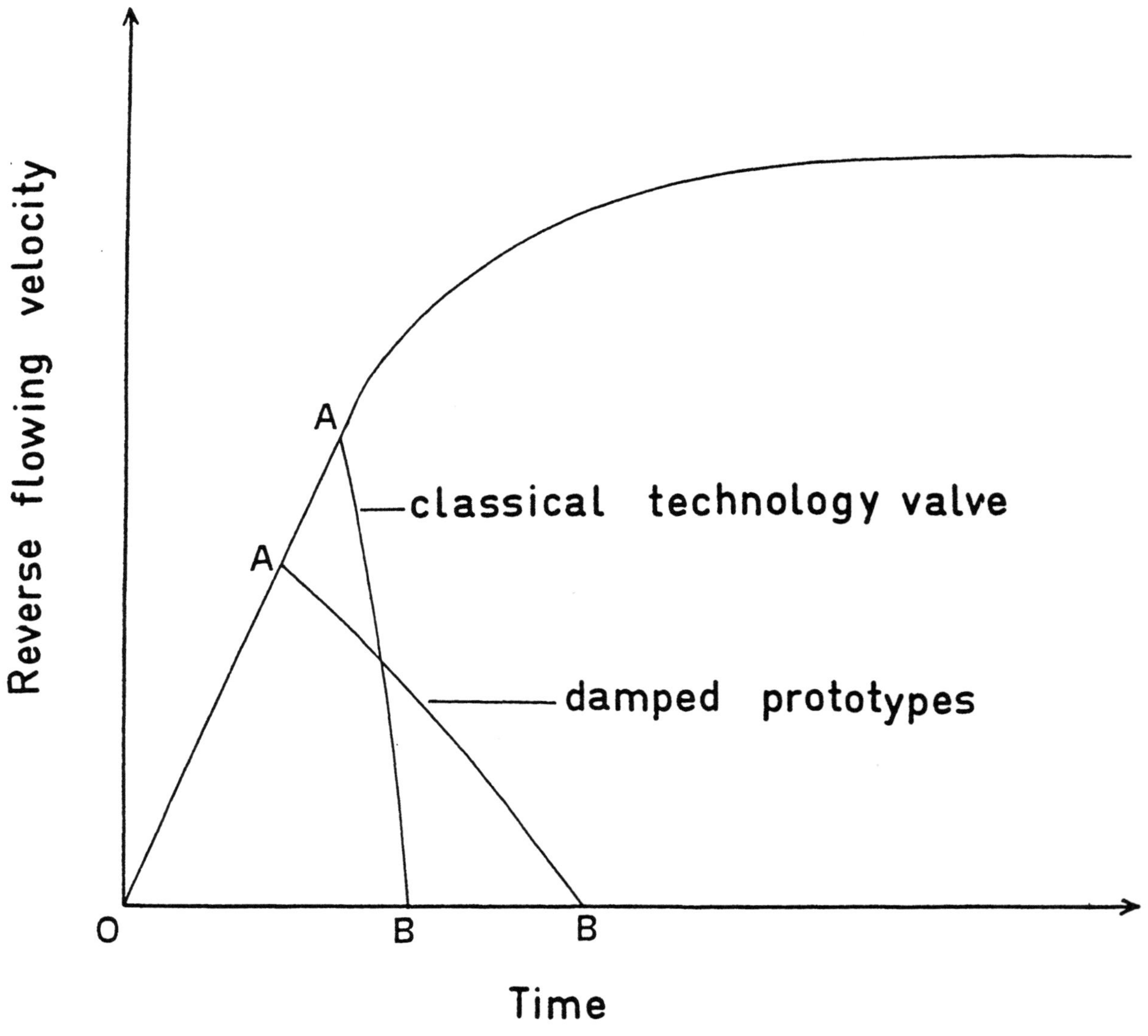

FIG 3 Reverse flow time history hypothesis

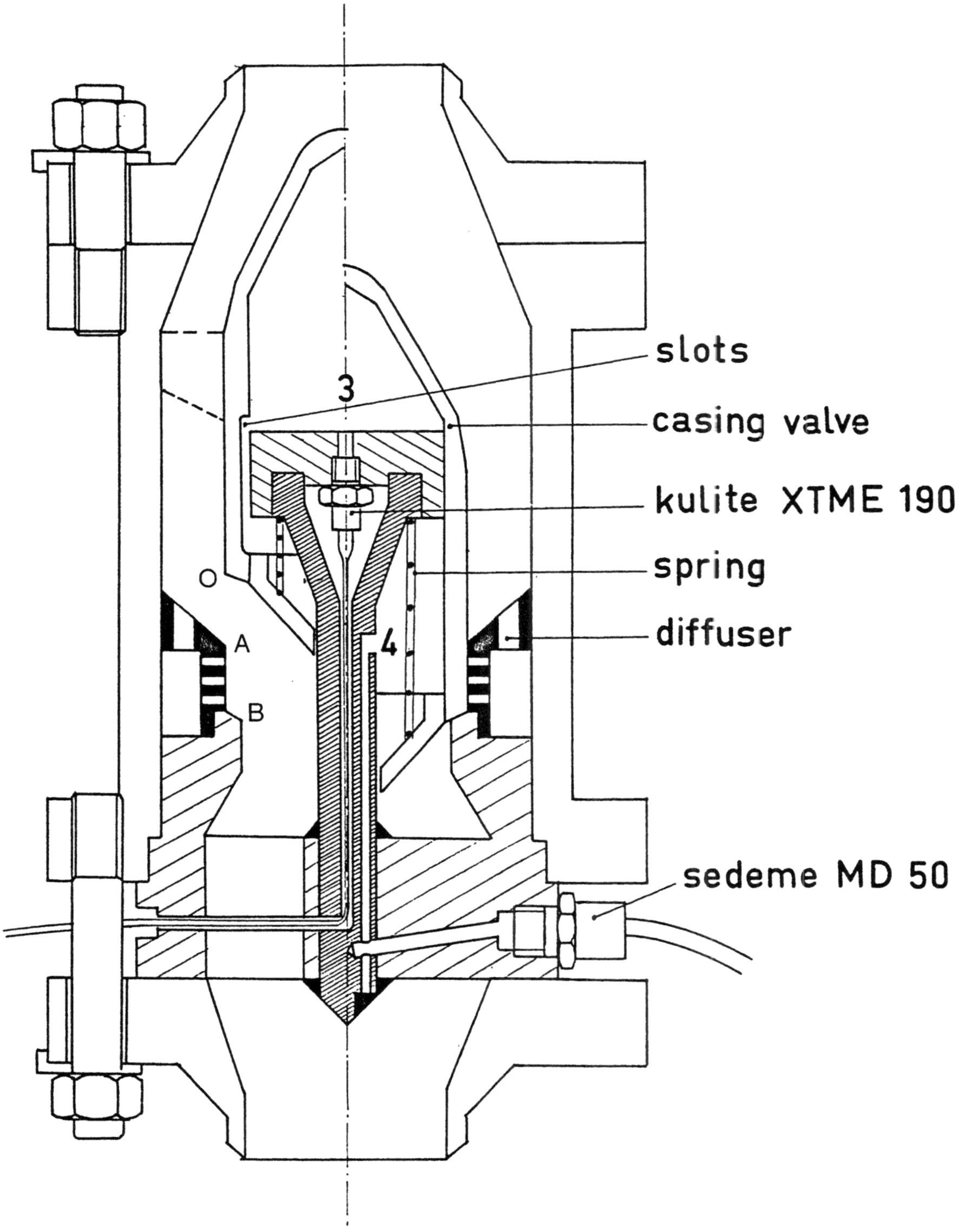

FIG 4 VANATOME Prototype

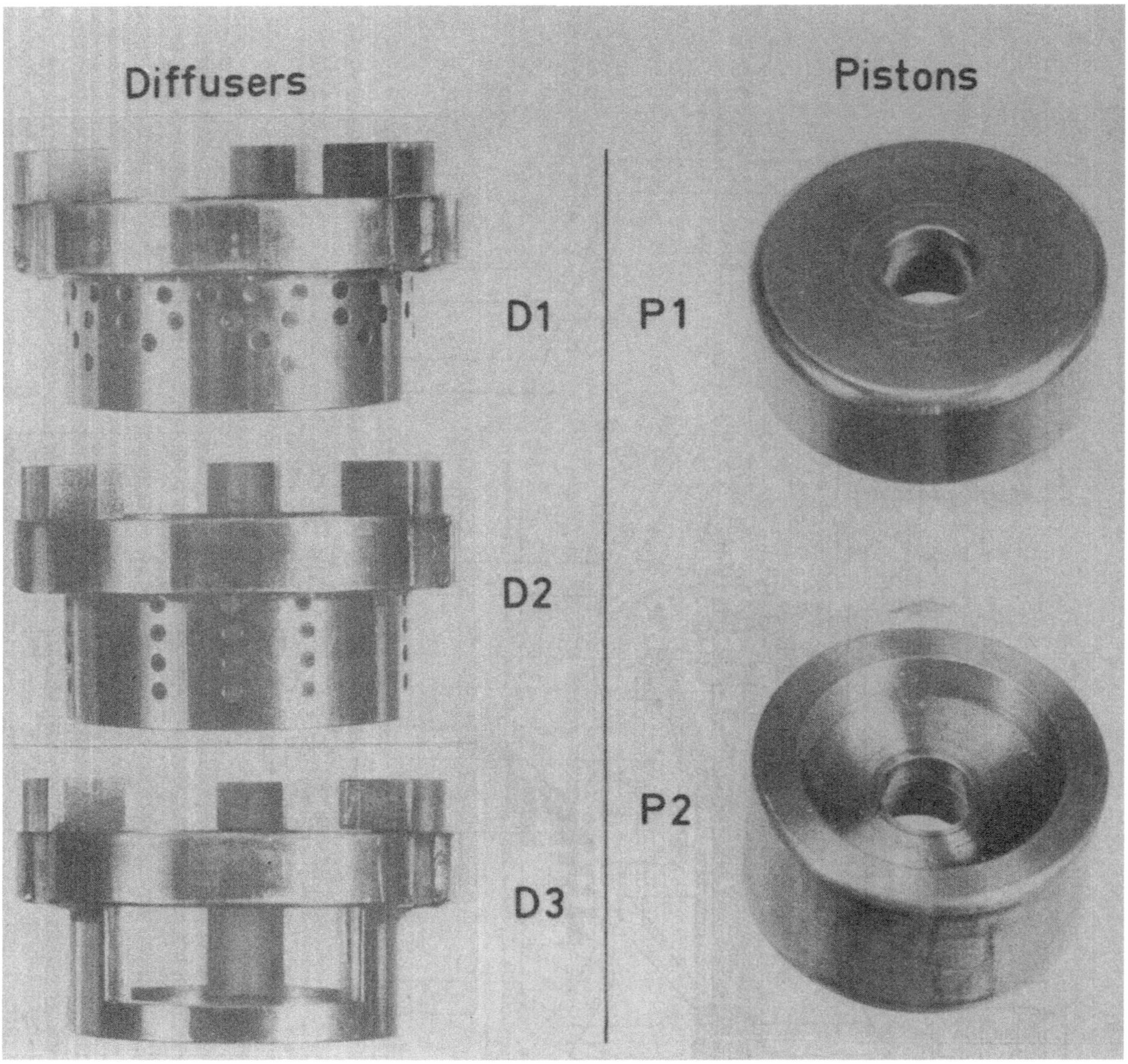

		P1		P2	
D1	test n°	13-14-15	17	7-8-9-10	18-19
	H, Δ Ø (mm)	16, 0.25	16, 0.15	22, 0.07	23, 0.15
D2	test n°	6		11	
	H, Δ Ø (mm)	16, 0.25		22, 0.07	
D3	test n°			12	
	H, Δ Ø (mm)			22, 0.07	

H: piston height, ΔØ: casing piston diametral clearance

FIG 5 Diffusers, pistons and tests numbering

DN 50 VANATOME Prototype (lift 35 mm)

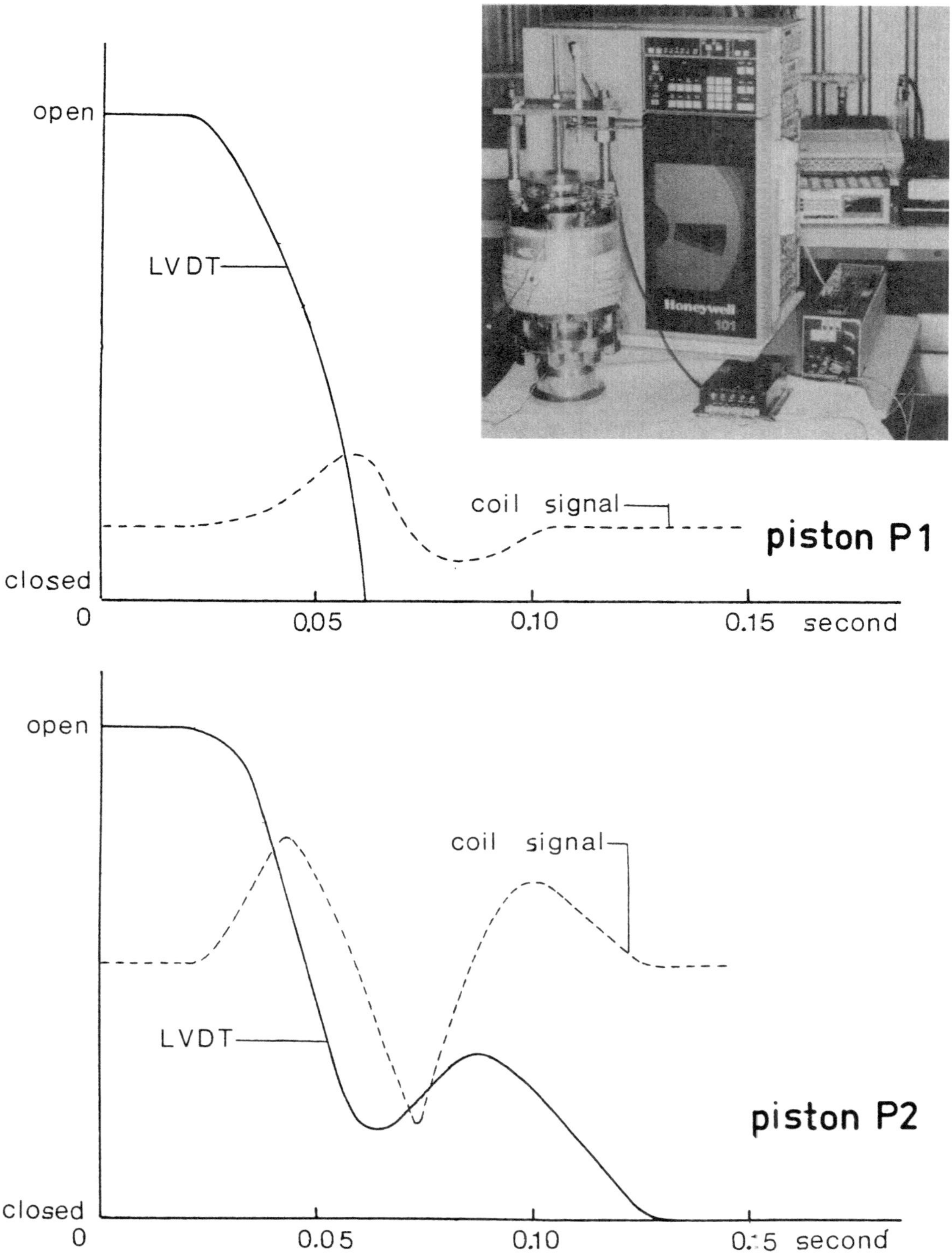

FIG 6 Coil and LVDT signals
closure tests out of water

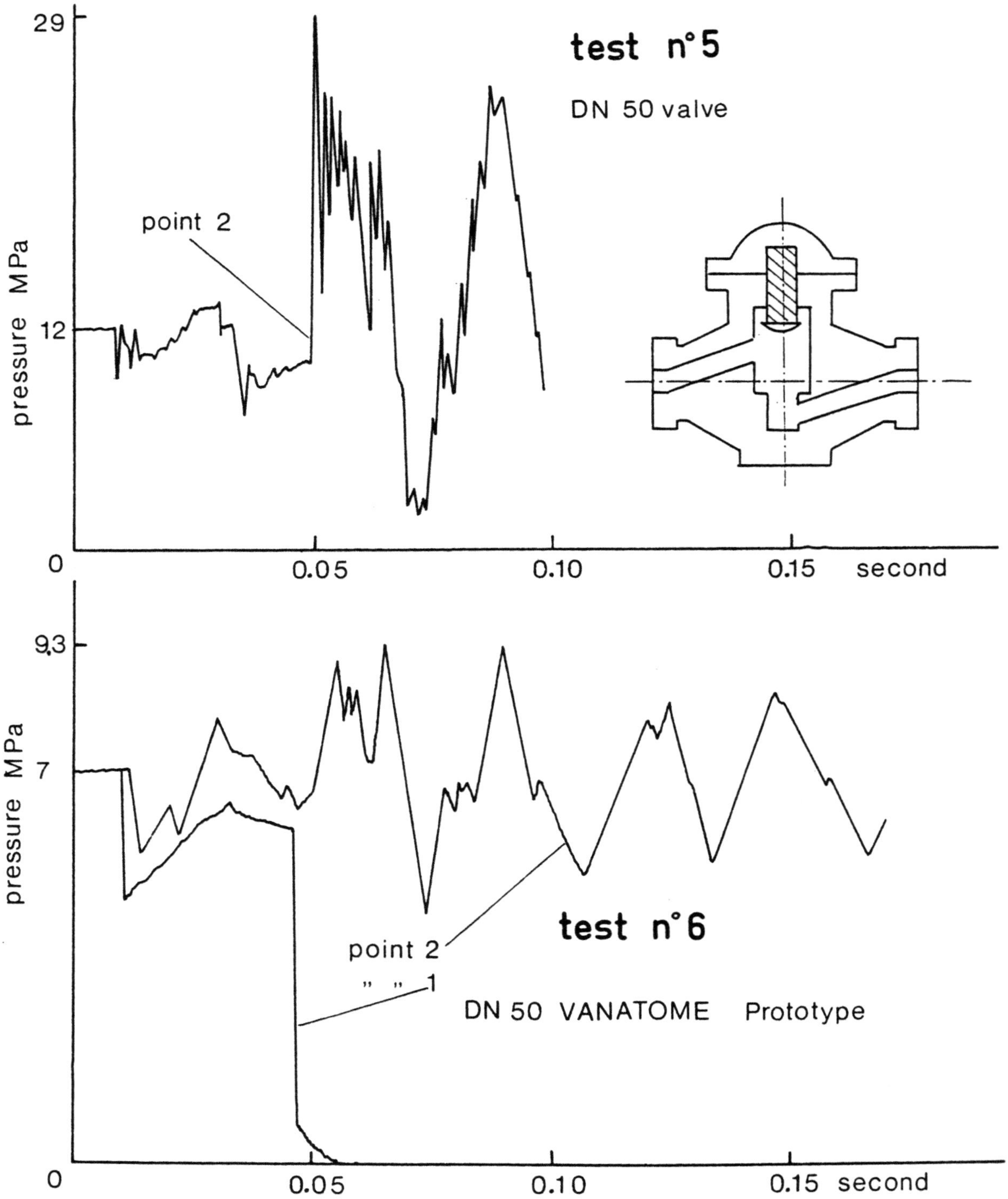

FIG 7 Pressure transients

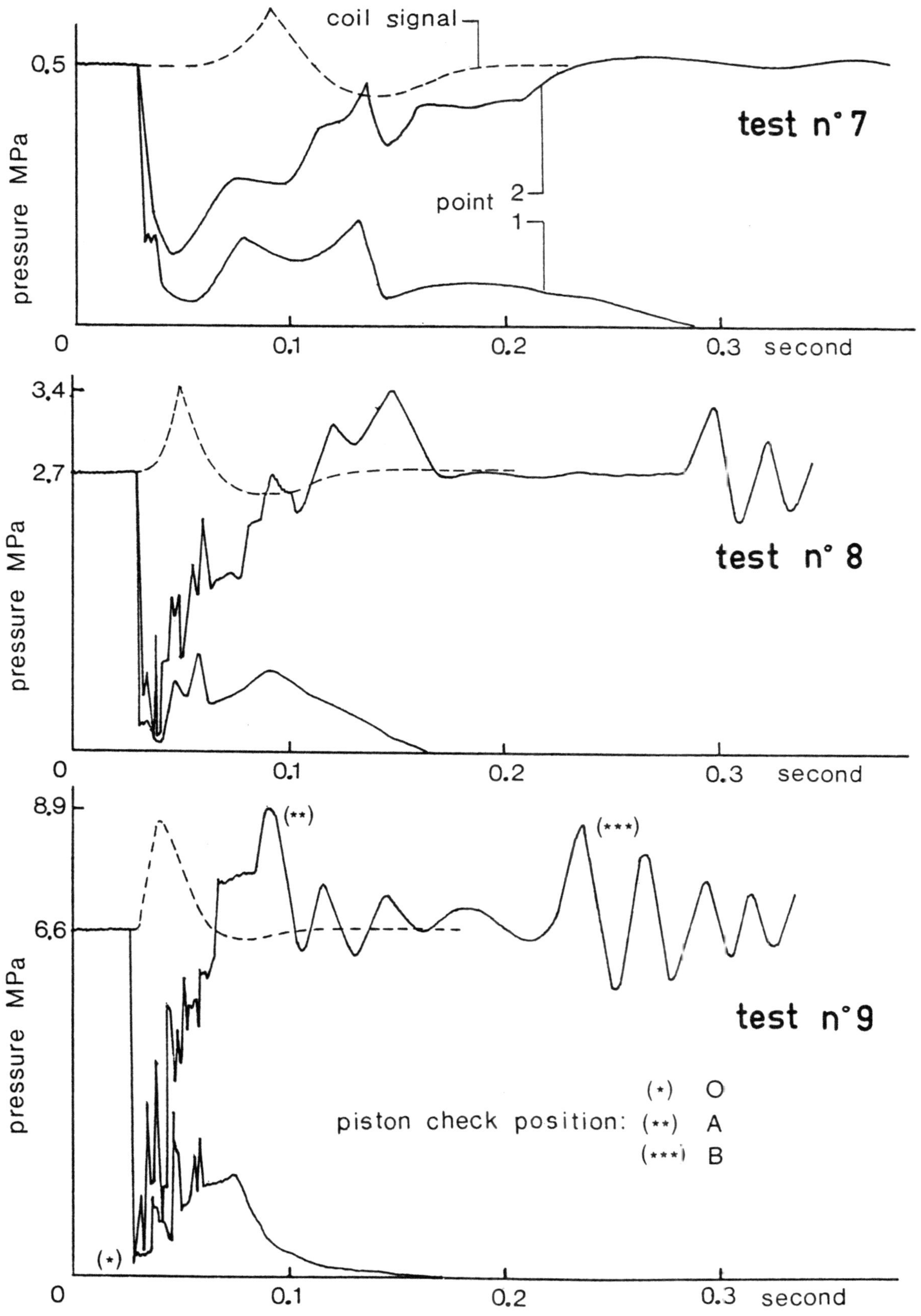

FIG 8 DN 50 Prototype pressure transients

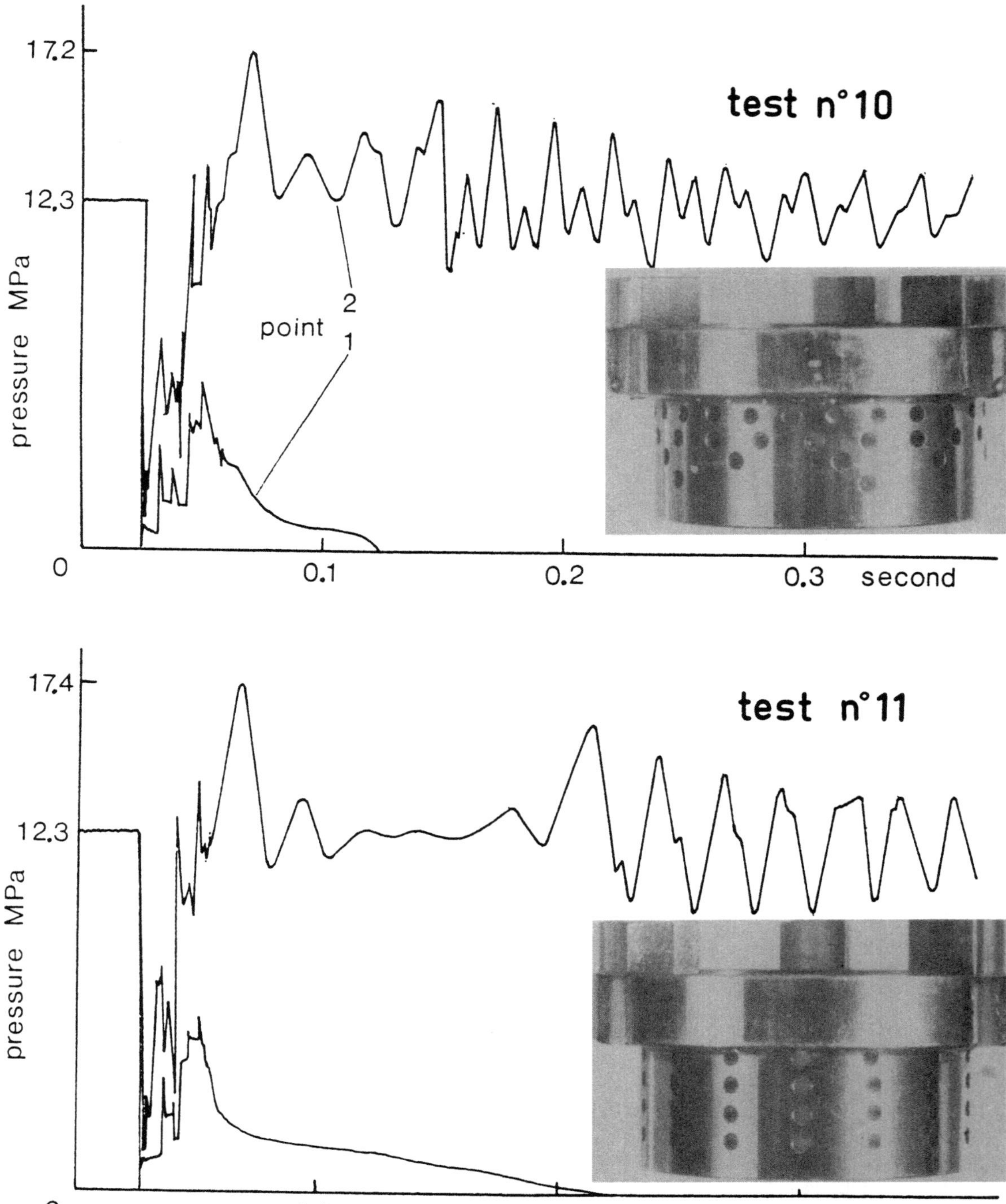

FIG 9 DN50 Prototype pressure transients
(influence of the diffuser)

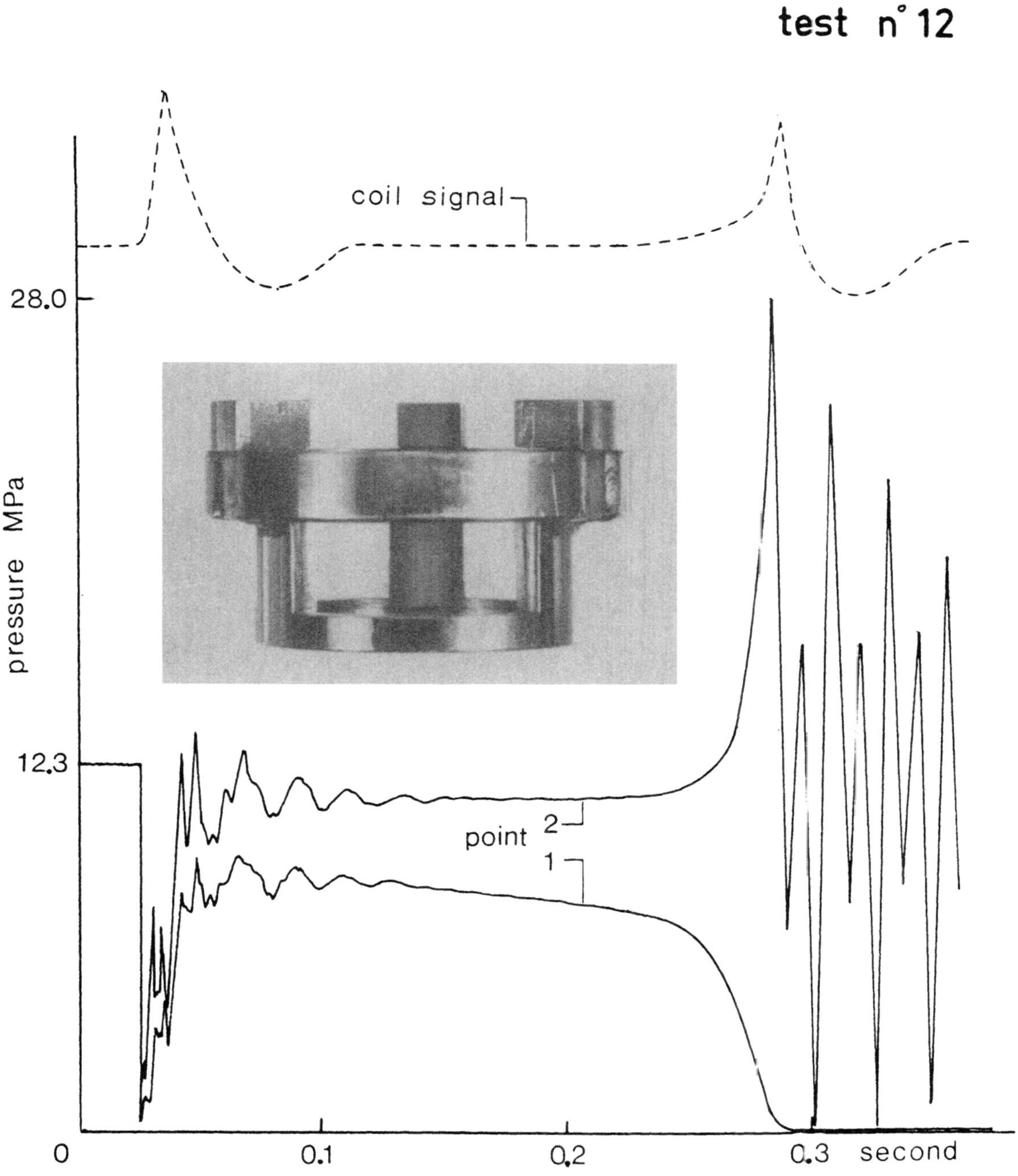

FIG 10 DN 50 Prototype pressure transient
(influence of the diffuser)

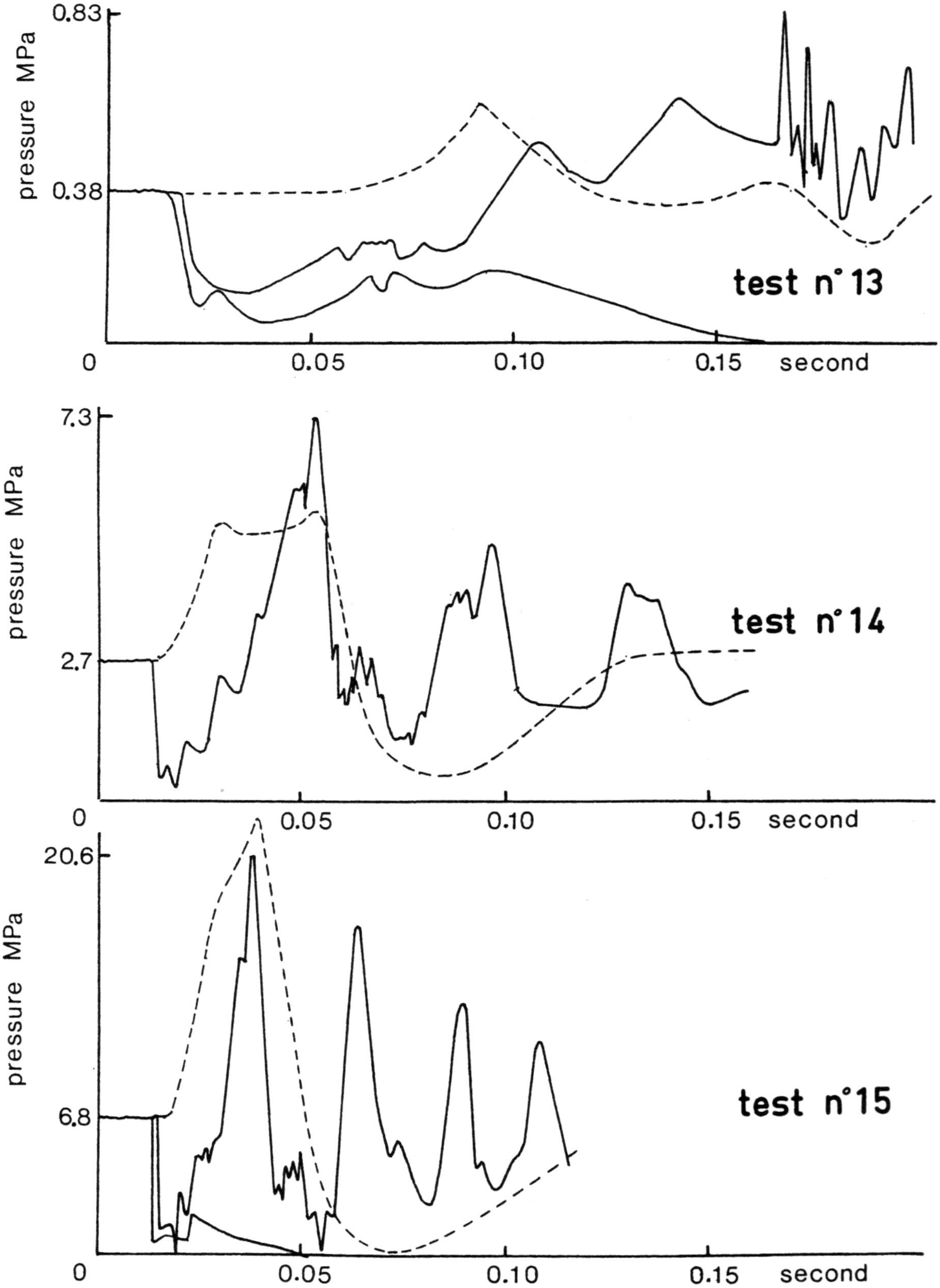

FIG 11 DN 50 Prototype pressure transients

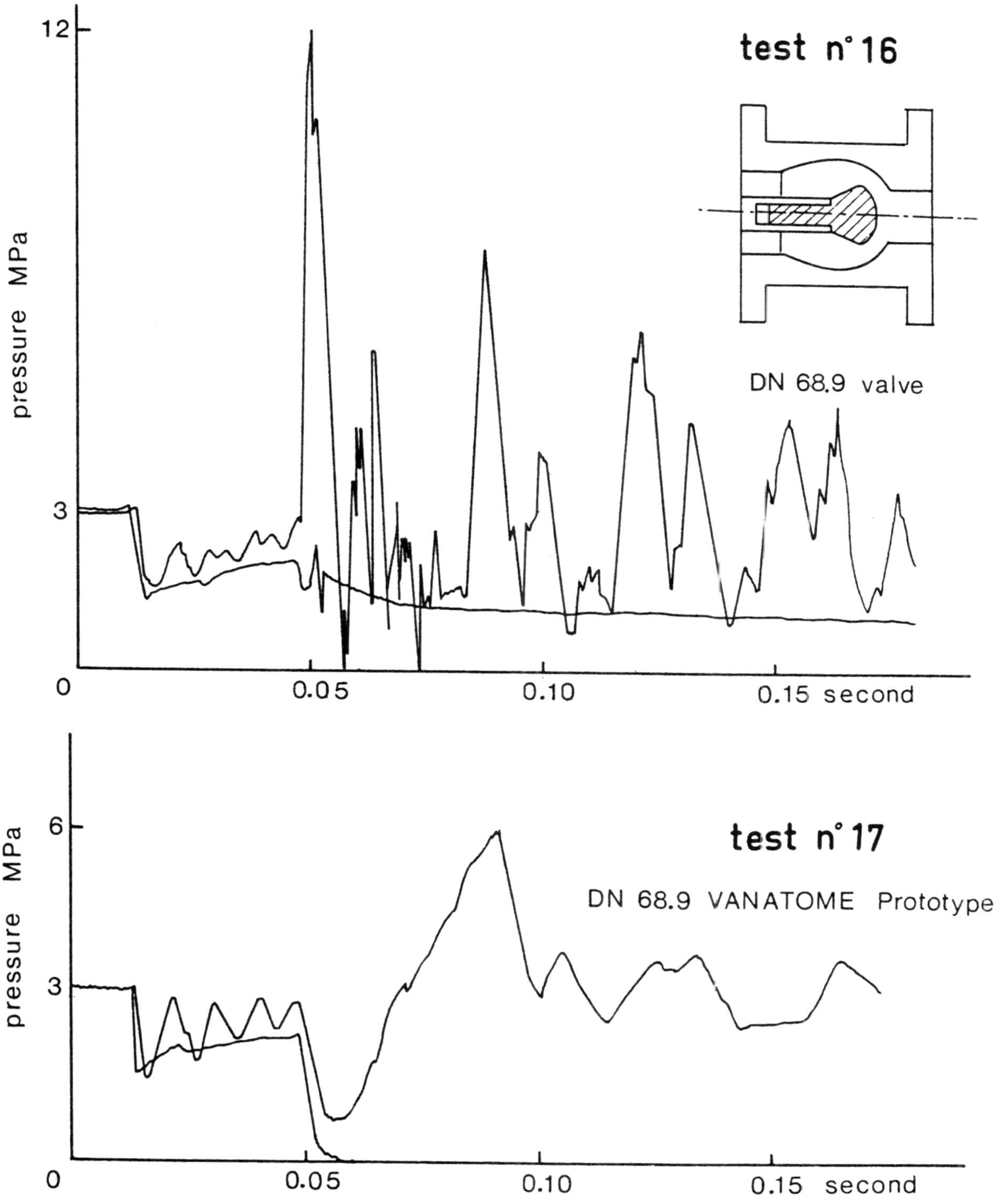

FIG 12 Pressure transients

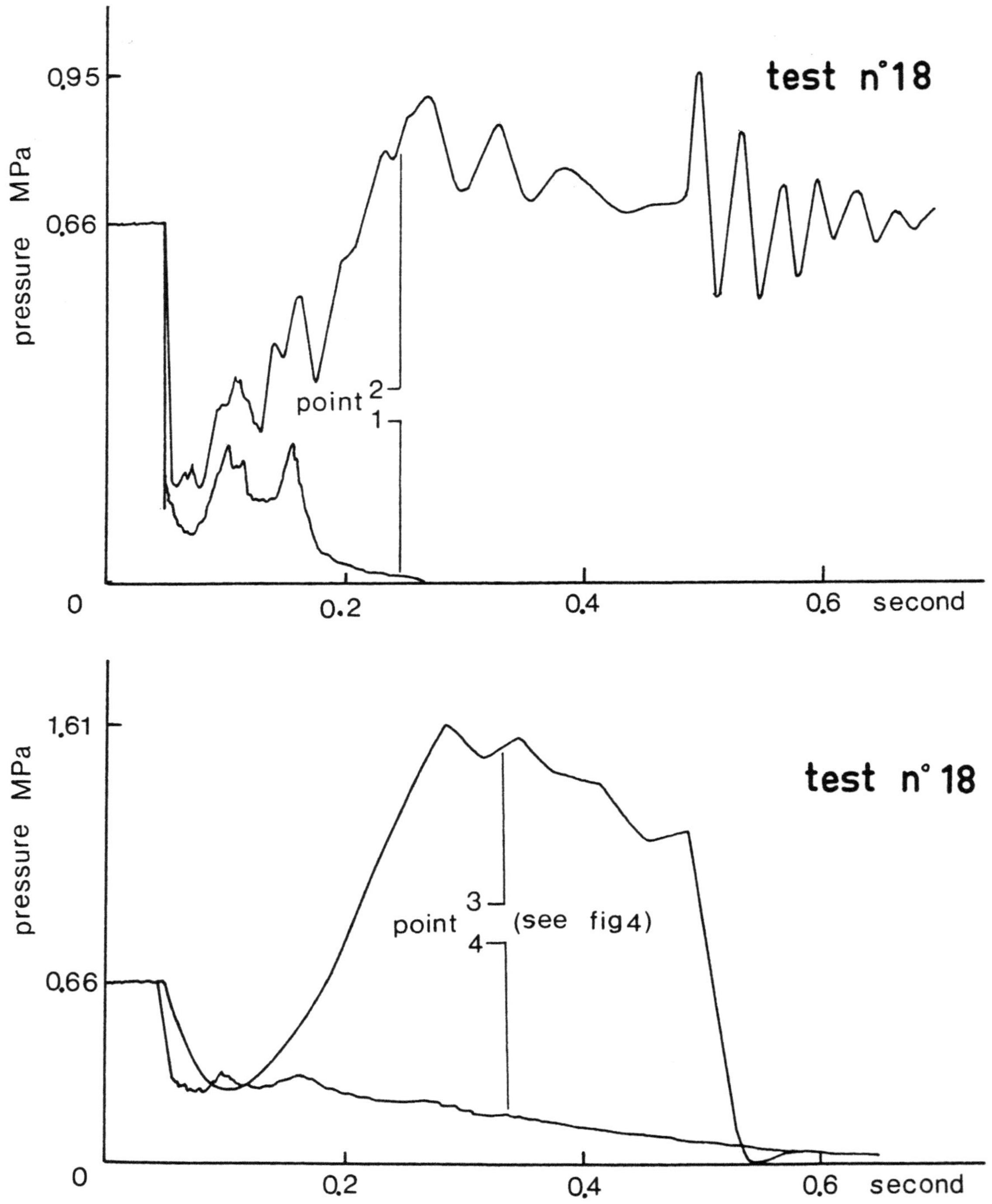

FIG 13 DN 68.9 Prototype pressure transients

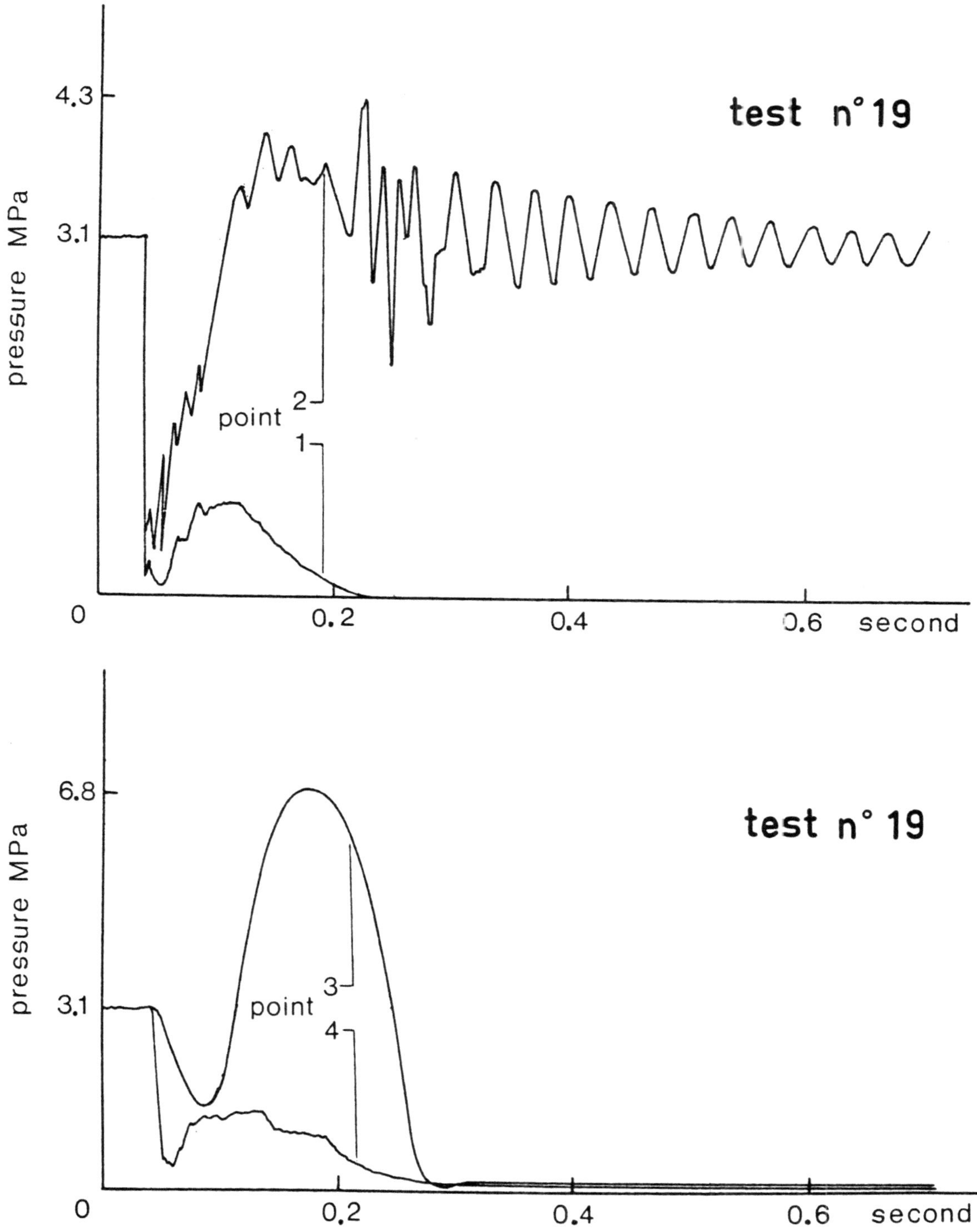

FIG 14 DN 68.9 Prototype pressure transients

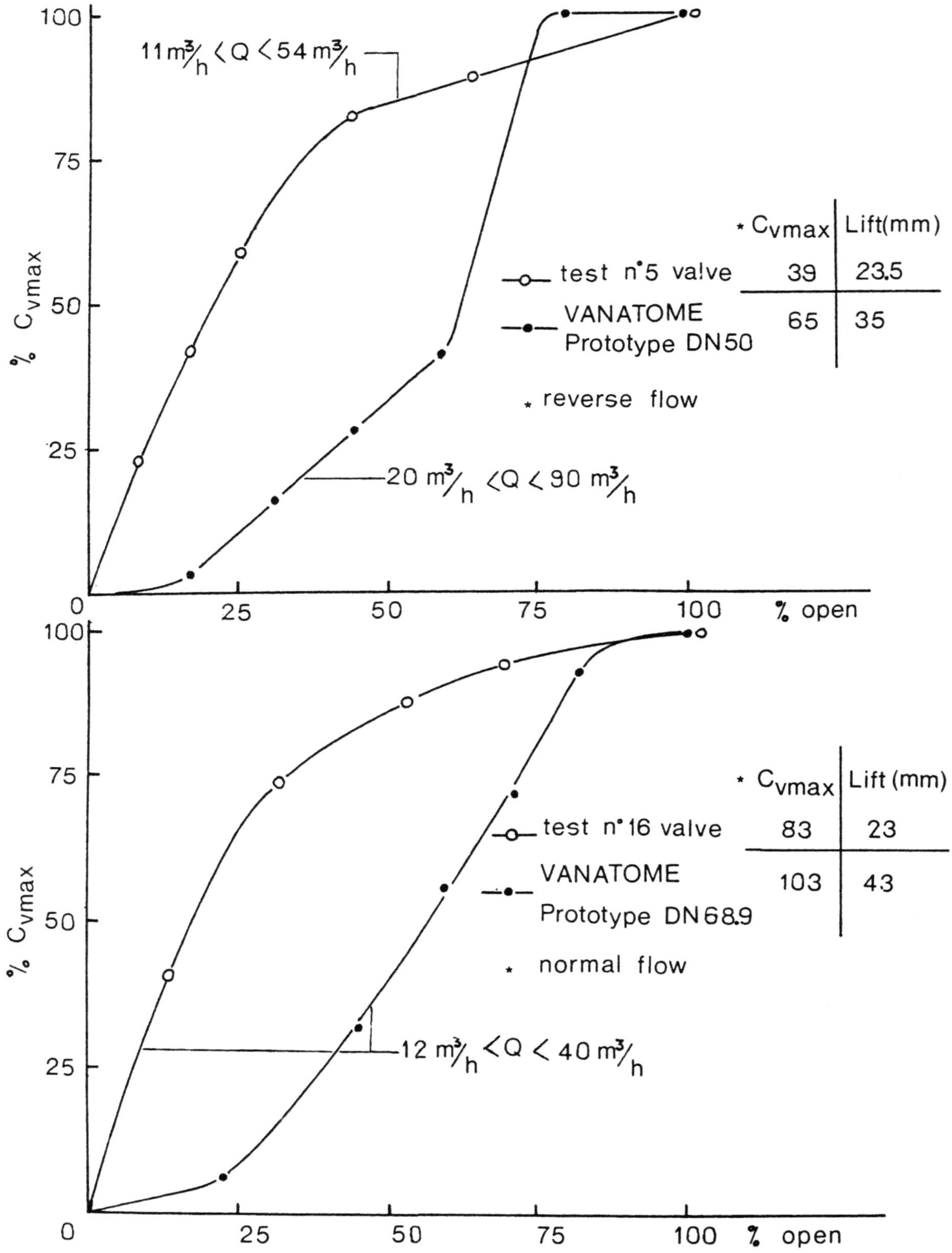

FIG 15 Flow coefficients C_v

2nd International Conference on

Developments in Valves and Actuators for Fluid Control

Manchester, England: 28-30 March 1988

PAPER H2

CHECK VALVE CLOSURE BEHAVIOUR; EXPERIMENTAL INVESTIGATION AND SIMULATION IN WATERHAMMER COMPUTER PROGRAMS

A.C.H. Kruisbrink
DELFT HYDRAULICS, Industrial Hydrodynamics, the Netherlands

SUMMARY

A standard of comparison for undamped check valves with respect to the closure behaviour is the dynamic characteristic. The pressure surges due to valve closure and the degree of check valve slam are directly related to this characteristic.

A general method is proposed to perform dynamic tests and to elaborate test results into dynamic characteristics. The method is based on results of an experimental investigation on the influence of the initial flow on the closure behaviour of check valves (initially fully open) and the influence of the valve itself on the flow during closure. Measurement data are processed taking time shift effects into consideration.

Recent developments are to present the dynamic characteristic in a dimensionless form. Theoretically the curves of geometrically similar check valves coincide. Moreover the curves of spring loaded check valves coincide when applying different springs.
The usefulness of the dimensionless characteristic is investigated for several spring loaded nozzle check valves up to diameters of 800 mm for various springs.

Based on the dynamic characteristic a method is presented for simulation of check valve closure behaviour in waterhammer computer programs.
The method is valid for all types of undamped checkvalves and may be useful to system engineers in the selection of check valves.
Examples of applications are included.

NOMENCLATURE

A	valve disc area (m^2)
c	propagation speed of pressure waves (m/s)
dv/dt	characteristic fluid deceleration (m/s^2)
D_i	inner valve diameter (m)
F_h	hydrodynamic force on moving element(s) (N)
F_o	net spring and gravitational force on moving element(s), valve just fully open (N)
F_s	net spring and gravitational force on moving element(s), valve just closed (N)

Held in Manchester, England. Organised and sponsored by BHRA, The Fluid Engineering Centre and co-sponsored by the British Valve Manufacturers Association and the Institution of Mechanical Engineers Process Industries Division

K_f flow coefficient (-)
L_f distance from flowmeter to check valve seat (m)
L_k distance from pressure tap to check valve seat (m)
Q fluid flow (m^3/s)
t time (s)
t_b moment of initiation of dynamic test (s)
t_o moment the fluid velocity equals the critical velocity (s)
t_R moment the maximum back flow velocity at the check valve is reached (s)
T_o net spring and gravitational moment on moving element(s), valve just fully open (Nm)
T_s net spring and gravitational moment on moving element(s), valve just closed (Nm)
v fluid velocity (m/s)
v_b initial fluid velocity (m/s)
v_o critical velocity (m/s)
v_R maximum back flow velocity (m/s)
x valve disc position (m)
ρ_m density moving element(s) (kg/m^3)
ρ_f fluid density (kg/m^3)
τ time shift of flowmeter (s)
θ valve disc angle
' measured signal

1. INTRODUCTION

Check valves are generally used in pipeline systems of industrial plants to prevent reverse flow. They are often located at the discharge side of pumps, to protect the pumps and driving motors from reverse rotation.

Undamped check valves are simple devices consisting of a casing around a moving element. The motion of the moving element is controlled primarily by the fluid passing through (see figure 1: situation 1).
An ideal check valve will close in a decelerating flow without the occurence of reverse flow. In practice however there will always be some degree of reverse flow depending on the inertia of the moving element(s), strength of springs, counterweights, magnitude of the deceleration, etc, resulting in waterhammer and check valve slam phenomena.

At the instant of closure the reverse flow is suddenly reduced to zero resulting in a pressure rise downstream of the moving element (see figure 1: situation 2). The high pressure accelerates the closure of the disc in its seat, resulting in check valve slam [ref 1]. The greater the back flow velocity the greater the pressure rise and the phenomenon of check valve slam will be. Therefor the back flow velocity is a direct, test rig independant, measure for such performances when dealing with undamped valves.

The abovementioned phenomena may be coupled with a damage to the valve and/or surrounding pipeline system, indicating the importance of a proper selection of check valves.

2. DYNAMIC CHARACTERISTIC

In the Dynamic Characteristic the maximum back flow velocity is represented versus a characteristic deceleration [ref 2].
The horizontal axis represents a characteristic fluid deceleration $|dv/dt|$ during valve closure. The vertical axis represents the maximum back flow velocity $|v_R|$ that usually occurs immediately before the moving element(s) are seated. Figure 2 shows a dynamic characteristic of an 800 mm nozzle check valve [ref 6].

The ideal check valve is one that closes without the occurrence of back flow, without aggravating the associated pressure changes and with a minimum of slamming. This performance should be achieved independant of the rate at which is decelerated. The dynamic characteristic of an ideal check valve should coincide with the horizontal axis.

3. MEASUREMENT OF DYNAMIC CHARACTERISTICS

The characteristics are obtained by performing dynamic tests whereby the flow through the valve is decelerated. Both the fluid deceleration and the maximum back flow velocity are measured. Typical test results are shown in figure 3, showing the flow, the valve disc position and the up- and downstream pressures as a function of time [ref 6].

From the flow signal a characteristic deceleration and the maximum back flow velocity are determined. Hereby the other measuring signals are relevant as will be shown in section 5. Each dynamic test supplies data for one point of the dynamic characteristic.

4. TEST FACILITIES

4.1 GENERAL

Dynamic tests can be performed in any test rig in which the flow through the check valve can be decelerated. However to obtain dynamic characteristics series of measurements have to be performed whereby the deceleration must be controllable, in such a way that the range in decelerations overlays the practical range of about 0 - 10 m/s^2.
In order to get unambiguous results the deceleration should be as constant as possible. Available test rigs designed for this purpose are described by Ellis [3]. In addition at DELFT HYDRAULICS since 1986 a test rig is operational for the dynamic testing of large check valves (see section 4.2).

Any standards with respect to the quality of test rigs and test procedures are not available yet.

4.2 TEST RIG AT DELFT HYDRAULICS

Specifications

The specifications of the test rig are:

Test medium	: water
Flow capacity	: 2 m^3/s
Valve size range to be tested	: 300 - 800 mm (12" - 32")
Length of measuring section	: 14 metres
Maximum deceleration for an 800 mm check valve	: 20 m/s^2

Principle of operation

Figure 4 shows a schematic diagram of the test rig.
Eight centrifugal pumps supply water from the basement reservoir to the head tank. The water level in the head tank is fixed by means of an overflow at a height of 24 m. Excess water flows back through the return lines into the basement reservoir. Under steady state conditions the flow is controlled by a 600 mm control valve.

To perform a dynamic test the static head downstream of the check valve must be increased rapidly to a constant and controllable level. Therefore an air reservoir is connected to the circuit via a high pressure tank.
Under steady state conditions before starting a dynamic test the air reservoir is isolated from the high pressure tank by means of a fast acting (300 mm) air valve. The air reservoir is pressurized and the water level in the high pressure tank is controlled by a vent valve.

A dynamic test is initiated by quickly opening of the air valve. The flow in the test-section will be decelerated due to the high air pressure. The back flow of water is supplied by the high pressure tank.

After the check valve is closed the air valve is shut off. Consequently the pressure in the high pressure tank will be reduced, the check valve will reopen and the next test can be carried out, after the air reservoir is pressurized to a different level.

Measuring equipment

The flow is measured with a fast-acting electro-magnetic flowmeter with a flow range from -2 to 2 m^3/s and a frequency range of 40 Hz.
The flow meter has a time shift which is constant for the whole frequency range. Therefore apart from high frequencies the flow signal will be measured without deformation.

The up- and downstream pressures are measured with piezo-electric pressure transducers with a pressure range of 100 bar and a frequency range of 40 kHz.

5. PRELIMINARY STANDARDISATION INVESTIGATIONS

In general the flow and pressure measurements do not take place at the check valve but at some distance from it. Therefore the real flow and pressure variations at the valve are measured with some time delay. Apart from this there is a time shift in the measuring equipment. To obtain the maximum back flow velocity from these signals corrections are needed (see section 5.1).

According to the definition of the dynamic characteristic the deceleration should be constant during a dynamic test. However in practice the deceleration always changes to some degree due to changes in boundary conditions (test rig properties) and changes in check valve resistance (check valve properties).
Also in a pipe line system where a check valve is used the deceleration will not be constant. For example in case of pumpfailure pump inertia effects will influence the deceleration.
Therefore it is necessary to define a relevant or characteristic deceleration which can be used both in obtaining dynamic characteristics from measurements and in the use of the characteristics in surge calculations [ref 6], e.g. in waterhammer computer programs (see section 8).

In behalf of the standardisation of dynamic tests and results experimental investigations are performed on:
- The influence of initial fluid velocity on maximum back flow velocity (see section 5.2)
- The influence of check valve on flow (see section 5.3)
- The influence of flow on check valve behaviour (see section 5.4).

The check valves tested are 800 mm spring loaded nozzle valves (translating type). The tests are performed in the test rig as described in section 4.

5.1 ANALYSES MEASUREMENT SIGNALS

In figure 5 real and measured signals are presented. The following moments are distinguished:

t_b : initiation of dynamic test
t_o : fluid velocity equals the critical velocity v_o
t_z : fluid velocity is for the first time zero
t_R : maximum back flow velocity v_R at the check valve is reached
t_{cl} : valve is just 100% closed
t_{max}: maximum back flow velocity at the flowmeter is reached

It is assumed that the fluid column between flowmeter and check valve can be considered to be be rigid after initiation effects have been damped out. This means that the flow through check valve and flowmeter is equal until the moment the maximum back flow velocity is reached. The time delay in the measured flow is caused by the time shift τ of the flowmeter.

In the time interval $[t_R, t_{cl}]$, which is of the order of milliseconds, the maximum back flow at the check valve is reduced to zero. The reduction of flow is coupled with a pressure rise at the downstream side and a pressure drop at the upstream side.

In the time interval $[t_R, t_{max}]$ the back flow through the flowmeter further increases to a maximum value at the moment t_{max}. When the distance between flowmeter and check valve is L_f m the pressure wave caused by valve closure reaches the flowmeter after L_f/c s. During this time interval the fluid at the flowmeter continues to decelerate.

The sudden reduction of the flow at the flowmeter after the moment t_{max} is reached may be coupled with high frequencies which exceed the frequency range resulting in a deformation of the measuring signal.

It is assumed that the frequency range of the pressure transducers is that high that the deformation in the pressure signal can be neglected. Moreover the shift of the transducers is neglected.
There will be a time shift in the measurement due to the fact that the transducers are located at some distance of the check valve. This time shift is L_K/c s if the distance from pressure tap to check valve seat is L_K m.

5.2 THE INFLUENCE OF INITIAL FLUID VELOCITY ON MAXIMUM BACK FLOW VELOCITY

The closure behaviour of check valves depends on the initial valve disc position. The smaller the distance/angle through which the disc has to travel the smaller the maximum back flow velocity will be [refs 1, 5].
Thus the most conservative results are obtained starting with a flow at which the valve is full open.

The investigation described here is related to these cases whereby the valve is initially 100% open, which means that the initial fluid velocity is greater or equal to the critical velocity (i.e. the fluid velocity whereby the valve is just 100% opened).

A consideration of the Navier-Stokes equations learns that demands for geometric similar flow patterns are equal (or sufficient high) Reynolds numbers and equal boundary conditions.
With respect to the latter the question arises in how far a change in initial fluid velocity $v_b > v_o$, which in fact is a change in boundary conditions, will influence the dynamic behaviour of the valve.

This influence was investigated by performing dynamic tests in which the initial flow is varied from once to twice the critical flow.
Some results are presented in the figures 6a and 6b .
The results of figure 6a show no influence of the initial flow on the dynamic characteristics. Here the measurement points lie within the accuracy of the measurements.
The results of figure 6b show a small influence of the initial flow on the maximum back flow for decelerations up to 2,5 m/s^2.

In all cases the most conservative results are obtained at an initial fluid velocity equal to the critical velocity.
It can be concluded that in principle the dynamic tests should be performed at an initial velocity which is equal to the critical velocity.

5.3 THE INFLUENCE OF CHECK VALVE ON FLOW

In general the magnitude of the deceleration of a fluid column is determined by the pressure difference per length unit over the fluid column dp/dx.
dp/dx will not be influenced by a resistance (check valve) as long as there are no changes in pressure loss over the resistance. The influence of the resistance can be neglected as long as the changes in pressure loss over the resistance are small compared to the total pressure difference over the fluid column of consideration.

The influence of check valve on flow is investigated by performing tests with and without check valve.
Some test results are presented in the figures 7a and 7b.
In order to be able to make a good comparison between the measurements with and without valve the flow signal is made dimensionless, according to:

$$\frac{v}{v_b} = f\left(\frac{t}{\Delta t}\right) \tag{1}$$

Where: v_b = initial fluid velocity (m/s)
$\Delta t = t_z - t_b$ (s)

When $t_b = 0$ this results in: $f(0) = 1$ and $f(1) = 0$. In general the function $f(t/\Delta t)$ will be dependant on test rig, initial flow and deceleration.
Therefore only measurements can be compared which agree more or less on initial flow and deceleration.
The figures show that the slope in the flow signal becomes more constant with increasing deceleration, thus indicating there is a dependency of deceleration in the test circuit of consideration.
The differences in the relative velocity v/v_b with and without check valve lie within the accuracy of the measurements, with the exception of $t/\Delta t <= .1$ due to initiation effects.

It can be concluded that in all cases the influence of the check valve on flow is neglectable until the moment the maximum back flow velocity is reached, which means that during valve closure the changes in pressure loss are comparatively small.

5.4 THE INFLUENCE OF FLOW ON CHECK VALVE BEHAVIOUR

The influence of flow on check valve behaviour in first instance can be found in the hydrodynamic forces on the moving elements. Assuming the relative fluid velocity with respect to the valve disc velocity is relevant, the hydrodynamic forces acting on the disc can be described with (translating type) [ref 4]:

$$F_h = \frac{\rho_f K_f}{D_i^2} (Q - \dot{x} A) \left| Q - \dot{x} A \right| \qquad (2)$$

Where: F_h = hydrodynamic force on moving element(s) (N)
ρ_f = fluid density (kg/m³)
K_f = flow coefficient (-)
D_i = inner valve diameter (m)
Q = flow (m³/s)
x = valve disc position (m)
A = valve disc area (m²)

In figure 8 some dynamic test results are presented of a translating type of check valve whereby the fluid flow and valve displaced flow are presented as a function of time. From these graphs it can be seen that during the greatest part of the valve travel the valve disc velocity differs from the fluid velocity, which means that the hydrodynamic forces are of importance during practically the whole valve travel. Although the hydrodynamic forces become comparatively small after the flow has reversed the ratio of hydrodynamic and spring forces remains more or less constant because the spring forces decrease when the disc moves to a more closed position.

6. STANDARDISATION OF DYNAMIC TESTS AND RESULTS

6.1 PERFORMANCE OF DYNAMIC TESTS

The dynamic tests can be performed in test rigs as described in section 4.

The measuring equipment sufficient and needed is:
- A fast-acting flowmeter with a sufficient frequency range.
  The time shift of the flowmeter preferably should be constant.
- Two pressure transducers located at the initial up- and downstream side of the check valve.

It is recommended to perform the dynamic tests at an initial velocity somewhat higher than the critical velocity in such a manner that initiation effects on the flow signal are damped out on the moment the valve starts to close (see section 5.1).

Definition: The critical velocity is the fluid velocity at which the valve is just 100% open, measured under steady state conditions.

6.2 DETERMINATION OF THE MAXIMUM BACK FLOW VELOCITY

The maximum back flow velocity is reached on the moment t_R when the pressure drop or build up starts. This moment can be determined from the upstream or downstream pressure signal. A correction for the time shift of flowmeter (τ) and pressure signal (L_K/c) yields (see figure 5):

$$v_R = v(t_R) = v(t'_R + \tau - L_K/c) \tag{3}$$

Where:
- v_R = maximum back flow velocity
- t'_R = moment the start of the pressure drop/build up is measured (s)
- τ = time shift of the flowmeter (s)
- L_K = distance from upstream/downstream pressure tap to check valve seat (m)
- c = propagation speed of pressure waves (m/s)

6.3 DETERMINATION OF THE CHARACTERISTIC DECELERATION

Based on the results described in section 5 the characteristic deceleration is defined as the average deceleration in the time interval $[t_o, t_R]$:

$$\frac{dv}{dt} = \frac{v_R - v_o}{t_R - t_o} = \frac{v_R - v_o}{t'_R - t'_o + \tau - L_K/c} \tag{4}$$

where:
- v_o = critical velocity (m/s)
- t^o = moment the fluid velocity equals the critical velocity (s)
- t'_o = moment the critical velocity is measured (s)

Notes:
- If the dynamic tests are performed at an initial velocity which is smaller than the critical velocity (valve is not 100% open) in equation (4) v_o, t_o and t'_o should be replaced by v_b, t_b and t'_b respectively.
- It is assumed that, apart from high frequencies, the flow signal is measured without deformation, which means that the time shift of the flowmeter should be constant over the whole frequency range. If not the flow signal needs to be corrected by means of a frequency analysis (e.g. Laplace transformation).

7. THE DIMENSIONLESS DYNAMIC CHARACTERISTIC

For a range of geometrically similar check valves the dynamic characteristic can be presented in a dimensionless form [ref 6]. By means of a dimensional analysis which is applied in a more general sense by Thorley [ref 7] this results in the following equation yielding for check valves with translating moving elements:

$$\frac{v_R}{v_o} = f\left(\frac{D_i}{v_o^2} \cdot \frac{dv}{dt}, \frac{\rho_m}{\rho_f}, \frac{F_o}{F_s}, \frac{x_b}{D_i}\right) \tag{5}$$

Where:
- V_R = maximum back flow velocity (m/s)
- v_o = critical velocity (m/s)
- D_i = inner valve diameter (m)
- dv/dt = characteristic deceleration (m/s^2)
- ρ_m = density moving element(s) (kg/m^3)
- ρ_f = fluid density (kg/m^3)
- F_o = net spring and gravitational force on the moving element(s), just fully open valve (N)
- F_s = net spring and gravitational force on the moving element(s) valve just closed (N)
- x_b = initial valve disc position (m)

When dealing with check valves with rotating moving elements the group F_o/F_s should be replaced by T_o/T_s, i.e. the ratio of the net spring and gravitational moments on the moving element(s) in just fully open and closed position.
Further the group x_b/D_i should be replaced by θ_b/θ_o, i.e. the ratio of the disc angles

in the initial and the fully opened position.
It is assumed that the number of Reynolds is sufficient high, so there is no Reynolds dependency.

When for a range of geometrically similar check valves the parameters ρ_m/ρ_f and F_o/F_s (T_o/T_s) are constant, and the initial valve position is choosen equal in the range, equation (5) can be reduced in the form:

$$\frac{v_R}{v_o} = f\left(\frac{D_i}{v_o} \cdot \frac{dv}{dt}\right) \qquad (6)$$

The usefulness of the dimensionless characteristic in the form of equation (6) is investigated by performing an experimental investigation in which v_o and D_i are varied. At all tests the valves where fully open and in a horizontal position.

Figure 9 shows dynamic characteristics of an 800 mm nozzle type check valve for three different springs. In figure 10 the same test results are presented in a dimensionless dynamic characteristic.
The mass and size of the used springs increases with increasing spring strength. Therefore the ratio ρ_m/ρ_f is approximately constant.
The ratio F_o/F_s will be constant for the different springs if the length of the springs in unloaded condition is equal. As there were small differences in spring length this requirement was not met completely.

The results show that the usefulness of the dimensionless characteristic is reasonably well although not all conditions are fulfilled.

In figure 11 a dimensionless dynamic characteristic is presented of a nozzle type check valve whereby both v_o and D_i are varied [ref 6]. The data are obtained using a mathematical model of the valve describing the valve motion as function of dv/dt. As a result all calculated points lie virtually on the same curve. The small deviations for the 800 mm valve are probably caused by non similarity in ρ_m/ρ_f and F_o/F_s.

The dimensionless dynamic characteristics might be useful for several reasons:
- The offer the opportunity of model testing as frequently used for eg. pumps, turbines and airfoil sections. Check valves are manufactured up to diameters of 2 m. Scale laws make it possible to test smaller models.
- The required data needed in surge calculations e.g. waterhammer computercodes can be reduced to data of valve types. This is comparable with standard pump and valve characteristics.
- The dimensionless dynamic characteristics can be used in the selection of springs in a particular application.

8. THE USE OF (DIMENSIONLESS) DYNAMIC CHARACTERISTICS IN WATERHAMMER COMPUTER CODES

8.1 MODELLING OF CHECK VALVE CLOSURE BEHAVIOUR

The closure behaviour of undamped checkvalves can be simulated in waterhammer computer programs with the aid of the (dimensionless) dynamic characteristic.

In section 5 it is shown that during valve closure the undamped check valve does not influence the flow up to the moment the maximum back flow is reached. The test results show that the moment the maximum back flow velocity is reached and the moment the valve is 100% closed practically coincide. These results are of twofold interest:

1. In most cases it will be reasonable to assume that the changes in pressure section loss and the pressure loss over the valve are small with respect to the total pressure difference over the system. In these cases the valve can be simulated as a component without resistance as long as it is (partly) open.
2. During valve closure only the moment of closure is relevant. This moment can be determined with the aid of the dynamic characteristic and the calculated flow time function (see figure 12). Hereto, in accordance with the definition (see section 6.3), the characteristic deceleration $dv/dt = (v-v_o)/(t-t_o)$ has to be calculated in every time step. The corresponding maximum back flow can be determined from the

dynamic characteristic. When the actual flow is greater than the maximum back flow the simulation is continued with the valve open; when the actual flow equals the maximum back flow the valve has to be closed within one time step.

In this way an arbitrary check valve can be modelled. All needed is the dynamic characteristic and the critical velocity. It is evident that also the dimensionless dynamic characteristic can be used. It should be noted that this model can be used for the first closure of the valve. In general when the valve reopens it will not be 100% open. Therefor the boundary conditions at which the dynamic characteristic is measured are not valid here.

8.2 APPLICATIONS

DELFT HYDRAULICS has implemented the abovementioned model of the check valve into their waterhammer computerprograms.
Some simulations have been made to compare the results obtained with the "real" check valve as described in section 8.1 and an ideal check valve, in fact a diode.

In figure 13 a pipeline is considered consisting of a pump with a check valve. Downstream of the check valve an airvessel is placed. The pump is rated at 33,7 m head when delivering 0,278 m^3/s at a speed of 1430 rev/min and having a power requirement of 115 kW. A pumpfailure after one second is simulated.

Due to pumpfailure the head at the pump decreases. In the case of the ideal check valve the valve closes at the moment the fluid velocity reaches zero. The real check valve closes somewhat later when the fluid velocity equals the maximium back flow velocity.
The results obtained with a real check valve show a pressure rise built up of two parts after the valve is closed. The first part is the Joukowsky pressure variation due to a reduction of the maximum back flow to zero within one time step. The second part is generated because the fluid column between the air vessel and check valve continues to decelerate until the pressure wave due to valve closure reaches the air vessel. During this short time interval the back flow velocity increases and exceeds the maximum back flow velocity at the valve. This results in a further increase of the pressure.
In case of the ideal check valve there is no Joukowsky pressure varation. The pressure rise is caused by the decelerating fluid column after valve closure.
As a result the total pressure variation in the simulation with the real check valve is more than twice as high as in the case of the ideal check valve.

As a conclusion can be drawn that the use of an ideal check valve in waterhammer analysis is not sufficient for high risk systems like this.

9. CONCLUDING REMARKS

With respect to the closure behaviour of undamped check valves the Dynamic Characteristic is a measure for check valve slam and pressure surges.

A general method is proposed for performing dynamic tests and how to determine the characteristic deceleration and the maximum back flow velocity in order to obtain unambigious dynamic characteristics.

The method is among others based on experiments in whereby the influence of the initial flow and the influence of the check valve on the flow is investigated. The most conservative dynamic characteristic is obtained at an initial flow which is equal to the critical flow (valve just 100% open). In all cases the influence of the tested

nozzle check valves on flow is neglectable until the moment the maximum back flow is reached. Howver, for the sake of completeness the experiments ought to be repeated for some swing check valves (rotating type).

The usefulness of the dimensionless dynamic characteristic is investigated for different type of valves whereby the spring strength and the diameter are varied and appears to be well although not all conditions are fulfilled.

Based on abovementioned test results a method is presented in which check valve closure behaviour can be simulated in waterhammer computerprograms. In the model the (dimensionless) dynamic characteristic is used, which means that the model can be used for all types of check valves if the characteristics are available.
With the aid of some computer simulations a comparison is made between an ideal check valve and the model. The results obtained with an ideal check valve are far too optimistic and cannot be used for high risk systems.

References

1. THORLEY, A.R.D.
   Can we develop a safe check valve?
   Procs. Internatinal Conference on Developments in Valves and Actuators for Fluid Control,
   BHRA, Oxford, U.K., September 1985

2. PROVOOST, G.A.
   The dynamic characteristic of non-return valves
   Procs. 11th IAHR Symposium of the section of Hydraulic Machinery, Equipment and Cavitation; "Operating problems of pump stations and power plants", Amsterdam, September 1982

3. ELLIS, J. and MUALLA, W.
   Selection of check valves
   Procs. 5th International Conference on Pressure Surges, Hannover, F.R. Germany, 22-24 September, 1986

4. PROVOOST, G.A.
   To present the dynamic property of undampened check valves
   Paper presented at the ASME Winter Annual Meeting, Boston, Massachusetts, U.S.A., November 13-18, 1983

5. PROVOOST, G.A.
   The dynamic behaviour of non-return valves
   Procs. 3rd International Conference on Pressure Surges, Canterbury, England, March 25-27, 1980

6. KRUISBRINK, A.C.H., LAVOOIJ, C.S.W. and KOETZIER, H.
   Dynamic behaviour of large non-return valves
   Procs. 5th International Conference on Pressure Surges, Hannover, F.R. Germany, 22-24 September, 1986

7. THORLEY, A.R.D.
   The state of the art
   (to be published)

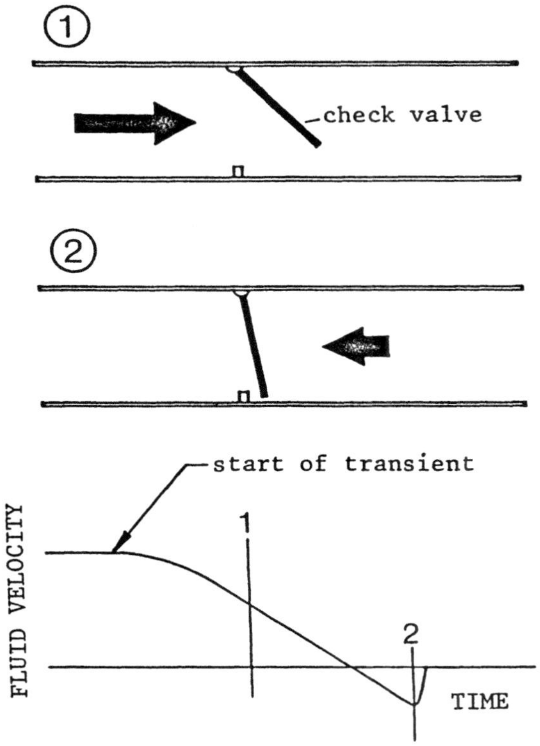

Figure 1. Check valve slam

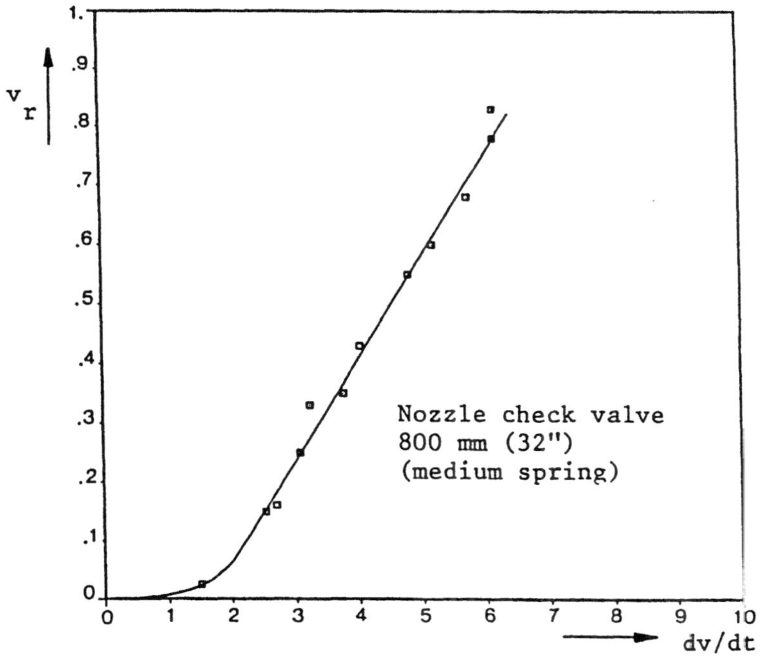

Figure 2. The dynamic characteristic of a 800 mm (32") Nozzle check valve (medium spring)

a.

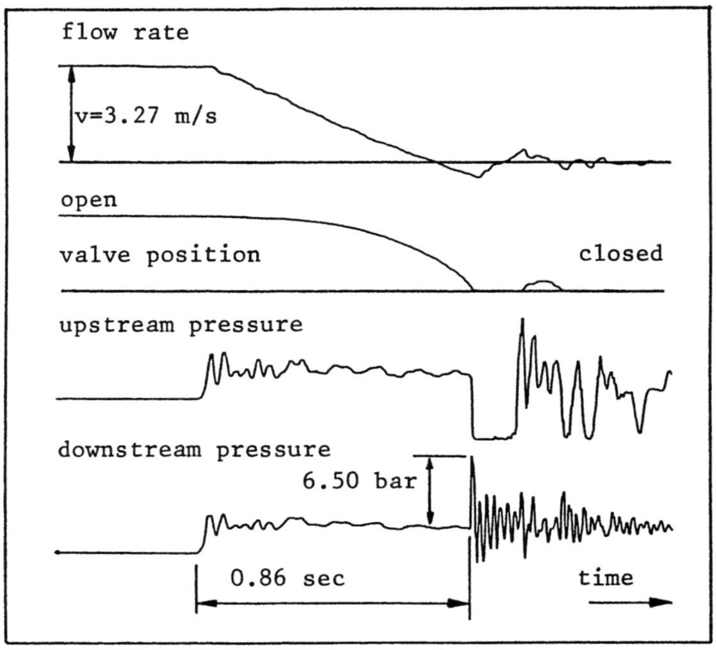

b.

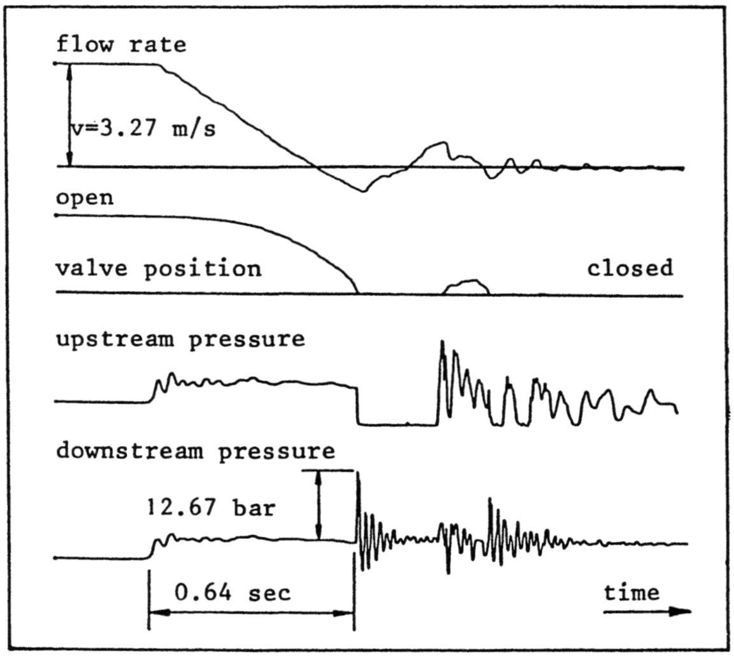

Figure 3. Dynamic test results of an 800 mm (32")
Nozzle check valve (medium spring)

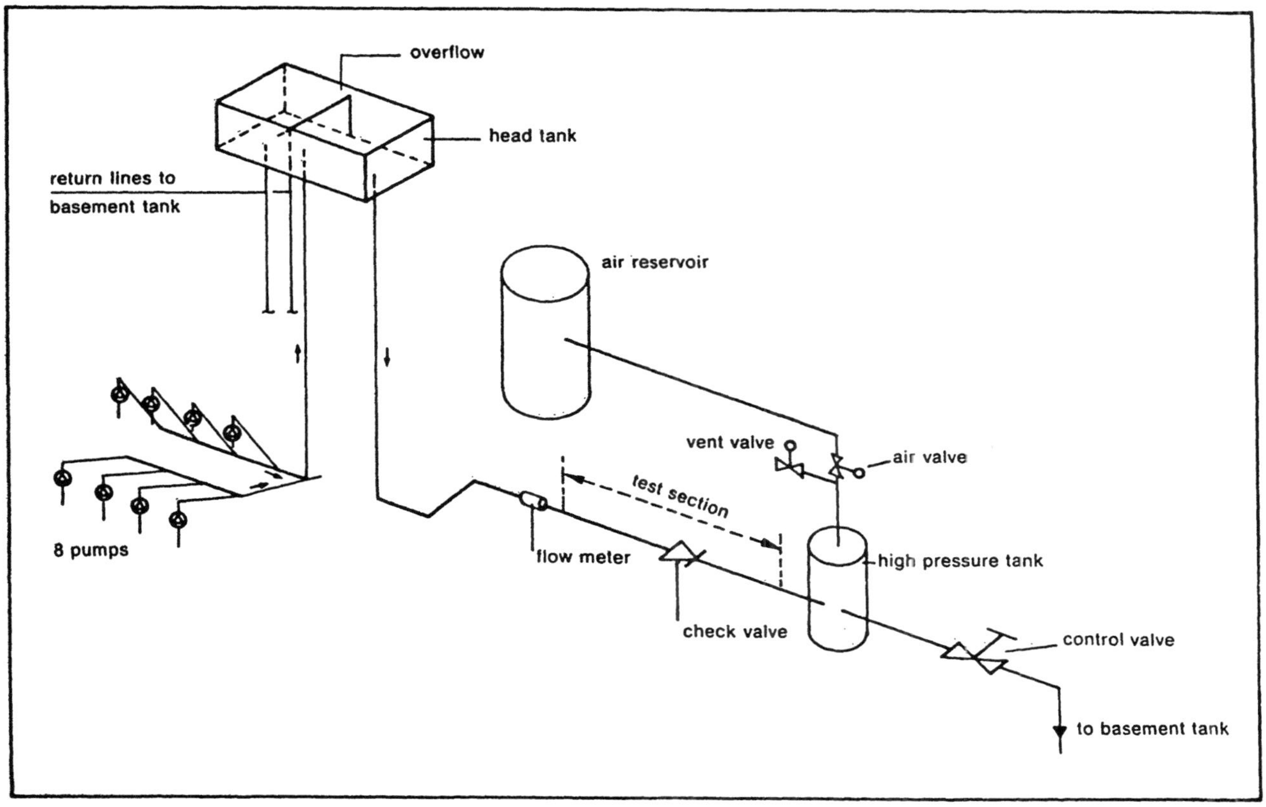

Figure 4. Schematic diagram of the test rig for large check valves

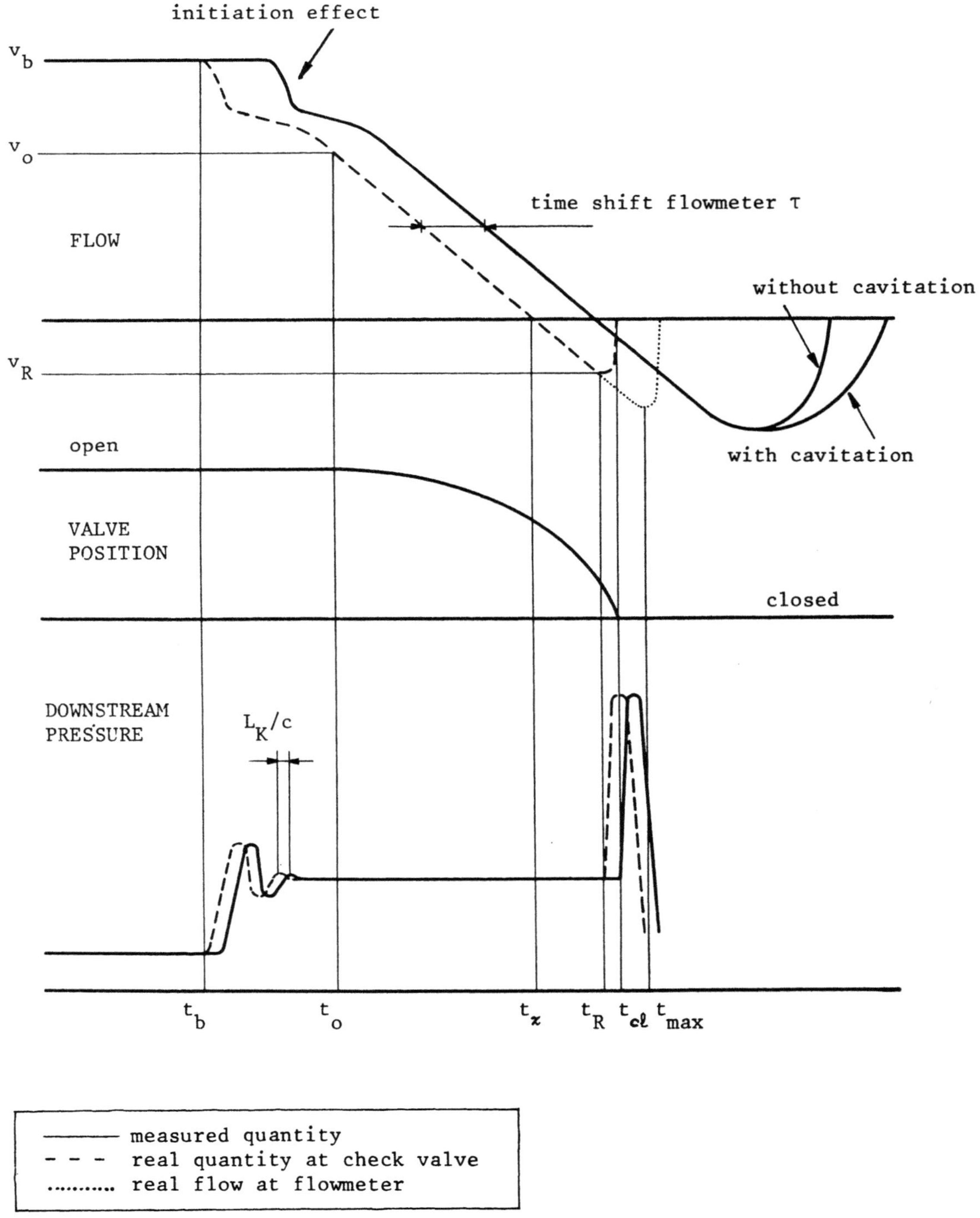

Figure 5 Analyses measurement signals

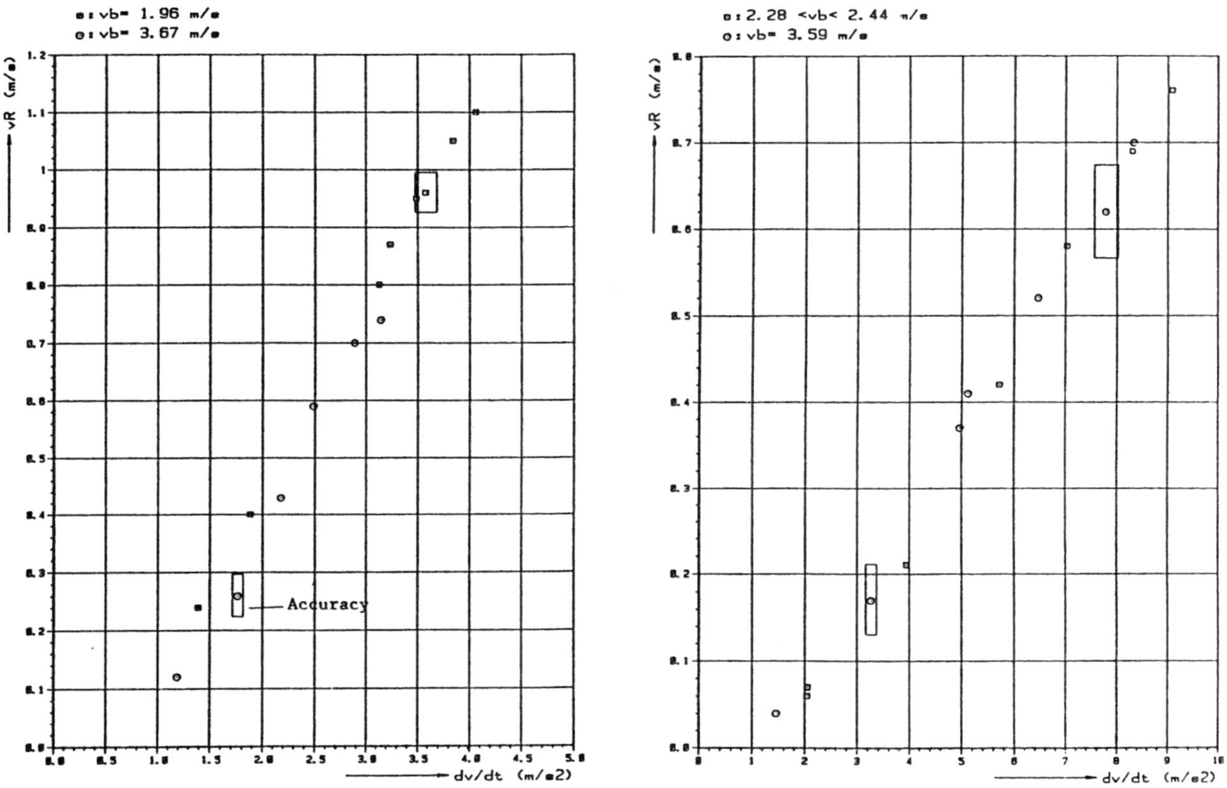

Figure 6 The influence of initial fluid velocity on maximum back flow velocity

a.

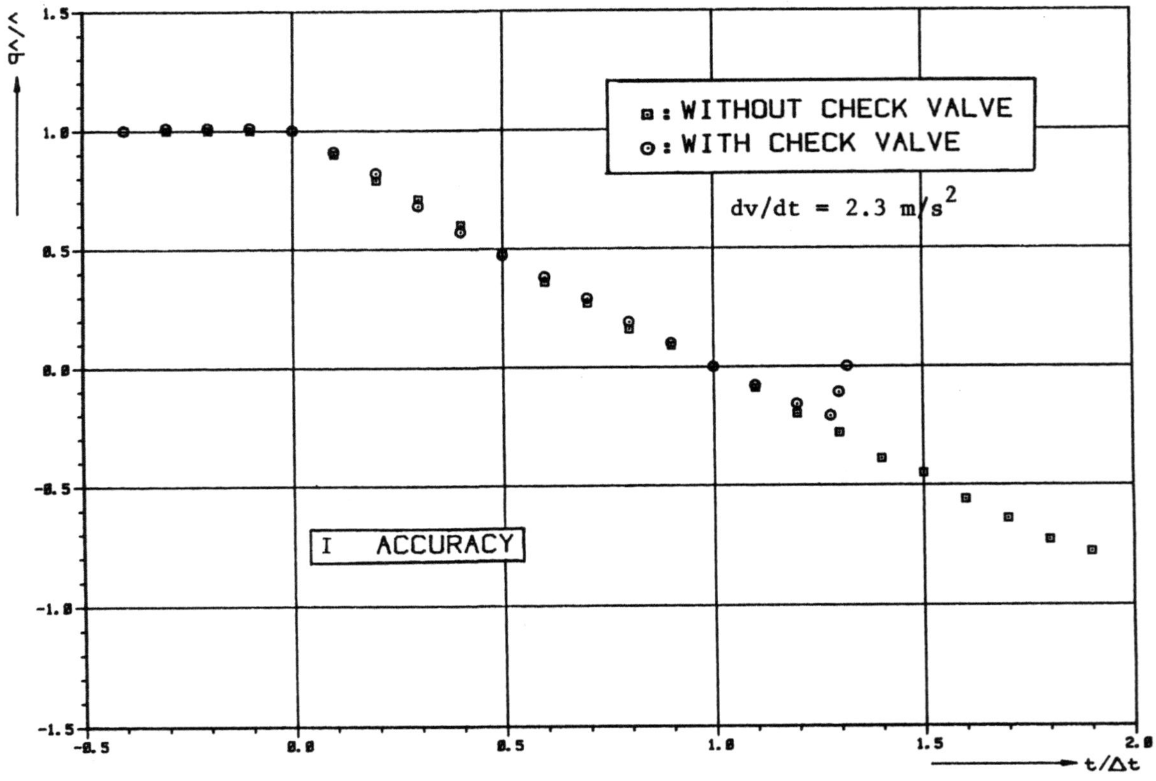

b.

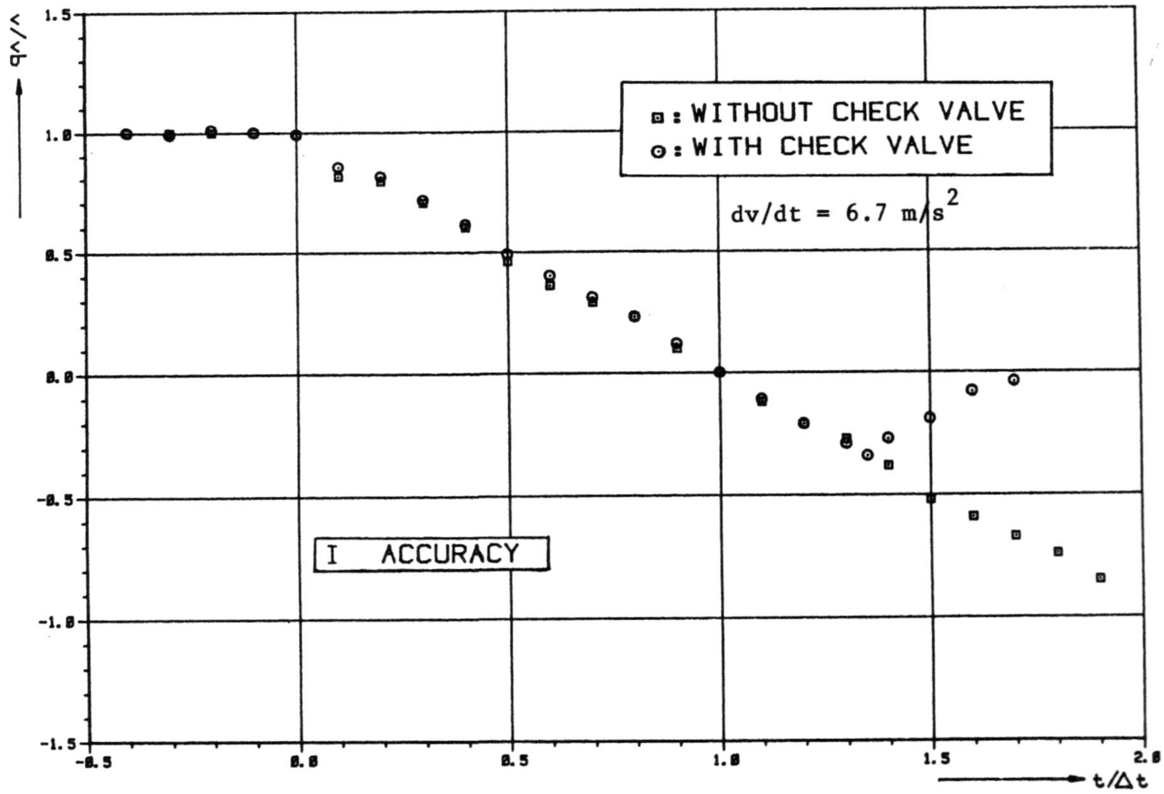

Figure 7. Influence of check valve on flow

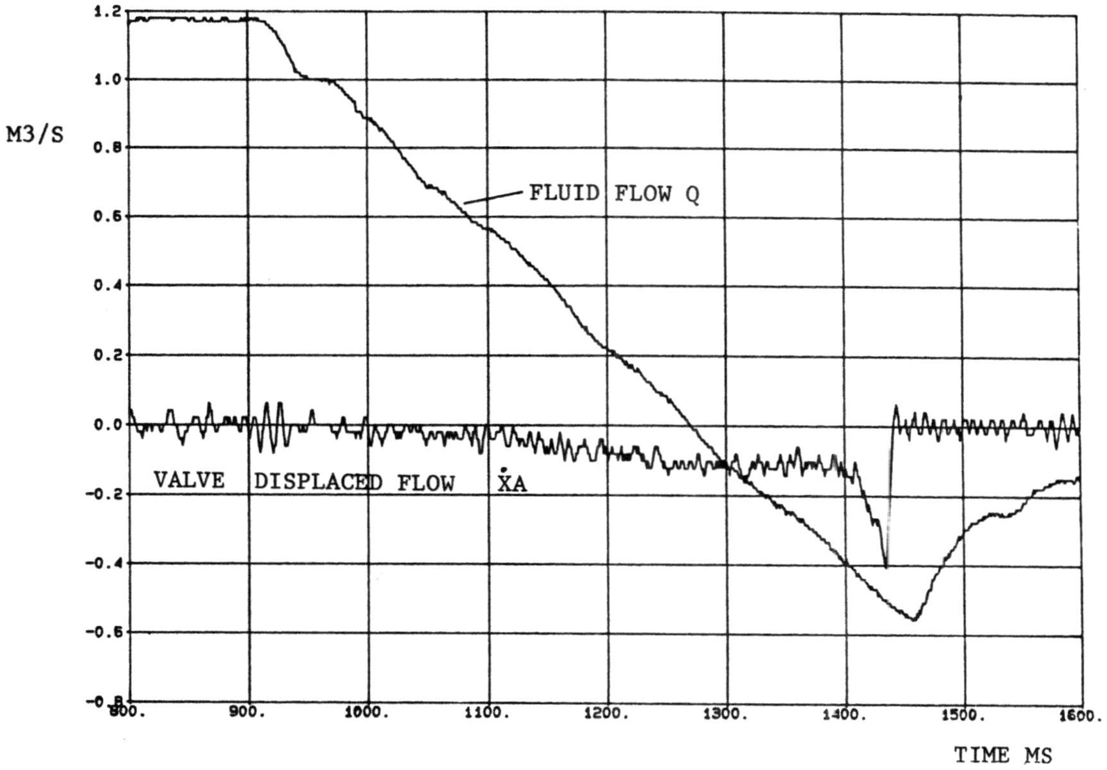

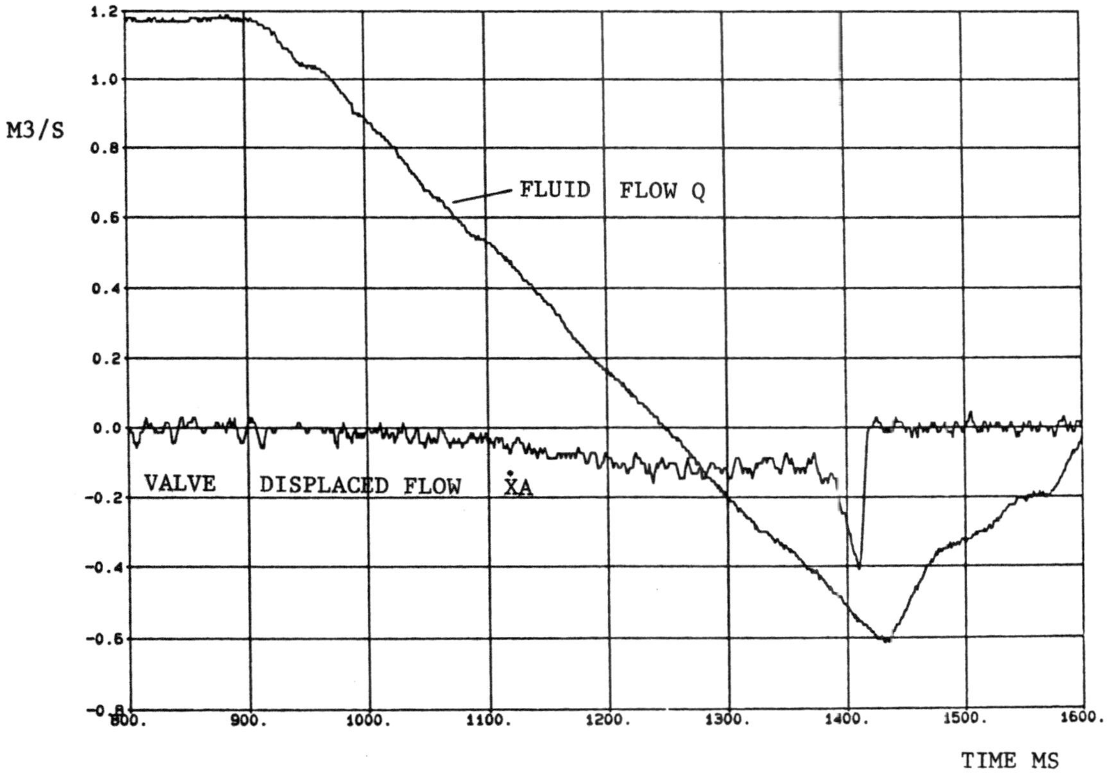

Figure 8 Influence of flow on check valve behaviour

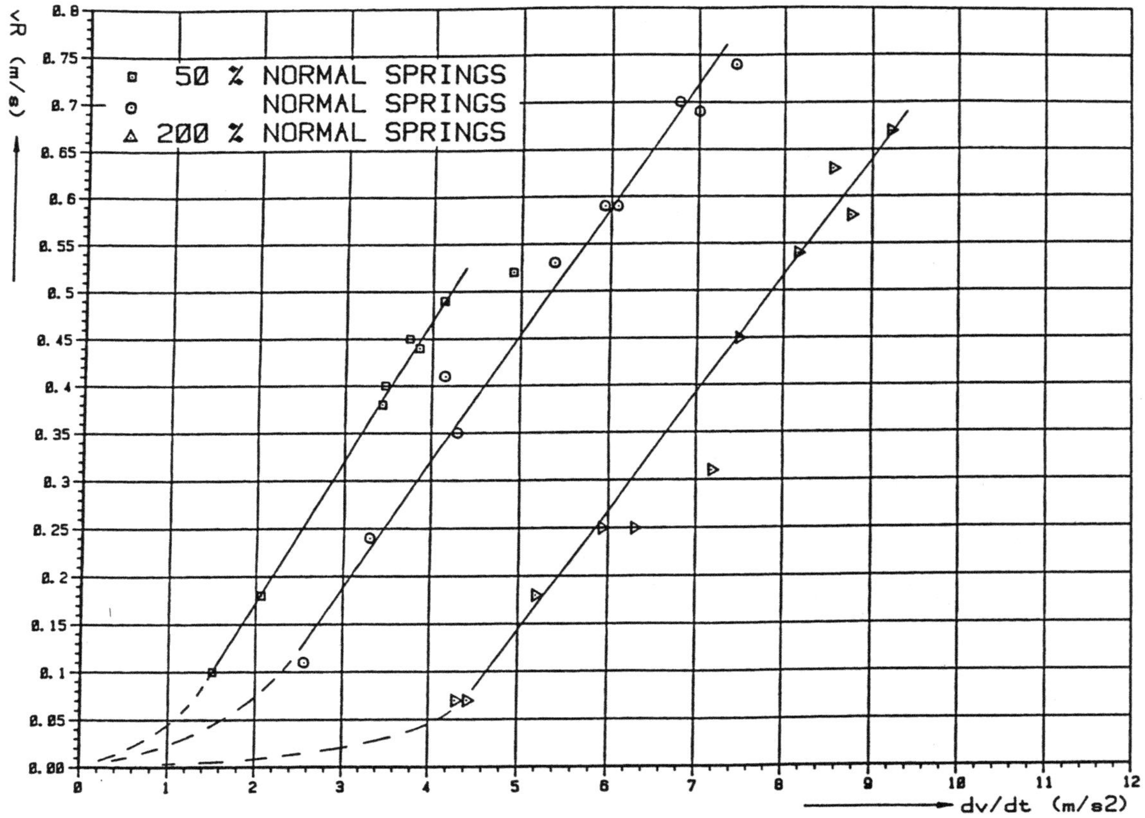

Figure 9 Dynamic characteristics of a 800 mm Nozzle check valve for three different spring stengths

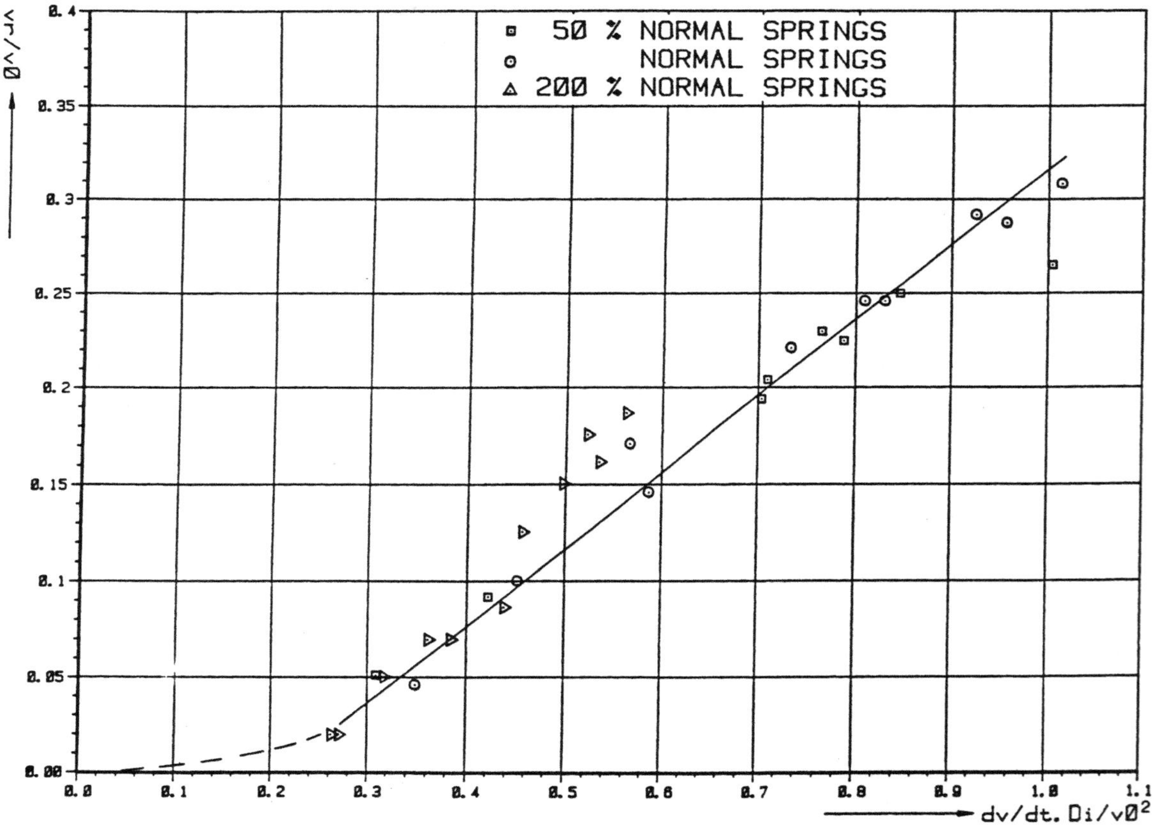

Figure 10 Dimensionless dynamic characteristic of a 800 mm Nozzle check valve

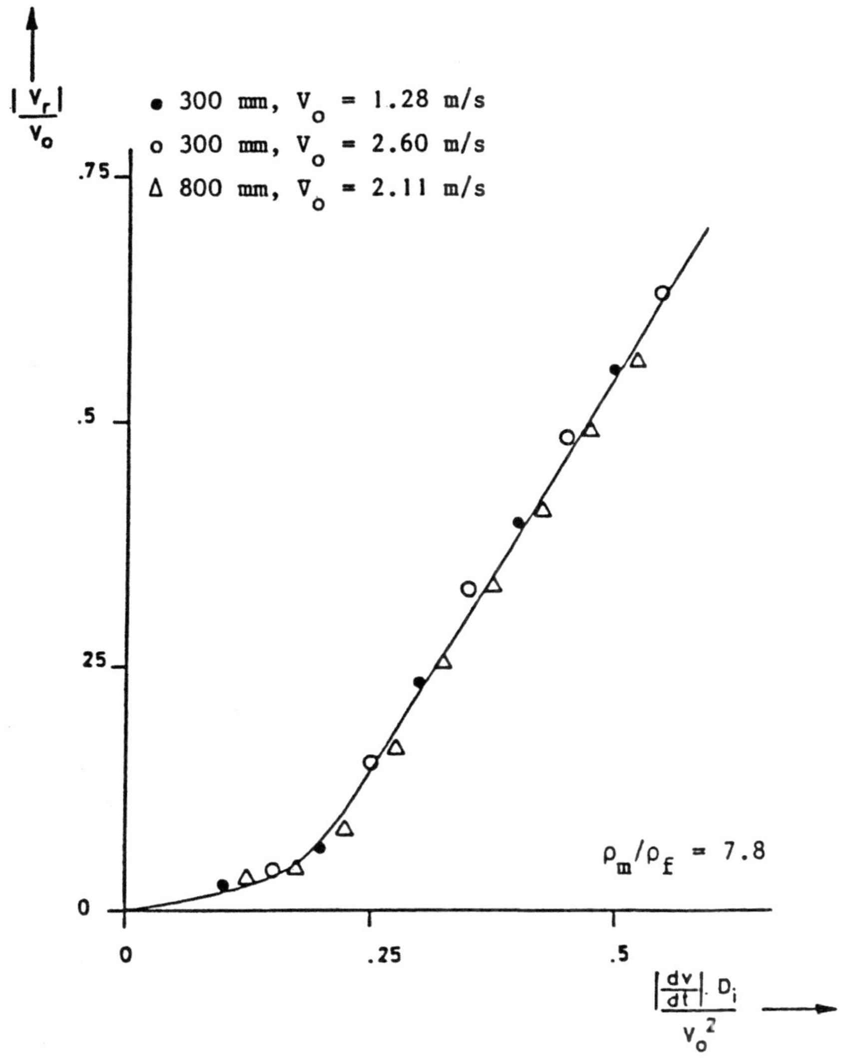

Figure 11. Calculated dimensionless dynamic characteristic of a nozzle type check valve

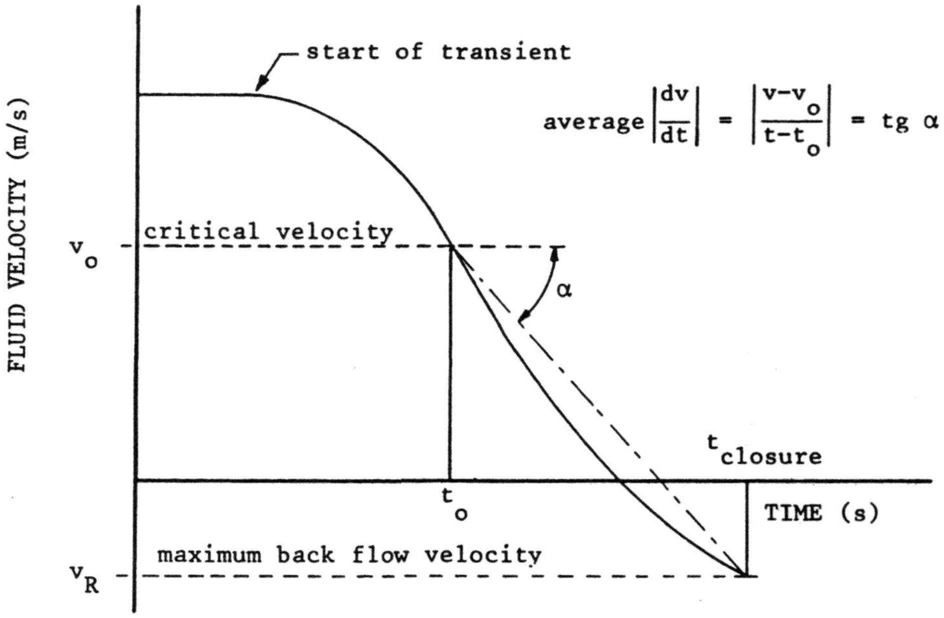

Figure 12 Determination of the characteristic deceleration; Modelling of check valve closure behaviour.

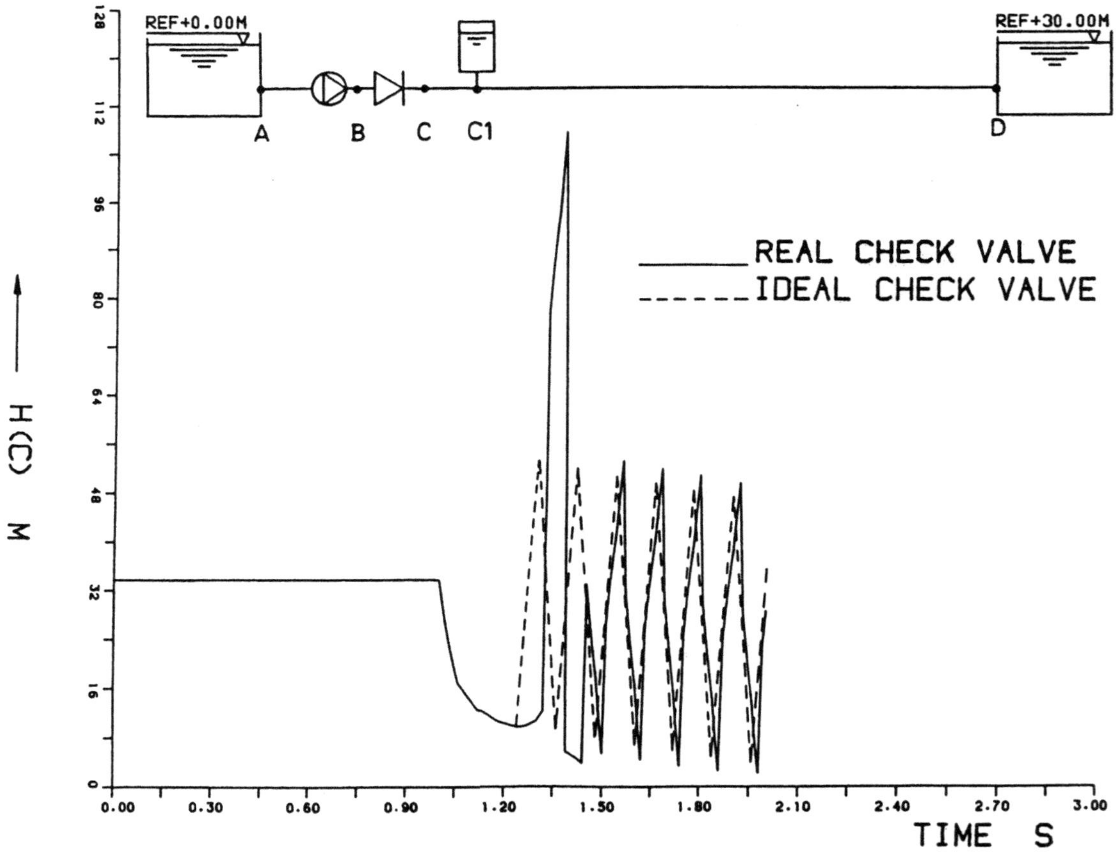

Figure 13 Computersimulations; Pumpfailure

2nd International Conference on

Developments in Valves and Actuators for Fluid Control

Manchester, England: 28-30 March 1988

PAPER H3

INNOVATIONS IN NON-INTRUSIVE TESTING AND TRENDING OF MOTOR OPERATED VALVES AND CHECK VALVES

A. G. Charbonneau
MOVATS Incorporated

SUMMARY

Two significant problem areas recently noted in nuclear power plants are the inability of motor operated valves to perform as intended and premature failures of check valves in service. The purpose of this paper is to provide a status of testing and trending techniques which minimizes system and component down time, and reduces radiation exposure.

Because of the great difference in techniques in testing motor operated valves and check valves, this paper will be divided into two major sections. Section I will describe advances made in testing of motor operated valves. Section II will describe innovations in predictive maintenance techniques for check valves.

SECTION I

A. BACKGROUND

Many studies have shown motor operated valve (MOV) failures to be a common cause of problems at power and process plants. Such failures reduce the safety and reliability of plant systems and sometimes lead to costly down time. Plant betterment programs and design changes have resolved some MOV problems. However, such efforts have been of limited success because they are often directed at correcting the symptoms, rather than the causes of failures.

The need to identify and resolve the underlying causes of MOV failures led to development of the MOVATS® Diagnostics Test System. These portable test systems are used to obtain dynamic "signatures" of critical operator parameters. The signatures are analyzed to identify degradations that have led (or could lead) to operator failures. The signatures can also be used to confirm the accuracy and adequacy of current torque and limit switch settings and verify that the MOV will function under worst case differential pressure conditions.

Held in Manchester, England. Organised and sponsored by BHRA, The Fluid Engineering Centre and co-sponsored by the British Valve Manufacturers Association and the Institution of Mechanical Engineers Process Industries Division

Full MOVATS® diagnostic testing is accomplished using temporarily installed transducers mounted on the valve operator. Access to the MOV may be restricted during normal plant operation, and it is often inconvenient or impossible to remove the valve from service to allow transducers to be attached. For this reason, diagnostic testing is performed infrequently (usually during plant outages).

Unfortunately, operability of a MOV can be affected by maintenance activities, environmental factors, and random component failures. Current periodic testing techniques (primarily valve stroking or stroking while monitoring motor current) are not capable of identifying degrading operator condition in time to prevent failures.

MOVATS Incorporated recognized that there is a need for a MOV testing system that can be used at the Motor Control Center (MCC) during periodic testing. Such a system could be used for trending and identification of significant operational changes and verification that the MOV has sufficient capacity to respond to the maximum differential pressure condition. Using such a system, full in-situ diagnostic testing would be required only after extensive maintenance activities (such as motor or gear replacement).

Evaluation and Field Test

Motor operated valve motor control centers (MCC) typically contain circuit breakers, motor contactors, power circuits (AC or DC) and control power wiring to the operator. Therefore, it should be possible to monitor most operator torque and limit switch actuations, motor supply voltage, current, power factor, and power directly from the MCC.

Torque and limit switch actuations are monitored by present MOVATS® equipment at the valve operator. The same equipment and methods can be used at the MCC without modifications. To monitor dynamic performance of the MOV, however, various electrical parameters associated with the motor must be related to mechanical forces generated in the operator.

Extensive field testing was performed to select the most suitable parameter for monitoring operator performance from the motor control center. Tests were conducted at power plants using a system to apply various loads to valve stems. During the testing, the parameters of interest were monitored as mechanical load was increased.

For a variety of reasons, motor current, power factor, and input power were all found to be unacceptable as primary parameters for the test system. Results from motor current tests revealed that motor current does not vary linearly with increasing torque, and line voltage variations greatly affect the current required to produce any given torque.

Power factor measurements were taken as the line voltage and stem load were varied. Results showed power factor to change linearly with changing stem load though there is considerable scatter in the measured values and also indicated that power factor does not remain constant as line voltage changes. Similar problems were identified when motor input power was the primary test parameter.

As a result of such tests, motor current, power factor, and input power were all eliminated as primary indicators of operator and valve condition.

During the series of tests as previously noted, it was determined that motor input power could be used as an indicator of the mechanical performance of an operator if certain modifications were made. In particular, motor losses were modeled and subtracted from the input power. This modified input power parameter of the motor is defined as "Motor Load."

Motor load is a measure of the power being delivered by the motor shaft. Accurate measurement of this parameter is considered to be the most accurate and dependable method of testing motor operators remotely from the MCC.

Typical test results using the "Motor Load" parameter are shown in Figures 1 and 2. These graphs show that "Motor Load" varies linearly with stem thrust and that the "Motor Load" values are not affected by voltage changes as large as $\pm$10%.

B. THE MOTOR LOAD SYSTEM

The Motor Load System developed by MOVATS Incorporated (called the (Motor Load Unit or MLU) is a self contained unit designed to measure motor load. Voltage sensing leads for each phase and a clamp-on ammeter measuring current through a typical leg of the motor windings are utilized.

The output of the unit is displayed on a built-in panel meter and is also available as an analog signal using the amphenol connector on the rear panel of the unit. The analog output is connected to a modified recording oscilloscope for data collection and analysis. The oscilloscope is utilized during baseline testing of the valve during outages, and when motor load baseline signatures are obtained. The baseline signatures are transferred to a computer database. When subsequent maintenance is performed on the valve, such as adjusting packing, the motor load unit is connected during valve stroking test to determine operability. The signatures obtained are then transferred to the computer and compared to the baseline values.

C. TYPICAL TEST RESULTS

Figure 3 shows a typical plot of the motor load parameter as a valve is opened. This figure also shows the change in motor load readings that would result from increased mechanical drag (due to packing adjustments, lack of lubrication, etc.). Similar recognizable changes in the motor load parameter would result from mechanical degradations such as gear wear, valve binding, or a bent stem.

SECTION II

In the second portion of our report, we will discuss advanced ultrasound techniques for predictive maintenance measures on check valves.

A. BACKGROUND

Recent significant failures of check valves at nuclear power plants have brought to light an important area previously receiving little attention.

In studies following the 1985 failure of several check valves simultaneously at San Onofre Unit 1, it was revealed that the valves which failed were only marginally satisfactory under ideal flow conditions and were not adequate under the reduced flow to which they were subjected in the months preceding the failure. These reduced flow conditions caused instability in the valves which led to their premature failure.

Further studies by the Institute of Nuclear Power Operations (INPO) have indicated that a frequent cause of instability of check valves that leads to valve failures is design misapplication.

Misapplication can occur both in the original design phase or through subsequent plant modifications. Lengthy operation of the valves at other than design flow rate, such as that noted at San Onofre, could also be considered a form of misapplication.

Correct orientation and location of check valves is also important. Valves correctly sized and typed, but located too close to upstream or downstream obstructions may have unstable operating characteristics.

Since the San Onofre Unit 1 event, at least one valve company has issued a service bulletin describing the instability problem of their valves and have published velocity ranges for reliable operations. Unfortunately, the velocity range developed by the valve manufacturer under ideal laboratory test loop conditions may be irrelevant when applied to a check valve in a real world application. A significant contributing factor to the instability of check valves, and in some cases failure, is the installed location. Ideally, check valves should be located at least 10 pipe diameters downstream of pumps, control valves, elbows and other obstructions that can cause increased turbulence and an unstable environment for the valve.

In the past, check valve performance has been monitored only during plant outages by disassembling the valves and closely inspecting for wear of hinge and disc components. Acoustic monitoring and radiography have also been used, but with very limited success.

Presently, the only means of quantifying the operational stability of check valves is through in-plant testing.

In Reference 1, it is noted that,

"It is not possible to develop a mathematical expression to relate the stability of a check valve to the different valve and flow characteristics. The relationships are too complex and dependent upon too many other variables. The present state-of-the-art is to gain understanding about valve stability by physical testing."

B. ANALYSIS OF OPERATING CHARACTERISTICS

The characteristics of the ideal check valve under the full range of flow conditions is illustrated in Figure 4 by plotting disc position versus flow rate.

The "free swing flutter" area, shown in Figure 3 displays the characteristics of a valve with insufficient flow velocity to hold the disc firmly and continuously against the backstop. The resulting fluttering of the disc/hinge arm about the hinge pin is mainly caused by vortex shedding as the impingement forces are not sufficient to lift the disc to the backstop. Excessive free flutter can lead to worn or broken hinge pins and prying action on the interfacing sections of the disc stud.

As the flow rate increases to the optimum operating condition, the vortex shedding forces causing flutter decrease, and the major factor in backseat tapping becomes the forces generated by localized turbulence or system flow fluctuation. Backseat tapping can be very destructive because the disc or stud impacts the open stop. Testing has shown tapping occurring when a disc is positioned within 7° from the backstop.

The phenomenon referred to as galloping can be expected to occur at high velocities during which the fluid lifting force is strong enough that it interacts with the vibrations of the disc and generates a fluid dynamic instability. A maximum flow rate is usually given by the swing check valve vendors to prevent galloping. Galloping can be the most destructive phenomenon because the disc is on the open stop, characterized by short high velocity excursions from the stop and high impacting energy.

Innovative portable test equipment called Checkmate™ using ultrasonic technology has been applied successfully to various types of check valves used in the nuclear power industry. The system utilizes a pair of temporarily installed ultrasonic transducers (UT) that are placed in contact with the valve casting and held with straps or other mounting devices. The ultrasonic signals from the transducers are then "focused" on the check valve disc. For a given operating condition, the reaction of the disc to that flow condition can be accurately monitored and easily analyzed. Initial testing can be completed in approximately thirty (30) minutes and enables the users to immediately determine if the valve disc is unstable for that particular test flow condition.

When the initial test indicates that a valve disc is oscillating or fluttering, a signature of the disc's movements is electronically recorded. Subsequent computer-assisted analysis is performed on that signature to validate and quantify the following operating characteristics of an unstable valve.

1. Disc location (accuracy of ± 4 degrees)
2. Amplitude and frequency of disc flutter
3. Disc velocity and acceleration
4. Energy levels of disc tapping on the open stop

By quantifying the four operating characteristics listed above, the Checkmate™ user can identify the valve misapplications such as incorrect type, size or location.

In independent testing sponsored by EPRI, Electrical Power Research Institute, the effectiveness of the ultrasonic techniques were verified and the superiority over other techniques demonstrated. Quoting from the draft EPRI report summary,

> "The use of the ultrasonic method to detect check valve disc motion offers a big advantage over the other techniques discussed thus far in its ability to provide directly usable quantitative results rather than indirect and mostly qualitative results."

Further in the report, in the Test Results Section,

> "After reviewing the test data, some observations can be made about the Checkmate™ System. The system can be used to determine disc position and quantify disc movement with an accuracy suitable for the intended use of the data. The system also identifies tapping of the disc against the backstop and provides information which can be used and is necessary for computation of wear rates and fatigue life. For many valves (including the valve used during this test), the Checkmate™ System can determine if the disc is detached from the hinge arm. Techniques used for this determination required a reasonably accurate drawing of the valve, but such drawings are generally available at plant sites."

The check valve program progresses as follows. The disc position and the frequency and amplitude of hinge/disc vibration are determined from the Checkmate™ signatures and included in the Field Service Report. When the testing indicates that a valve disc is oscillating excessively, a Checkmate™ Signature Analysis Report is generated. This analysis involves further review of test results to quantify the extent of disc stud degradation and the rate of disc stud and hinge pin degradation. If the degradation rates are unacceptable, Engineering Design Review is performed to identify design or other changes required to achieve a high degree of check valve reliability.

CONCLUSION

Nuclear power plants world-wide have experienced failures of both motor operated valves and check valves. These failures have caused or worsened plant transients, challenged safety systems and caused prolonged loss of power generation capability.

Two techniques have recently been developed to allow the current conditions of motor operated valve and check valves to be assessed. These techniques allow identification and correction of degradations prior to valve failure. The techniques also allow long standing design or application problems to be diagnosed and corrected, thereby increasing valve reliability. Both techniques are currently being used in U.S. nuclear power plants.

REFERENCES

1. Rahmeyer, William J.: "Application of Check Valves with Unsteady and Non-Uniform Flow Conditions," Utah State University.

2. Chiu, C. and M. S. Kalsi. "Plant Availability Improvement by Eliminating Disc Vibrations in Swing Check Valves," ASME Paper No. 86-JPGC-NE-6.

3. Chiu, C. and M. S. Kalsi. "Failure Analysis of Swing Check Valves," - San Onofre Nuclear Generation Site - Unit 1." NRC Docket No. 50-206, April 4, 1986, Revision 2.

4. Institute of Nuclear Power Operations. "Check Valve Failures or Degradations," - Significant Operating Experience Report 86-03.

4. MOVATS Incorporated, Marietta, GA "Checkmate™."

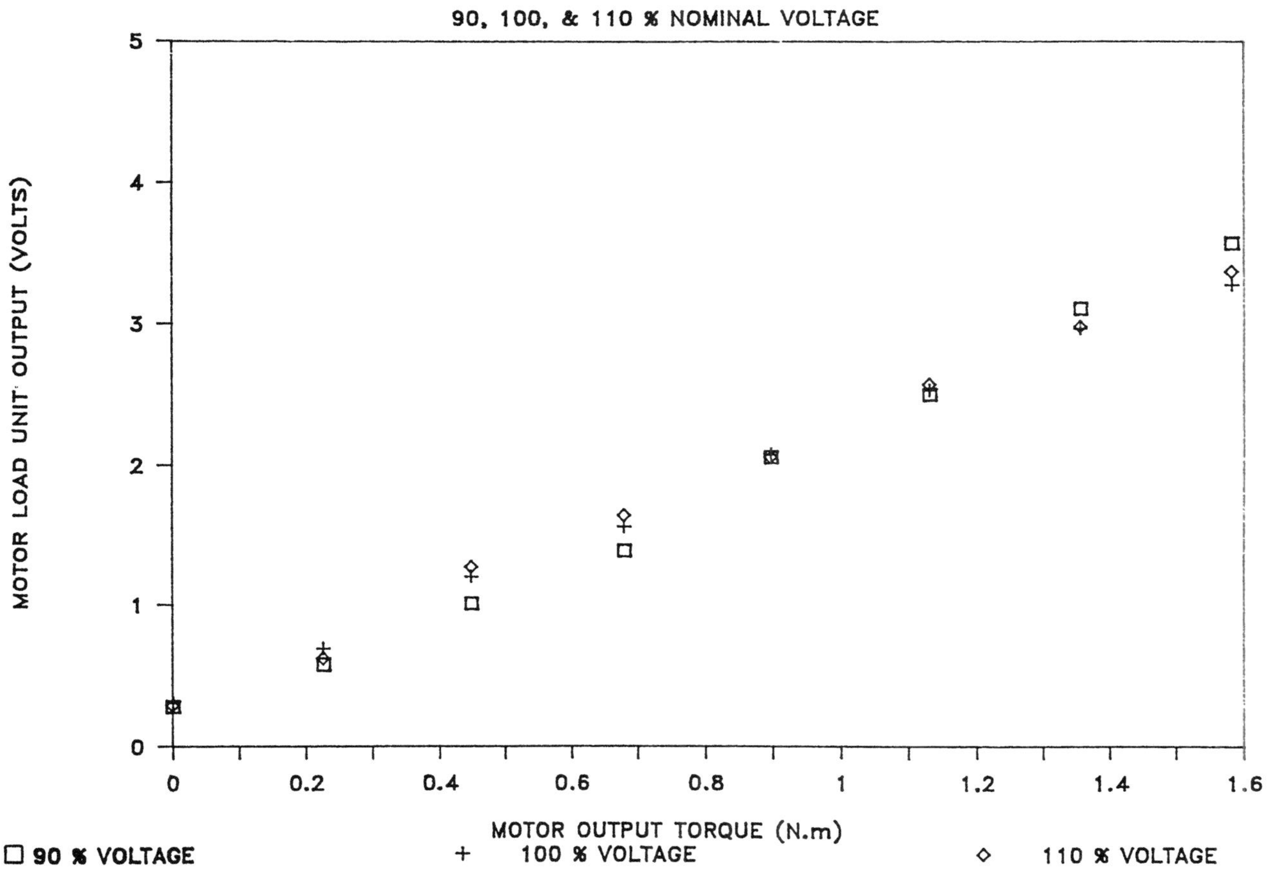

FIGURE 1: MOTOR LOAD UNIT OUTPUT VS MOTOR TORQUE

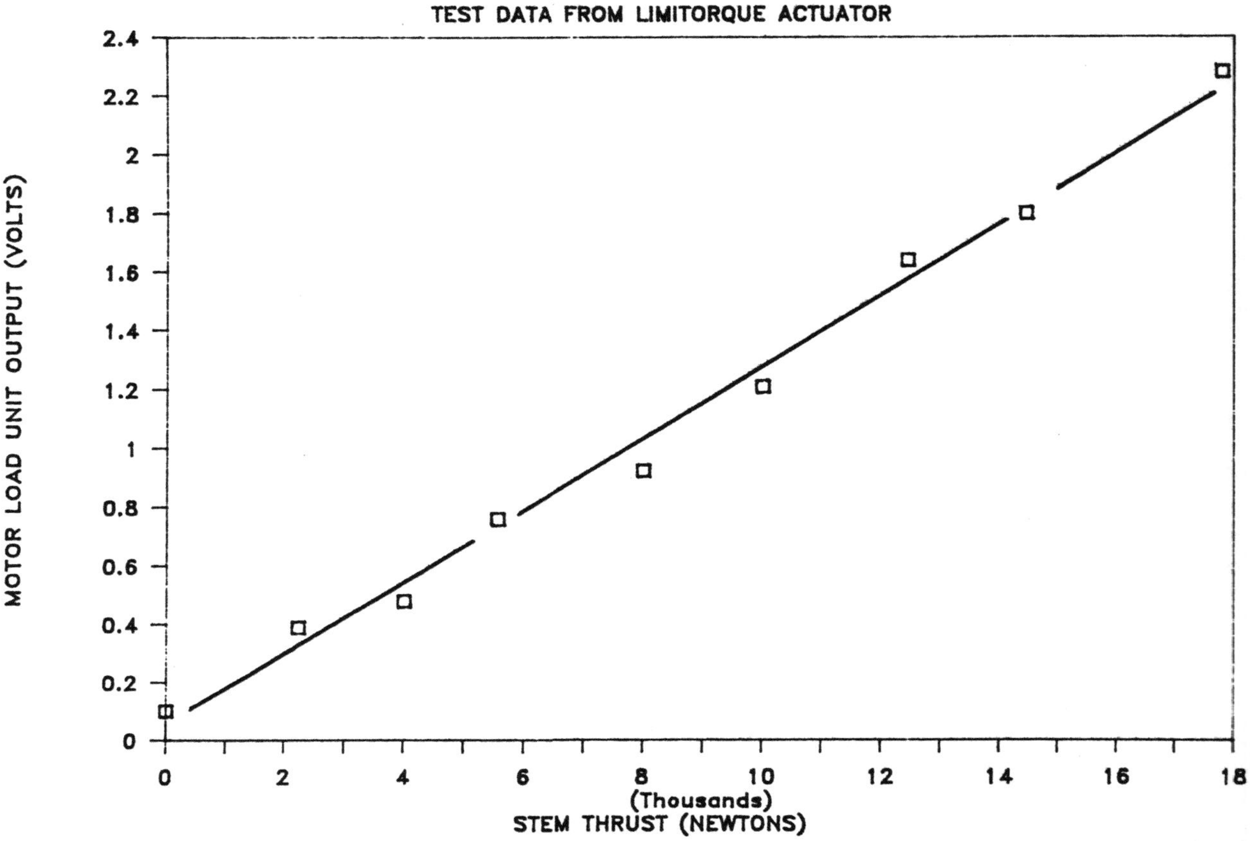

FIGURE 2: MOTOR LOAD VS. STEM THRUST

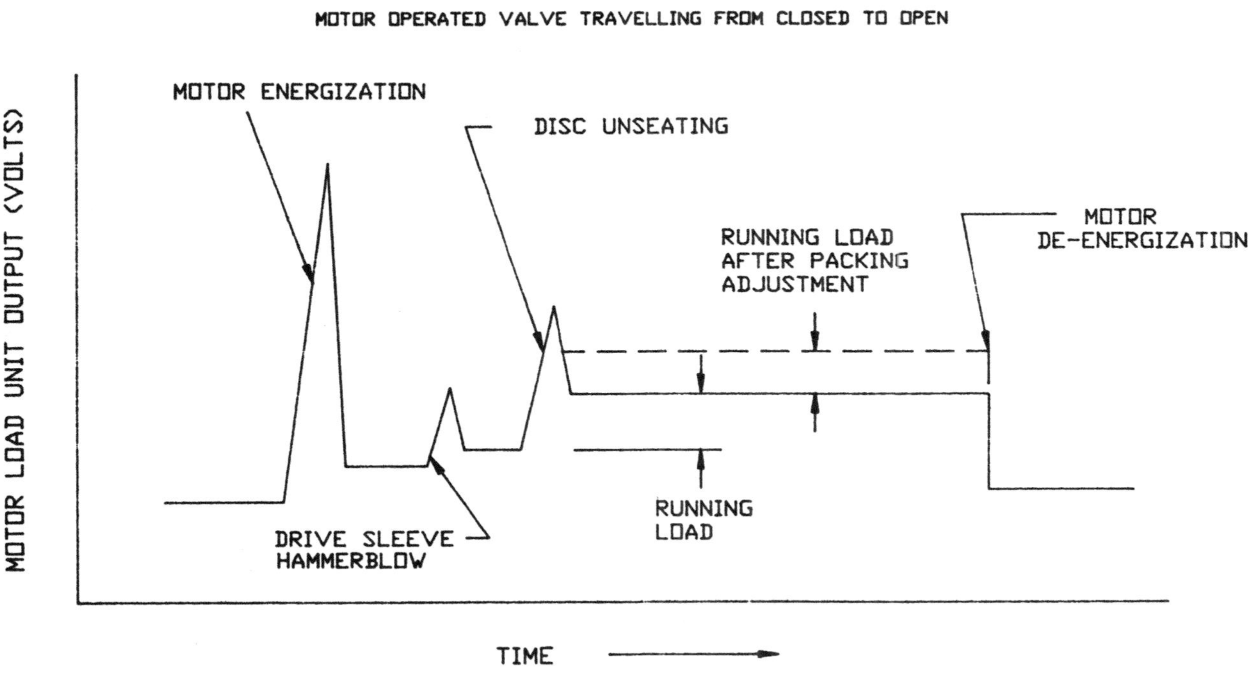

FIGURE 3: TYPICAL MOTOR LOAD MEASUREMNETS

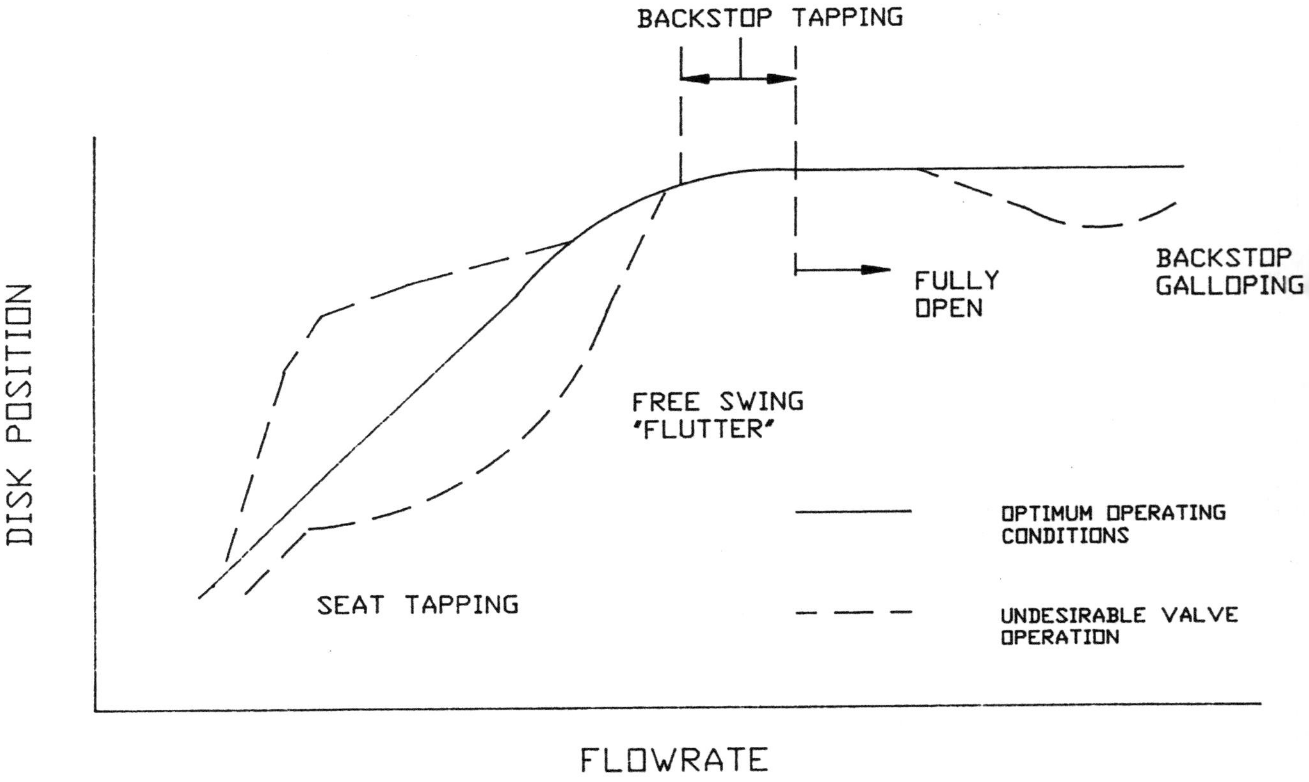

FIGURE 4: FLOWRATE VS DISK POSITION

2nd International Conference on

Developments in Valves and Actuators for Fluid Control

Manchester, England: 28-30 March 1988

PAPER J1

QUALIFICATION OF FRENCH PRESSURIZED WATER NUCLEAR POWER STATIONS VALVES.

Caulier R. - Grenet M. - Panet M.

ELECTRICITE DE FRANCE, FRANCE

Summary

This article presents the tests to which the valves of the French PWR power stations are subjected to in order to qualify under normal, incidental and accidental conditions.

1 - Introduction

A 900 or 1300 MWe French pressurized water power station has about 12 000 valves : an important number of them must have a very high reliability for two major reasons :

- To maintain the reactor in a satisfactory safety condition ;
- To limit as much as possible power station downtime which on an industrial complex of this size is very costly.

To achieve this objective, ELECTRICITE DE FRANCE has started an important qualification test program on the loop for normal and incidental operating conditions as well as for accidental ones.

Before developing this subject further however, it should be recalled that obtaining a high quality

Held in Manchester, England. Organised and sponsored by BHRA, The Fluid Engineering Centre and co-sponsored by the British Valve Manufacturers Association and the Institution of Mechanical Engineers Process Industries Division

equipment, results from a global approach integrating other stages :

- Conception,
- Manufacturing,
- Power Station commissioning,
- Maintenance and periodic tests during operation,
- Experience feed-back.

A brief comment will be given on each of these stages.

2 - Conception

In this stage the manufacturers must respect a certain number of formalized rules : Besides the applicable codes and general regulations (e.g : Conception and Construction Regulations for Mechanical Materials, and nuclear islands of PWR power stations : RCCM), a technical specification sheet adapted to the valves imposes a strict framework for the supplier. Finally, specifications and technical data sheets appropriate to the operation completes this item.

It should be underlined that establishing the latter implies that the functional studies and previous installations are as precise as possible.

This is particularly true for safety valves and control valves.

Operating or environmental accidental conditions must be taken into account in an exhaustive way as of the material conception.

3 - Loop trials

3.1 - Normal and Incidental conditions

For more than 10 years, ELECTRICITE DE FRANCE has carried out systematic trials on several hundred valves.

It concerns endurance trials directed at showing that not only is operability maintained, but that leakage remains acceptable after several hundred manoeuvres under pressure, temperature and flow.

Several test types co-exist :

- Qualification : test carried out at maximum authorised conditions for the valve (pressure and temperature) ; in general, 1500 open/close cycles are made.

  The accelerated aging function is completed by a thermal cycle (20 double shocks of 225·C amplitude). Generally speaking, the valves ordered were chosen amongst the qualified valves.

- Validation : trials carried out under real operating conditions : one can thus carry out several hundred thermal shocks on the same test valve.

  The isolating valve system of the main primary circuit has been subjected to this type of test ;

- Conformity : a trial aimed at verifying that valves manufactured in series are capable of guaranteeing the performances of the qualified valve.

- Investigation and perfecting.

  These trials have enabled valve weaknesses to be detected and to significantly improve their quality.

3.2 - Accidental Conditions

Amongst these are several hundred valves which must stay operational during and/or after an accident of internal or external origin, in order on the one hand, to bring back and maintain the reactor in a safe shutdown condition and on the other hand, to limit the radiological consequences of such an accident to values acceptable on the peripherary of the site.

A specific trial program has been implemented to this effect.

3.2.1-Envisaged scenarios

Scenarios can be of different natures and have diverse origins : we mention in particular :

- Sudden flow condition changes through the valve during a pipeline rupture.

- Loads transmitted by the piping in perturbed conditions (earthquakes for example).

- Accident environment : increase in temperature, pressure, humidity or irradiation.

The corresponding situations taken into account are the following :

- High energy pipeline rupture (RTHE),
- Safe shutdown earthquake (SSE),
- Loss of coolant accident (LOCA),

knowing that each situation includes other less serious types of accident.

3.2.2 -Test Sequences

By taking into account the above mentioned load situations, the following four test types have been defined.

We would point out that in all these trials, the criteria to be verified is that the complete valve movement is obtained, which is indicated by limit switches at opening and closing.

3.2.2.1-High energy pipeline rupture

This has three consequences :

- Sudden change in flow conditions through the valve

- Loads transmitted by the piping on the valve ends

- Environmental change.

A real accident simulation can only be made with regard to its simultaneous consequences, these are taken into account in three different trials.

Only the first one is treated in this paragraph, the other two being treated in paragraphs 3.2.2.2 and 3.2.2.4 respectively.

From the flow viewpoint, a piping rupture has two distinct effects, according to whether it is an isolation valve or a check valve.

For an isolation valve, the sudden downstream pressure loss means an increase in flow rate. It would therefore be advisable for the valves concerned, to verify that it is possible to section the fluid stream

With the test valve initially open this simulation was achieved by firstly establishing the flow rate through the intermediate of another valve, then closing the test valve (figure 1). The test facility used is the CUMULUS loop (figure 2)of the DIRECTION DES ETUDES ET RECHERCHES at Renardières (near Fontainebleau at about 50 km from Paris).

For a check valve, the sudden rupture of the piping, this time in the upstream direction, means chronologically, firstly flow reduction,followed by a sudden reclosing of the valve with a simultaneous increase in pressure and an important structural load.

The test facility (ECLAIR LOOP of ETUDES AND RESEARCHES figure 3),comprizes a water reservoir maintained under pressure by a nitrogen blanket linked to the piping on which the test check valve is fitted.

This valve is kept in the fully open position using a fine wire.

The trial was started, the circuit being pressurized (120 Bar) by the provoked rupture of a membrane down-stream of the relief valve.

- As far as the results are concerned, the main lessons learnt are the following :

Parallel seated gate valve seems more sensitive to this accident than others : Loads to achieve closure are however considerable and can lead to deterioration of the gates or even incomplete closure. Particular care must be paid to the servomotor sizing and mechanical clearances.

It is advisable to take into account valve motorization dimensioning, not only for the maximum upstream pressure (with zero downstream pressure) but also for the subsequent stress brought by the fluid velocity, without under estimating the friction coefficient effect under these high pressure conditions.

The extremely rapid closure of the check valve (the order of 10 milliseconds) and the sudden corresponding stoppage of flow, creates very important pressure increases in the piping which can reach five times the initial pressure.

One is then faced with the problem of the resistance of the downstream piping to these pressures, as well as that of the supports to the shearing effect initiated by the stoppage of the moving fluid mass. On the contrary, no difficulty on the check valves themselves has been observed.

3.2.2.2 - Operability under valve ends loading

A thermal expansion, an earthquake, generate a rupture in the piping and then mechanical stresses at the extremities of the installed valves under which certain materials must be able to function and maintain a good tightness.

This verification has been made on a test facility installed at the COMMISSARIAT A L'ENERGIE ATOMIQUE at Saclay and developed into the framework of a tripartite action COMMISSARIAT A L'ENERGIE ATOMIQUE/ELECTRICITE DE FRANCE/FRAMATOME (figure 4).

This is capable, either independantly or simultaneously, of producing the following loads at the valve end connections :

- Thrust load : traction of 300 kN (700 kN without bending)

- Bending moment : ± 400 kN.

- Torque : ± 400 kN.

It also permits the transmission of a longitudinal static load to the superstructure's centre of gravity (representing stress generated by an earthquake) of 150 kN.

Finally the internal pressure of the valves can reach 250 bar.

Valve diameters can vary between 50 and 500 mm.

Obviously, associated with the application of the stresses are measurements of deformation, constraints, displacement and leakage.

The stresses transmitted are regrouped at two levels :

- In a first phase one verifies the functioning of the valve in pressure at the minimum characteristic of its motorization, during the application of stresses corresponding to a stress equal to the elastic limit in the periphery of the piping ;

- In a second phase, and as a research and development exercise, stresses are brought to a level corresponding to the total plastification of the piping section.

The results obtained to date, show the very good behaviour of the valves. No opening/closing refusal has been noted and the effect of stresses on the leak level is negligable, except for spring operated valves.

3.2.2.3 Seismic Test

The motorized valves can be susceptible to an earthquake, on account of the eccentric mass constituting the motorization.

But the verification of their integrity and operability during this accident was preceded by vibratory aging trials which were divided into three successive phases :

- Initial resonance frequency research,
- Endurance test by frequency scanning,
- Endurance test on resonance frequencies.

Between each of these trials, the valve was subjected to operability trials.

The rectilinear and sinusoidal vibrations were generated by an electrodynamic exciter of a private laboratory (SOPEMEA) ; the trials were conducted successively on the three reference axes for each of the phases. The valve is rigidly fixed onto the test bench and fitted with accelerometers placed at sensitive points of its structure (figure 5).

Initial research of the resonance frequencies explores the band from 1 at 500 Hz in both directions with the excitation levels below :

- constant acceleration of 0.2 g between 1 and 26 Hz,

- constant displacement of 0.15 mm peak to peak between 26 and 57 Hz,

- constant acceleration of 1 g between 57 to 500 Hz.

The scan speed is constant and is one octave per minute.

The graphical restitution of the recordings enable one to determine by each measuring point :

- The resonance frequencies (Amplification coefficient more than 2),

- Amplification coefficient.

The endurance test by scanning, simulates the dynamic stresses to which the valves have been subjected, and by constituting an ageing of the apparatus, enables the weak points to be detected.

The valve is subjected during two hours to a scan in the frequency range 5 to 500 Hz with the excitation level given below :

- constant displacement of 0.15 mm amplitude between 0 and 57 Hz,
- constant acceleration of 1 g between 57 and 500 Hz,

. The endurance sequence on the resonance frequencies discovered during the first phase completes the aging previously started.

. The apparatus is then subjected to an excitation corresponding to the first three resonance frequencies, for thirty minutes.

. Finally the earthquake trial actually finishes the programme.

Two levels were envisaged successively :

- the dimensioned half-quake (DSD) corresponds to low intensity accidents or the aftershock of a major earthquake.

- the actual Safe Shutdown Earthquake (SSE).

The level of the specified response spectra attained is shown in figure 6. It takes into account on the one hand amplifications due to the floor height compared to to the ground and on the other hand the piping to which the valve is fitted.

In a 20 second duration trial, the frequency band explored goes from 0.5 to 50 Hz and the retained damping is 5%.

The hydraulic tables on which the valves are mounted enable bi-axial tests to be made.

In total, the seismic sequence appears as follows :

- first bi-axis OX/OZ or OY/OZ) :

  . five DSD level trials during which the operability of the valve under pressure is verified,

  . an SSE level trial, in which operability is only verified after the test ;

- the same trials for the second bi-axis.

. The lessons learnt from these vibration and earthquake trials are numerous ; in particular the following malfunctions were encountered :

- loosening of the boltings of the superstructures, resulting from :

  . insufficient torque load,
  . inappropriate blocking (for example by a split washer),
  . the absence of a rotational stop of the nuts or bolts.

- rupture of the copper feed pipeline of the pneumatic motorizations, mainly at the connections.

These malfunctions however, can generally be eliminated by simple techniques :

- applying a sufficient and controlled torque load by using a torque wrench,

- rotational stop of the nuts and bolts by using washers

- improving the air feed connections by :

  . changing to stainless steel piping,
  . respecting certain lay out rules for them,
  . using connections which reduce pinching of the pipes
  . stiffening the accessory supports fitted on the piping.

On the contrary, the dimensioning of the body and superstructures themselves (apart from connections) were shown to meet the functional requirements under these severe conditions.

3.2.2.4 - Thermodynamic accident trial

Following the Loss of Coolant Accident (LOCA), the environmental parameters of the reactor building were found to be considerably changed ; increase in pressure, temperature, humidity, irradiation, corrosive products,....

Moreover, such an accident can occur at the end of the lifetime of a power station and be consecutive with an earthquake. In order to simulate this life period, prior to the accident, an artificial aging was imposed on the tested valves ; the structure of the complete test sequence appears as follows :

. initial measurements and controls,
. trials at operational use limits,
. thermal aging (dry heat),
. humid heat trials,
. functional aging,
. aging irradiation,
. vibrating and seismic trials,
. accident irradiation,
. accidental and post-accidental environmental trials.

Without going into details of each test, one can as an indication, show the main parameters below :

. thermal aging test during 6 months (3800 h) at 95·C,

. functional aging relating to the validation trials mentioned above.

. aging irradiation can reach 25.10^4 Gray and accident irradiation 60.10^4 Gray ; these integral doses can be reduced as a function of the materials' implantation and the periodic replacement of sensitive elements ;

the accidental and post accidental environmental trials characterized by the steam pressure temperature profile of figure 7, to which is added a solution containing soda.

The valves were required to remain operational during and at the finish of this test program ;

Testing facilities used, besides those already mentioned of ETUDES ET RECHERCHES AND SOPEMA are those of the COMMISSARIAT A L'ENERGIE ATOMIQUE of Saclay irradiation and of Cadarache (Thermodynamic and Chemical accidents).

The main results so far obtained are the following :

- certain motorization components are susceptible to aging, which can be thermal, irradiation or functional (grease and lubricants of the electrical motorizations, membranes and seals of pneumatic motorization : they must be chosen with great care and their periodic replacement taken into account in maintenance operations. Concerning this, it has been found that Teflon behaves very badly to irradiation, and that ethylene-propylene membranes improved compared to their nitrile equivalents, vis à vis temperature and irradiation

- electrical connection tightness is a generic problem in the presence of pressurized steam, which can only be resolved by perfecting specific equipment taking into account this particular need,

- dimensioning the pneumatic servomotors must integrate the fact that abnormally elevated ambient pressure can significantly reduce, or even cancel out, the valve's functional capability.

3.3 - Retained qualification method.

Although it concerns qualification under normal, incidental or accidental conditions, in reality it is not possible to test all the valves used in the power station.

A first way to circumvent this difficulty is to qualify by testing the valves in the actual operating conditions encountered. As we saw previously such a method was used, for tests with both incidental and accidental conditions.

Such an expedient, however, is not enough to resolve all the problems.The only remaining method is therefore to qualify by analysis a family of products from a valve qualified by test, these representing a technological proximity vis à vis the test valve. Regrouping can thus be carried out on an analogy basis such as :

. valve type (butterfly, parallel seat, globe,...),

. similar diameters,

. similar operating conditions,

. similar superstructure behaviour (in the case of an earthquake qualification),

. same supplier...

It's a fact, however, that this qualification by analysis, which makes heavy demands on the engineers judgement, must be applied with considerable care in order to be valid.

4 - Manufacturing

It appeared necessary at this stage to specify a small number of valves as "sensitive", because of their importance for the safety and/or availability of the plant and on which inspections, particularly dimensional ones have been tightened up.

5 - Qualification Perenniality

Thanks to the efforts made at both the conception levels and manufacturing or test stages, the valve quality available to the exploiter has clearly improved over the last ten years.

In order to safeguard this improvement and avoid an uncontrolled drift of series manufactured materials compared to the qualified ones, a document control procedure has been set up.

Thus,"Reference dossiers" identifying all the important elements defining the valves (detailed drawings ; associated nomenclature, special manufacturing procedures, inspection and supply).They can only be changed with the clients agreement.

This vigilance is extended into the exploitation phase, during which envisaged modifications are subject to a planned procedure with the designers and manufacturers of the equipment.

6 - Power station commissioning trials

These are carried out in a systematic way without fluid. They are then retested with fluid, and under nominal conditions, during circuit trials.

7 - Maintenance

The maintenance programs and periodic trials enables one to make certain of the operability of the valves which is essential for the plant's safety.

8 - Experience feed-back

Regular meetings between designers, buyers and users allow experience feed-back to be taken into account.

The collection and filing of the most significant malfunctions are achieved with the help of a card index.

Amongst other things, this enables one to assign the part of the power stations non-functioning caused by by the valve system, compared to other causes.

9 - Conclusions

The introduction of nuclear technology has represented a challenge to the valve industry and operators of the electric power stations ; some specific problems present themselves: it is obvious that a steam leak in a coal fired power station does not have the same consequence as its equivalent in a radioactive circuit; a manoeuverability failure can in the same way initiate or aggravate an incident, a priori, minor : This was the case with the Three Miles Island power station in the USA in March 1979; likewise highly improbable accidental scenarios had to be integrated: earthquake, primary piping rupture, ...

We can say today, that the challenge has been largely taken up. ELECTRICITE DE FRANCE, who today benefits from a unique experience of about 200 reactor years in PWR technology, possesses a data base sufficiently large and reliable to quantify this fact (figure 8): the number of outage days due to the valves, dropped from more than fifteen days per year and per plant in 1979, to an average of 2.5 over the last five years.

The Reliability Data's Reception System (SRDF), which enables the defect level of the valves to be made, appears to follow a similar evolution.

Here again, the role assumed by ELECTRICITE DE FRANCE as main contracter, from conception to final product utilization is an important factor in assuming the quality and performance of this material.

Prudence, remains however indispensable : to maintain in an operating state, in all the French nuclear power stations, several tens of thousands of valves under normal conditions, and under particularly severe but hypothetical conditions a few hundred of them, remains a performance to be repeated continuously, and demands from the side of all the intermediates, a constant vigilance. Nothing is more harmful in this regard than "habit" which leads to imperceptible but perhaps fatal mistakes. Today, the development of a "Safety Culture" seems in the end, the only way to guarantee the required results.

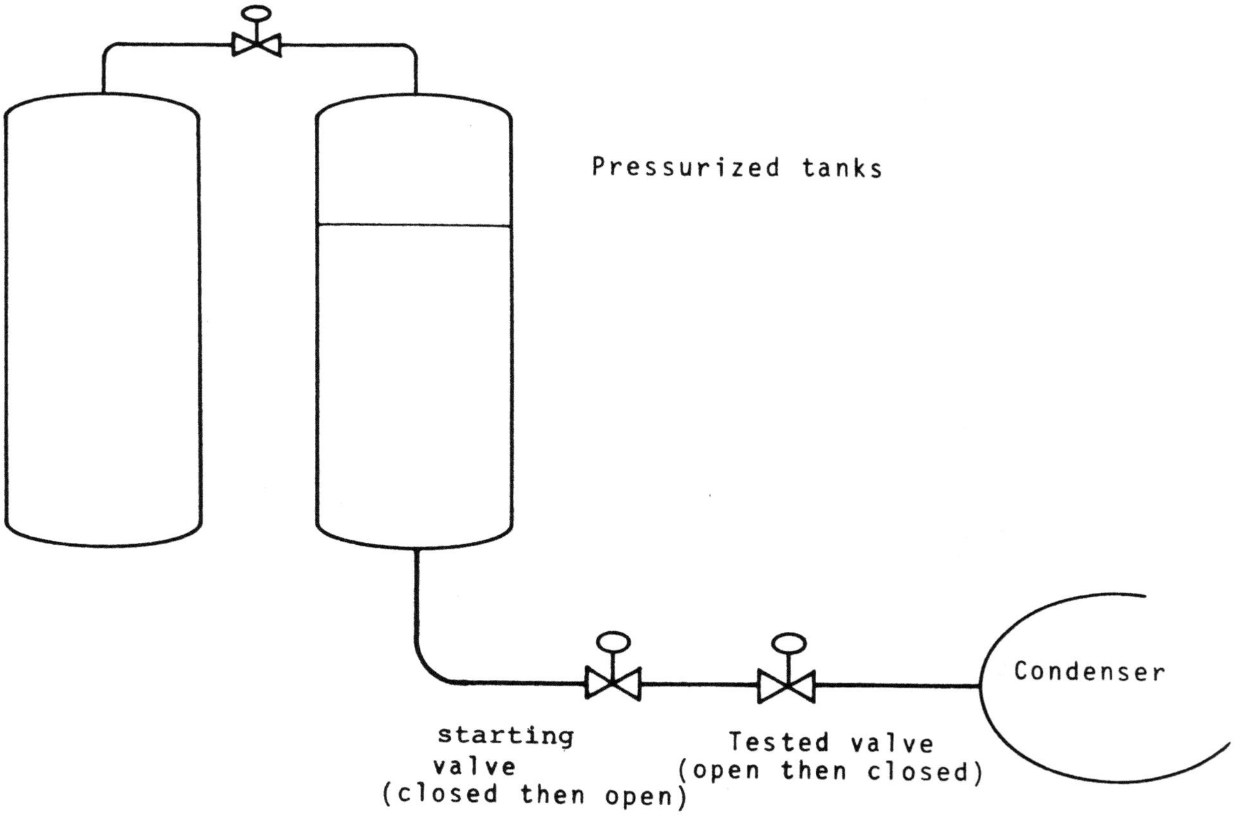

FIGURE 1

HIGH POWER PIPE BURST VALVES TEST

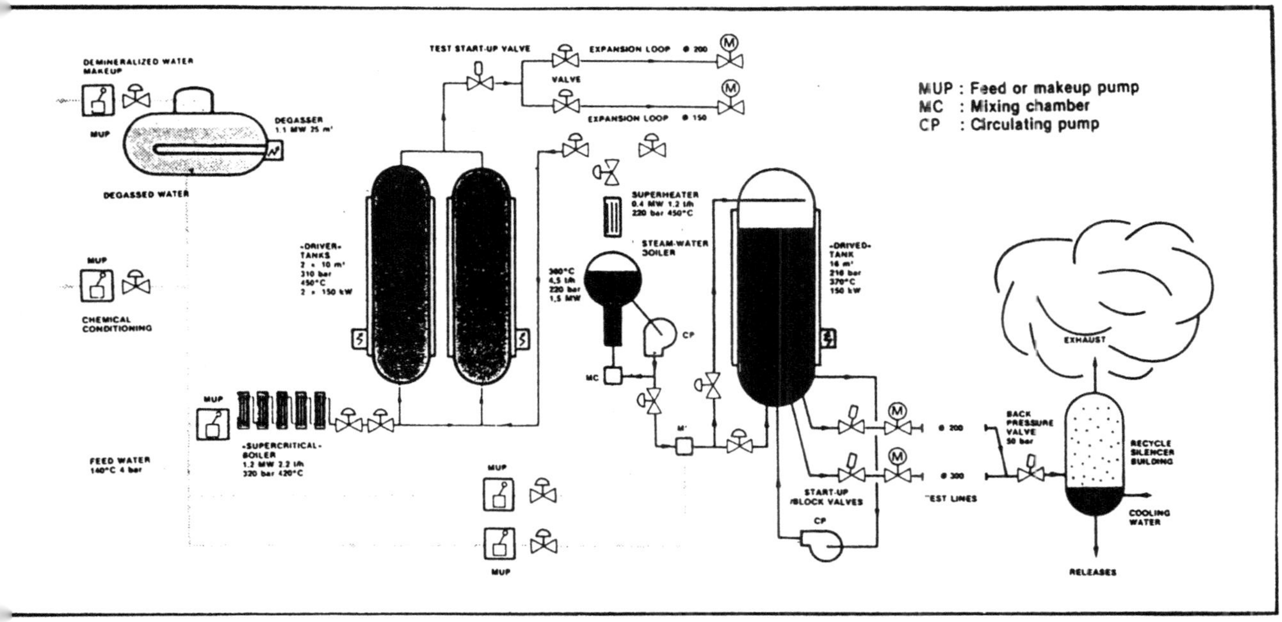

<u>Possible tests</u> :

CUMULUS loop provides, for example :

- a flowrate of 250 kg/s at 190 bar for 20 seconds in water,
- a flowrate of 180 kg/s at 100 bar for 5 seconds in steam.

This test facility is largely open to designers and manufacturers.

<u>FIGURE 2</u>

<u>THE CUMULUS LOOP</u>

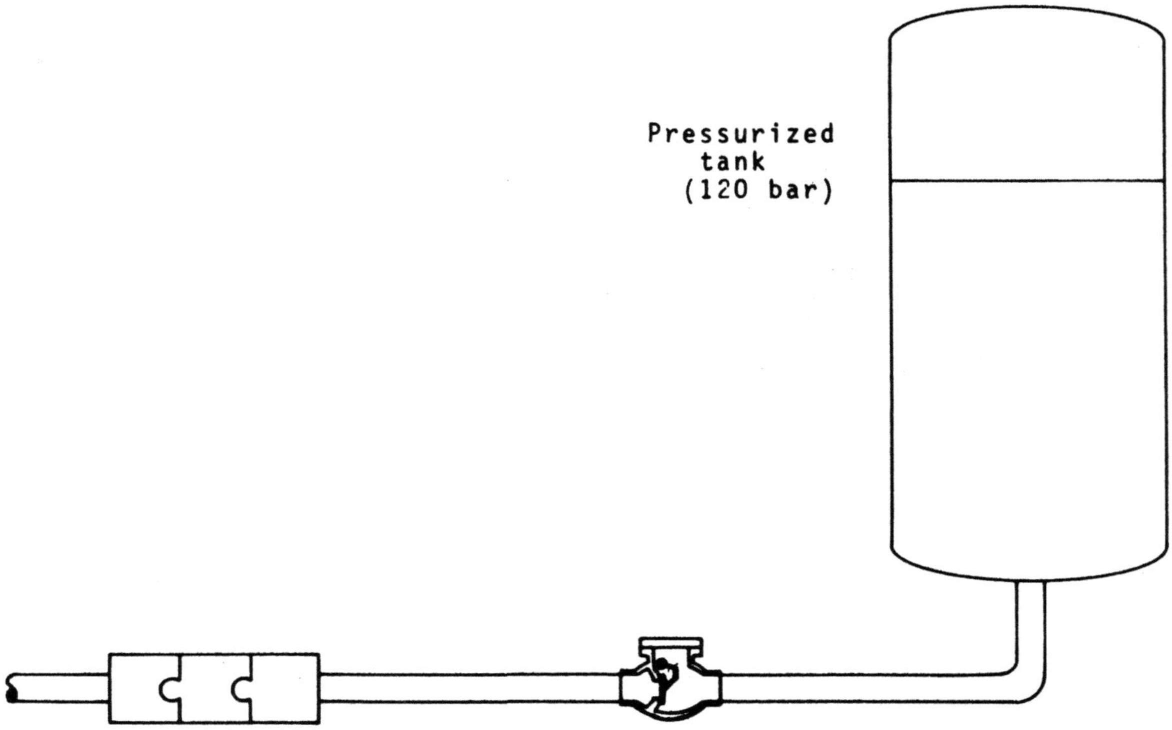

FIGURE 3

THE ECLAIR LOOP

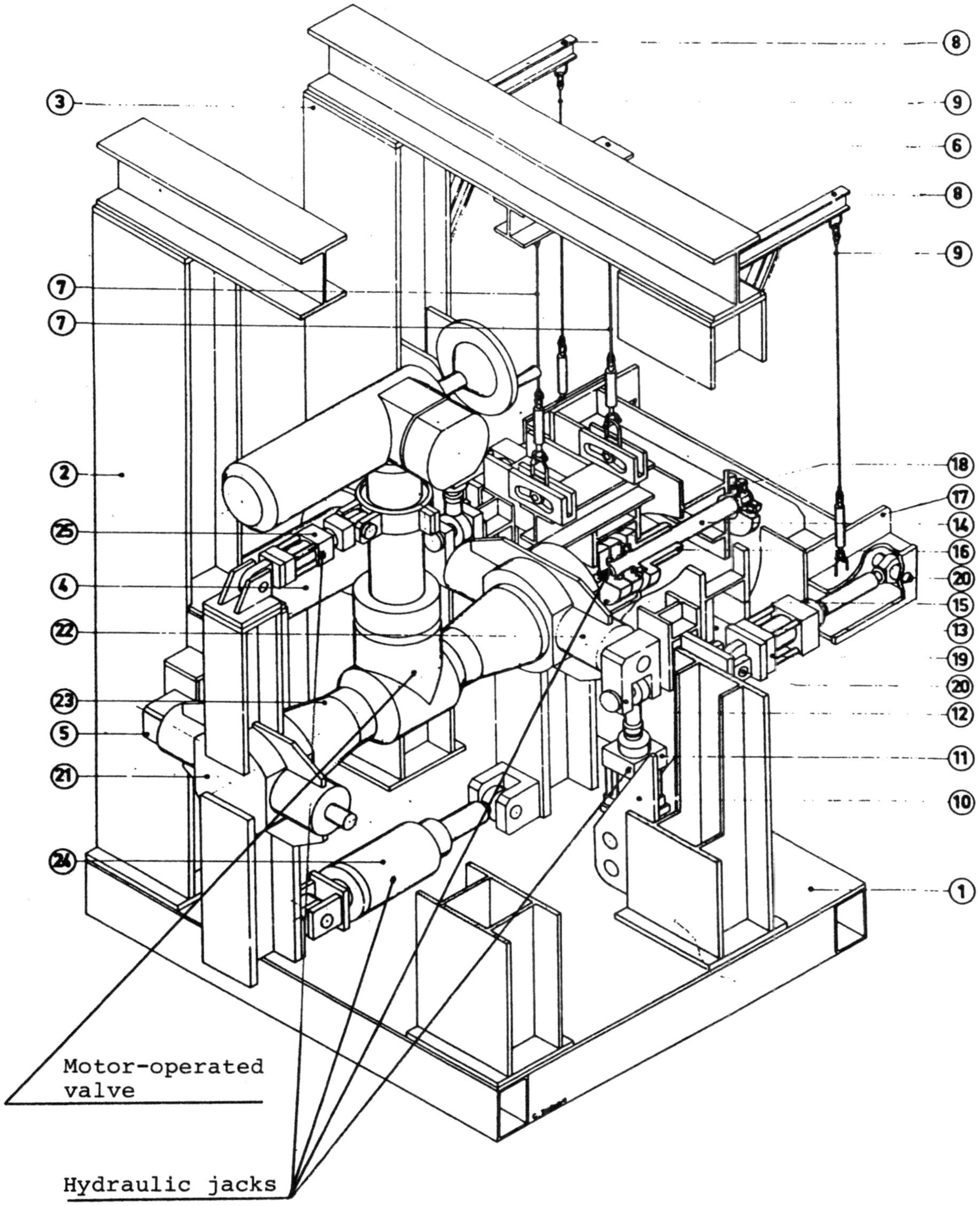

FIGURE 4

THE ESOPE FACILITY

FIGURE 5

SEISMIC TEST AT SOPEMEA

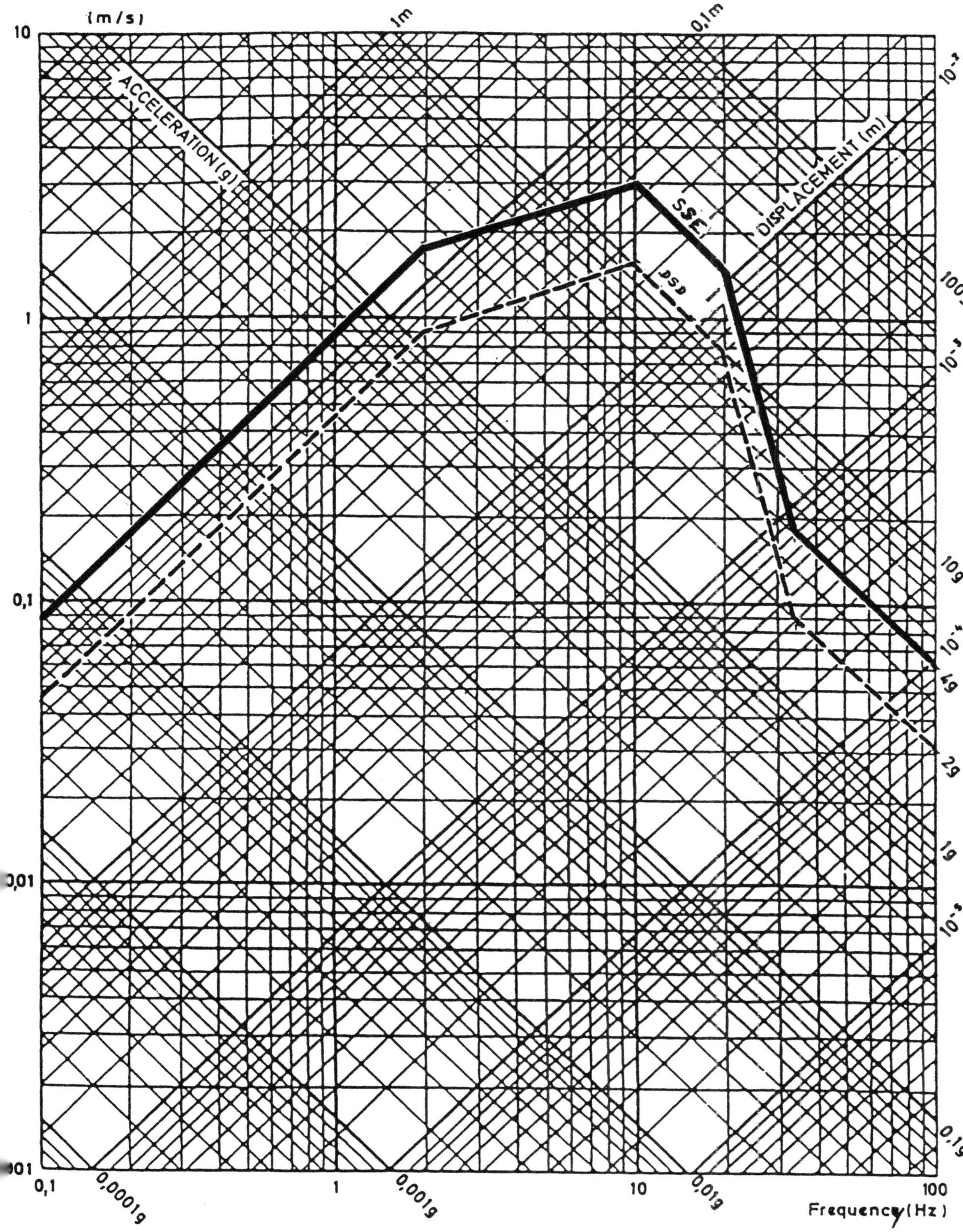

FIGURE 6

SPECIFIED RESPONSE SPECTRUM

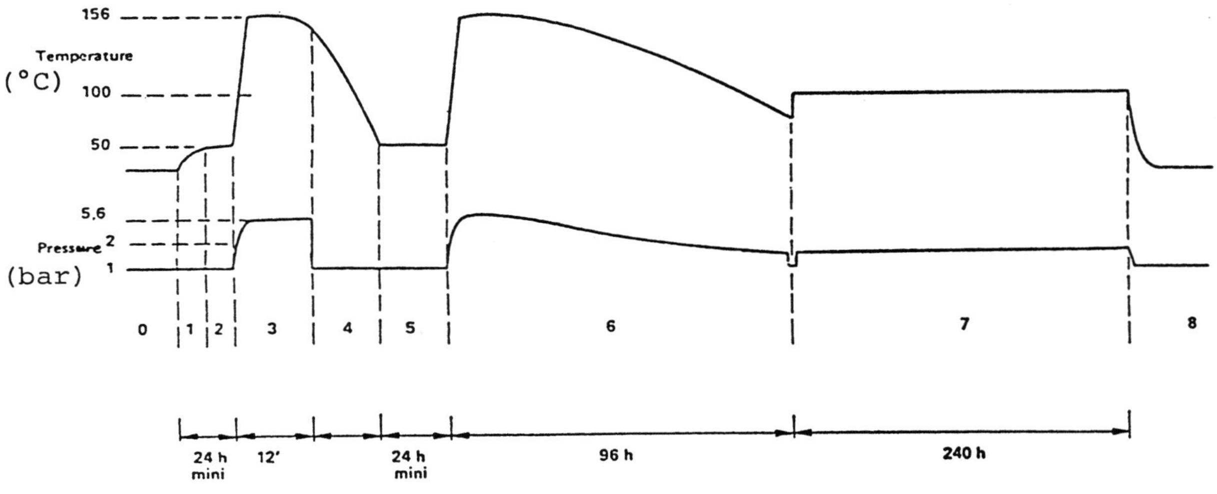

FIGURE 7

PRESSURE/TEMPERATURE CURVE ACCIDENT

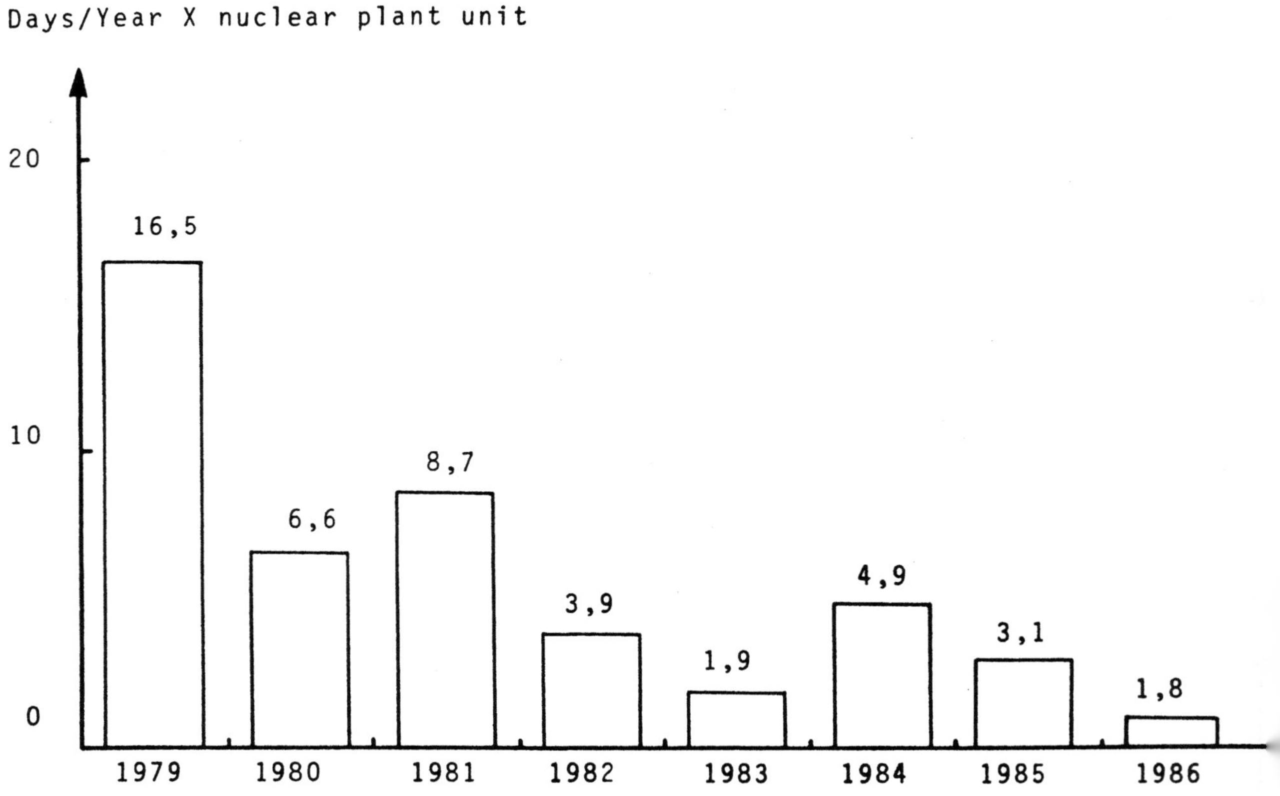

FIGURE 8

UNAVAILABILITY DUE TO VALVES

2nd International Conference on

Developments in Valves and Actuators for Fluid Control

Manchester, England: 28-30 March 1988

PAPER J2

LUBRICATION AND WEAR OF VALVE HEADGEAR - A CASE STUDY

D Aldham and T D Johnson
Central Electricity Generating Board
Operational Engineering Division (Midlands Area)
Scientific and Technical Branch
Bedminster Down
Bristol BS13 8AN

Summary

CEGB plant is normally fitted with parallel slide valves to isolate the turbines from the high pressure steam produced by the boiler. When new, the plant is generally on load continuously and so valve operations are infrequent. Later in life, valve duty is more onerous with several operating cycles per day. The valves are typically 5" venturi bore designed for operation at up to 18 MPa pressure with 13.5 MPa differential.

In recent years a number of failures of valve headgear have been experienced; a common feature being the wear and subsequent shearing of the aluminium bronze drive screw threads. Problems have also been encountered with the thrust bearing assemblies.

Details of these failures and mechanisms involved are given. In order to give guidelines for future operation of this type of valve a programme of wear and friction testing of the relevant materials combinations was undertaken.

This has led to revised guidelines for valve actuator selection and for lubrication requirements and has given confidence in a 3 year major maintenance interval.

1. Introduction

The work described in this paper was initiated after a parallel slide boiler stop valve failed to close under fault conditions. At the time of the fault the 500 Mw(e) fossil fuelled unit was generating at 485 Mw(e) and the

Held in Manchester, England. Organised and sponsored by BHRA, The Fluid Engineering Centre and co-sponsored by the British Valve Manufacturers Association and the Institution of Mechanical Engineers Process Industries Division

differential steam pressure across the valve was estimated to be of the order of the maximum design pressure of 13.5 MPa with a steam temperature of 540°C. Attempts to close the valve manually were unsuccessful, for although the drive shaft could be rotated this was not translated into movement of the valve stem (see fig 1). Removal of the valve stem cover from the top of the gearbox revealed that the drive nut thread had stripped (see fig 2a) but it was not clear if there was any other damage. The valve was subsequently removed intact from the steam line and taken to the laboratory for controlled dismantling and a detailed examination.

2. Examination

The general condition of the valve internals was good. There were no indications that unusually high forces had been generated around the valve throat that could cause the drive thread to fail. Other potential high load generating interfaces (eg valve stem gland, stem stop/pillar interaction) were similarly in good order. The reduction gearbox complete with the drive nut and its sheared thread were wound from the stem thread. Only a very thin smear layer of grease was present on the stem thread and this was accompanied by quantities of what appeared to be drive nut thread wear debris, (see fig 2b).

The stem thread (material EN57; thread; $1^1/2$ in, 4 tpi class 3G ACME) was in poor condition, showing extensive adhesive transfer of aluminium bronze from the gearbox drive nut to the upper thread face (see fig 3a). Measurements of the thread profile clearly shows the effect of this transfer (see fig 3b) for an unworn thread is symmetrical about the centre of each tooth (see fig 3c), but the teeth in fig 3b have a raised (~0.1mm) and rounded lip on the upper face crests corresponding to the region of pick-up.

An etched section of the damaged thread is shown in fig 4 where it can be seen that the thread corner face has undergone plastic flow and the transferred aluminium bronze from the drive nut can be seen adhesively attached to the flowed surface. In contrast, the opposite corner of the thread is in good condition.

The reduction gearbox was in excellent condition, but the drive nut thrust bearings (see figs 5a & 5b) were found to be almost devoid of lubricant. The thrust races (nominally 92.5 mm OD) and rolling elements were badly pitted and worn, as well as being generally corroded. So poor was their condition that a nominally stationary thrust race had rotated against the drive nut shoulder (fig 5c).

The aluminium bronze (BS 1400 AB2) drive nut (same thread form to stem) was sectioned axially to enable the remaining thread to be examined, fig 6a shows one of the halves. It can be seen that the majority of the thread has been worn away from the valve throat side of the thread prior to it failing at the root. The thread failed over its complete length with the detached portion remaining intact (see fig 6b), although the remainder of the teeth were severely deformed in a direction away from the valve throat. Measurement of the thread profile shows that the tooth root section had been reduced by wear on the closing side of the thread from an initial 4.2 mm to ~1.0 mm before failure.

2.1 Condition of other similar valves

These findings initiated inspections of the head gear on a further 11 identical boiler stop valves. Varying degrees of lubrication and wear of both the drive nut and the thrust bearings were found. Fig 6c shows a section of the most worn nut where the tooth root section has been reduced by 2.2 mm and the thread deformed by ~6° in a direction away from the valve throat. This nut is clearly showing incipient signs of the same failure mode as described above. The associated, and similar in terms of design and materials, but far smaller ($^3/4$" nominal bore), bypass valves by the same manufacture were also examined and their headgear found to be wearing in the same manner.

2.2 Service Lives

This design of valve had suffered from a history of stem bending problems and improved stem designs had been progressively introduced. The failed valve had been uprated ~1 year before the incident but the drive nut had not been replaced nor any fresh grease applied, other than an assembly lubricant, during the rebuild. Its service history prior to refurbishment is unknown, but since then the valve was cycled ~118 times at a differential pressure of 0.7 MPa before failing under fault conditions. The 3 other valves associated with this unit, all of which were believed to have been similarly refurbished and subjected to identical duty, were examined and 2 of the drive nuts found to be ~50% worn, whilst the third was only ~12% worn (suggesting that additional maintenance had been carried out).

Valves on a further 2 units were also examined; one set of valves (4) had new unworn components fitted whereas the other set had drive nuts which were ~25% worn. Unfortunately a service history for the latter unit is not available, but it is thought that the valves had seen at least 120 cycles.

The bypass valves will have been subjected to the same duty as their associated main valve, but separate maintenance details were not available.

With regard to maintenance, the lubrication schedule for the valves calls for a relubrication interval of 3 months, but it is not clear what quantity of grease was added.

2.3 Previous Failures

Plant records showed that in the preceding 18 months there had been 3 failures leading to seized valves. All of these cases involved failed drive nut threads and damaged thrust bearings, but further details were not available. Similar failures had also occurred at a sister station equipped with the same installations.

3. Wear Testing

These failures of valve headgear led to three main requirements for information. Firstly, what lubrication was required to maintain low wear and friction, secondly, what lifetimes and refurbishment intervals would then be satisfactory and finally what coefficients of friction and hence drive nut efficiencies could be expected under the different lubrication regimes? In the time available, it was not possible to model the geometry but an existing rig was modified to simulate the sliding velocities and nominal contact pressure etc. The rig comprised a reciprocating carriage on which flat aluminium bronze specimens (simulating the drive nut) were mounted. This was driven, via a load cell, by a hydraulic actuator. There are two opposed hydraulically loaded stationary specimens of 12.5 x 12.5 mm simulating the valve stem. These were mounted in spherical seats so as to allow good alignment, see fig 7.

Tests were conducted both unlubricated and with two nominal types of lubrication: designated trace and good lubrication. From the operating conditions experienced by the valves a clay based mineral oil grease containing MoS_2 has been selected for plant use as this had a higher temperature limit than lithium soap based grease (150°C c.f. 120°C). This was used for the majority of the tests but a few comparative tests with the Li based grease and with antiscuffing spray were also undertaken. For the 'well lubricated' tests a rubber shield - which wiped the bronze specimen - was installed around the small specimen holders; these were packed with grease.

For the 'trace lubrication' ~0.5 g of grease was smeared over each large specimen, most of which was displaced during the first few test cycles. The ranges of test parameters studied is shown in Table 1, along with 'typical' plant conditions.

Prior to testing, specimens were degreased and weighed, installed in the rig and degreased again prior to any chosen lubrication conditions. Different sequences of experiments involved changing one variable (load, distance, lubrication) whilst maintaining the others at nominal values. Similar loads and lubrication conditions were always employed for both specimen pairs as this enabled a mean friction coefficient to be determined. In practice, the results indicate that the geometry of the lower specimen pair led to less full lubrication than for the upper pair.

3.1 Results

A total of 32 experiments were carried out. In all cases the aluminium bronze wore more than the steel, which often showed a small weight gain, due to material transfer. Wear rates were calculated from the weight changes assuming a steady progression of wear by utilising the relationship

$V = KLS$ where V is the wear volume (m^3), L the load (N), S the sliding distance (m) and K the 'specific wear rate' in units of m^3/Nm.

Examination of the results showed that there was no strong dependence of either load or sliding distance on the friction or wear rates. The results have therefore been grouped by lubrication type and are summarised in Table 2. The standard deviations refer to the individual specimen wear rates and to the two specimen pair mean (steady) coefficients of friction respectively. Friction coefficients often started at lower values but reached steady state early in each test. They may be biased a little high as the lower specimen pair had less satisfactory lubrication (and therefore an anticipated higher friction) than the upper pair.

The tests with Li based grease showed similar behaviour to the main series tests. Antiscuffing spray (a MoS_2 spray in an organic binder) was utilised in two tests, one with and one without lubrication. In both cases its effect was minimal, the wear and steady state friction being within the scatter band of the control tests. The lubricated wear values are very similar to those found by Sullivan and Wong (1985) for aluminium bronze on steel with boundary lubrication by aviation kerosene.

The surface and subsurface appearance of the bronze specimens was investigated. There was an increase in final roughness of the aluminium bronze as the lubrication quality deteriorated. This is shown in fig 8a, b and c and average roughness (Ra) values range from 1 μm for the well lubricated specimens, 4 μm for trace lubrication and 6 μm for no lubrication, compared with an original roughness of 0.7 μm.

There is thus a wear/friction/surface roughening correlation based on average values from a number of tests (there is considerable scatter) as shown by Rabinowicz (1980) This may well have benefit in plant applications as surface roughening is fairly readily observable and will indicate inadequate lubrication.

The surface appearance of the steel specimens showed some transfer of bronze, with more transfer in the absence of lubrication. This is shown in fig 9a, b and c.

Single tests were therefore made in which aluminium bronze was worn against itself under nominally similar conditions. The results, Table 3, show very similar wear and friction in the absence of lubrication and somewhat less wear and/or lower friction when lubricated.

A section through a worn, unlubricated bronze specimen shown in fig 10a shows significant subsurface plastic deformation to a depth of $\sim$80 μm. Similar but less pronounced effects were seen in the lubricated specimen, fig 10b. The corresponding steel counterfaces are shown in figs 10c and 10d.

The wear rates found in this study may be compared with the predictions made by Rabinowicz (1980) who has produced tables of wear coefficients for various materials' combinations and lubrication conditions. The steel and the bronze are considered partially compatible whilst, for bronze on bronze, the 'identical' materials tabulation was used. The derived wear rates are shown in Tables 2 and 3. There is very good agreement for the steel/bronze combination whilst the Rabinowicz method predicts more wear of bronze on bronze than was observed. However, as stated above, the bronze transfer onto the steel resulted in bronze on bronze at the sliding interface, in both series of experiments. The agreement with theory may therefore be considered somewhat fortuitous.

4. Comparison of Plant and Rig Wear

Transfer of aluminium bronze onto the steel is observed both in plant and in the rig tests. Similarly, gross wear of the bronze, with underlying surface disruption is found in both situations. However, no transfer is observed for well lubricated plant components or on the steel thread face loaded during valve opening (for which loads are expected to be lower). The explanation may well be that the loads actually experienced in plant are lower than those used in the tests, due to a lower than design (worse case) pressure differential across the valve.

Alternatively, 'good lubrication' may be better than achieved in the tests. Overall, therefore, the results tend to pessimism and so conclusions based on the rig tests will be conservative.

5. Discussion

The rig tests show coefficient of friction values rising from a mean of 0.13 with 'good' lubrication, to 0.28 in the absence of lubricant. Dependent on the degree of security required in a particular application there are therefore two options. One is to size the actuator etc to allow for dry running (although a lubrication regime is required in order to keep the probability of this to a minimum). The alternative is to size for good lubrication conditions, for which $\mu\sim0.13$ is reasonable. This is clearly the cheaper option but the plant specifier should be aware of the risk of the valve failing to close following a loss of lubrication caused, for example, by a gland leak.

For the fault case described in this paper the thrust bearing was also badly lubricated to the extent that the race slid against the bronze drive nut. A friction coefficient similar to that for unlubricated bronze stem material is therefore appropriate. The method given in BS 3580 has been used to calculate the combined effects of high frictional losses. For the actuator and gearbox fitted to these valves this leads to the limiting conditions for valve operation shown in fig 11. This shows that, for normal valve differential pressures (~0.7 MPa) the valve would close even with high friction at both interfaces. However, it would fail to close, as observed, against the design differential (13.5 MPa). This explains the absence of any warning prior to the incident.

The wear rate figures show that, as expected, aluminium bronze is a comparatively wear resistant material but differences in wear rate of ~ x20 have been found. Calculation (see Appendix) of the likely operational use of the valves shows that the high wear rate will lead to the types of problems described in this paper within a 3 year maintenance period.

The relevant valves were therefore inspected and a number of partially worn drive nuts were replaced. The valves were then lubricated with the clay based grease prior to use.

The test runs, at sliding distances of up to 1000m show that lubrication intervals can be lengthy provided that there is no process for actively removing grease (eg steam leak) or for grease contamination by hard debris. A decision was made initially to relubricate monthly but to consider 3 monthly intervals. Relubrication was recommended immediately if gland leaks were observed.

Since the programme was initiated ~ $2^1/_2$ years ago there has been only one failure. This was due to a large steam leak causing overheating and subsequent seizure of a drive nut onto a valve stem.

Finally, however it should be recalled that a contributory part of the initial failure was due to frictional resistance of the badly lubricated thrust bearings. These were inaccessible for relubrication without dismantling the gearbox, an alternative means of lubrication has now been adopted. Thus it is important to adopt a systems approach to studies such as this in order to ensure that all wear and friction locations are identified and assessed.

6. Acknowledgement

The assistance of several colleagues in particular N Jones and G Orchard is gratefully acknowledged. This paper is published by permission of the Central Electricity Generating Board.

APPENDIX

CALCULATION OF WEAR RATES AND DRIVE NUT LIVES

In a 3 year maintenance period, with daily operation, there will be ~1000 valve closures each of ~3.6 m sliding distance. For each closure the axial force on the valve stem will increase as the valve closes due to differential steam pressure forces but will have steady components due to the unbalanced steam pressure on the stem cross section and to the gland resistance. Overall, an average axial force of $\sim 3.8 \times 10^4$ N is reasonable. The thread area of the drive nut, resolved perpendicular to this axial force can be readily found from the nut dimensions. In this case it is $\sim 1.7 \times 10^3$ mm^2. Hence the resolved contact pressure is 22 MPa (see Table 1).

The equation in the main text can now be used in modified form:

$$\frac{V}{A} = \frac{KLS}{A}$$ where A is the resolved thread cross section area

$$\Delta x = K.P.S.$$ where Δx is the (resolved) depth of wear and P the contact pressure

For good lubrication $K \cong 1.5 \times 10^{-15}$ m^3/Nm and Δx (1000 cycles) $\cong$ 0.12 mm

Whilst for the unlubricated case $K \cong 27 \times 10^{-15}$ m^3/Nm and Δx (1000 cycles) $\cong$ 2.2 mm.

This can be compared with the 3.2 mm wear found on the failed nut and the 2.2 mm on another one.

An alternative method is to work backwards from the observed damage. The failed nut had lost ~$^3/_4$ of the thread volume ie. ~4000 mm^3. Taking load and sliding distance as above the calculated wear rate is then

$$K = \frac{4000 \times 10^{-9}}{3.8 \times 10^4 \times 3.6 \times 1000} \cong 29 \times 10^{-15} \ m^3/Nm$$

Which is in excellent agreement with the unlubricated wear rate found in the rig tests.

7. References

Rabinowicz E. 'Wear Coefficients - Metals' in Wear Control Handbook ed M B Peterson and W O Winer ASME, 1980.

Sullivan and Wong 'Wear of aluminium bronze on steel under conditions of boundary lubrication' Tribology International 18, 275, 1985.

BS 3580; 1964 'A guide to design considerations on the strength of screw threads, Appendix B'.

Table 1

Test and Plant Conditions

	TEST RIG	PLANT
Contact Pressure, MPa	20 - 110	Max thread pressure when closing depends on ΔP across valve. Can rise to 110 MPa for 13.5 MPa differential pressure
Sliding Distance, m	100 - 1000	Distance per valve closure ~ 3.6 m
Stroke (P → P) mm	30	"
Sliding Speed mms^{-1}	32 - 55	Actuator/gearbox gives 46
Temperature °C (due to frictional heating)	60 - 120°C	Normally ~ 100 - 120°C but can rise if gland leaks

Table 2

Wear Rig Test Results for Valve stem/Drive nut materials.

CONDITION	DRIVE NUT MATERIAL WEAR RATE ($\times10^{-15}$ m^3/Nm)		COEFFICIENT OF FRICTION		PREDICTED WEAR RATE ($\times10^{-15} m^3$/Nm)
	Mean Value	Standard Deviation	Mean Value	Standard Deviation	
Good Lubrication	1.5*	1.2*	0.13	0.02	1
Trace Lubrication	16	11	0.17	0.04	10
No Lubrication	27	14	0.28	0.02	50

* Upper specimens only
Mean for lower specimens 4; std deviation 2.6

Table 3

Wear Rig Test Results (Single Tests) for Drive Nut Material vs Self

CONDITION	WEAR RATE ($\times10^{-15}$ m^3/Nm)	COEFFICIENT OF FRICTION	PREDICTED WEAR RATE ($\times10$ $^{-15}m^3$/Nm
Good Lubrication	0.35	0.075	15
Trace Lubrication	3	0.16	150
No Lubrication	20	0.29	750

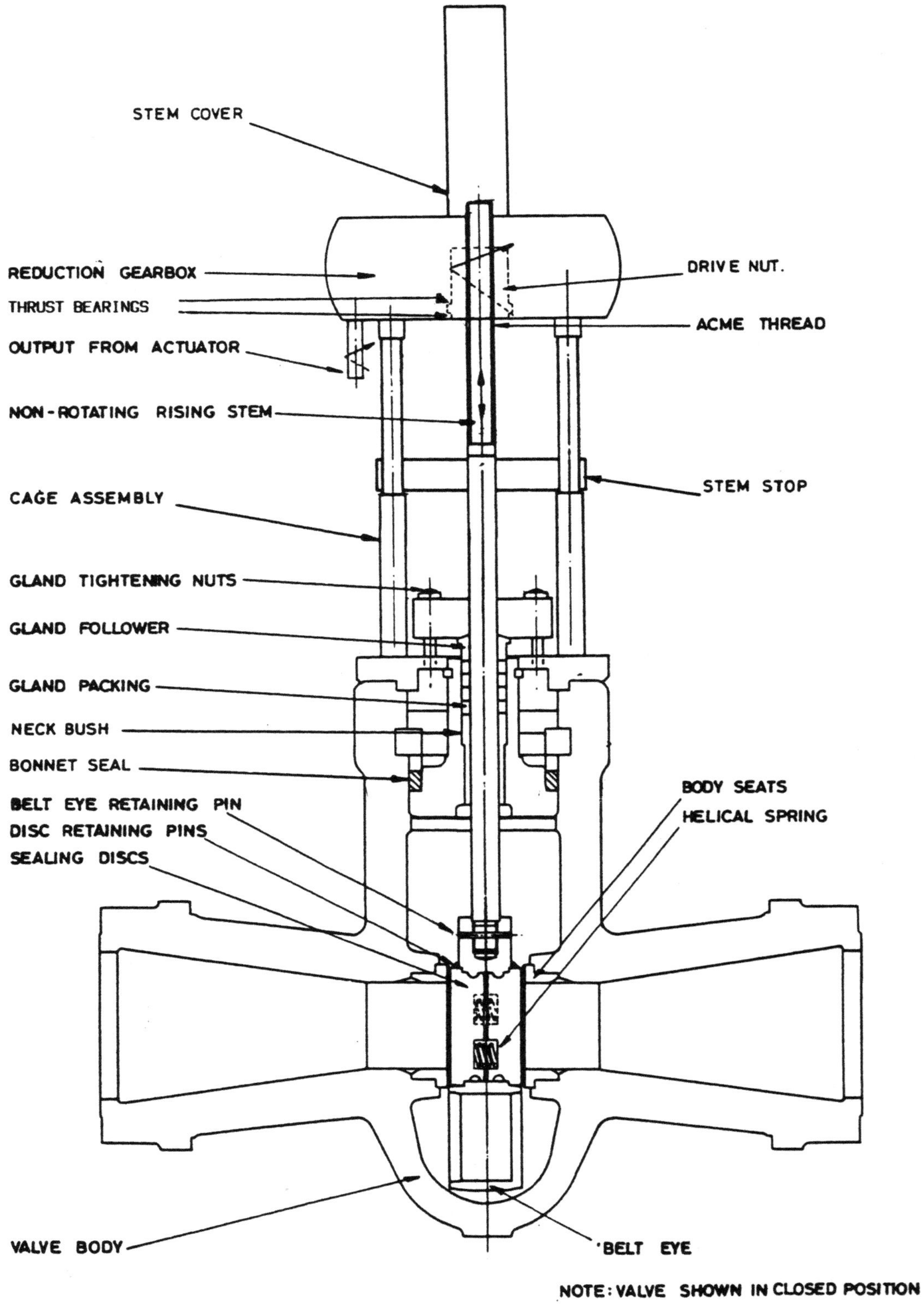

Figure 1: General arrangement of valve

a

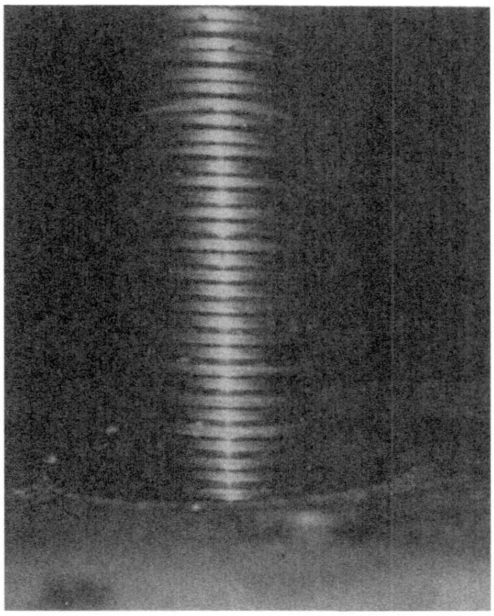

b

Figure 2: a) Stripped drive nut thread, b) Stem thread showing drive nut debris

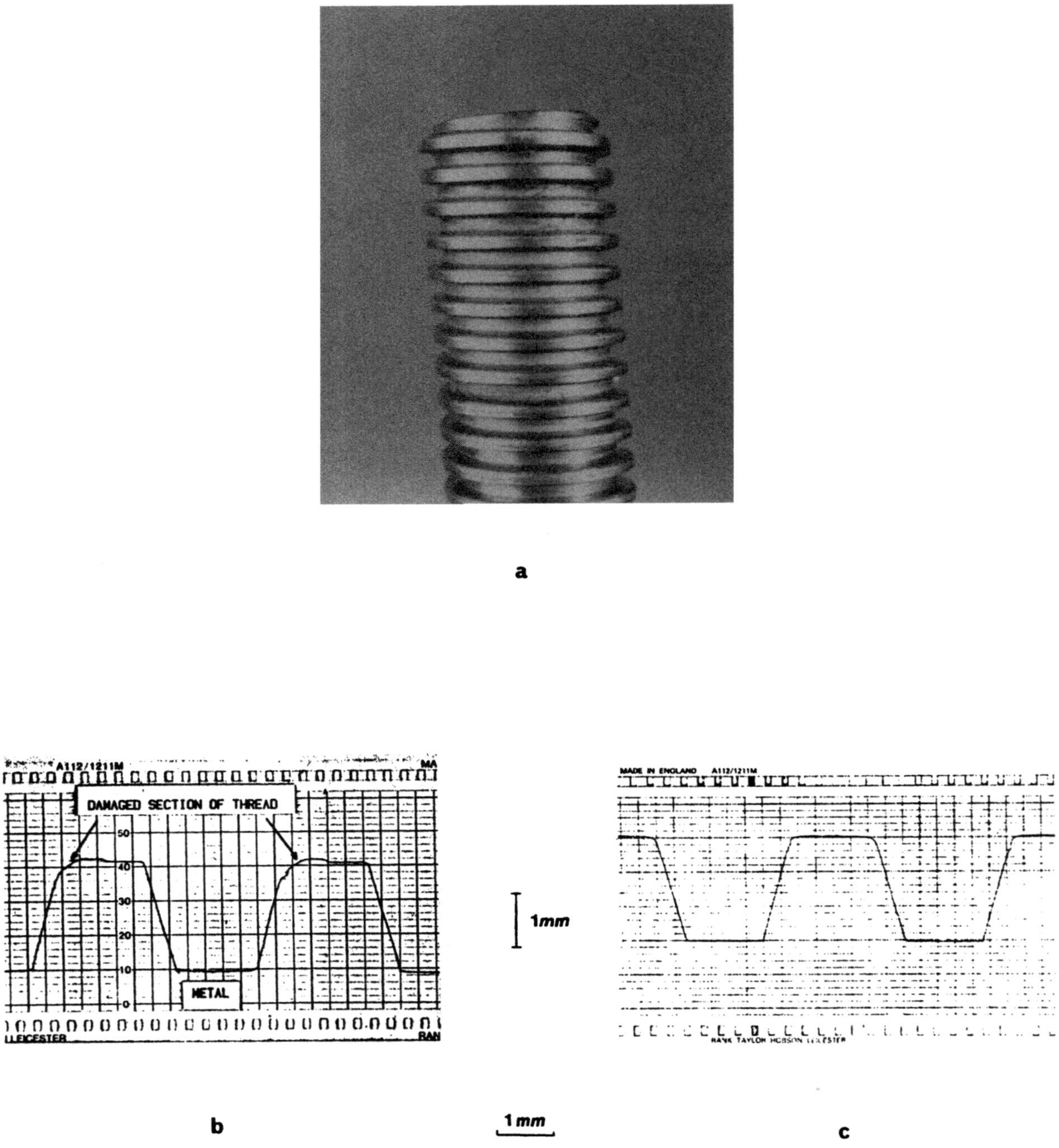

Figure 3: a) Stem thread showing transfer, b) Thread profile of (a), c) Unworn thread profile

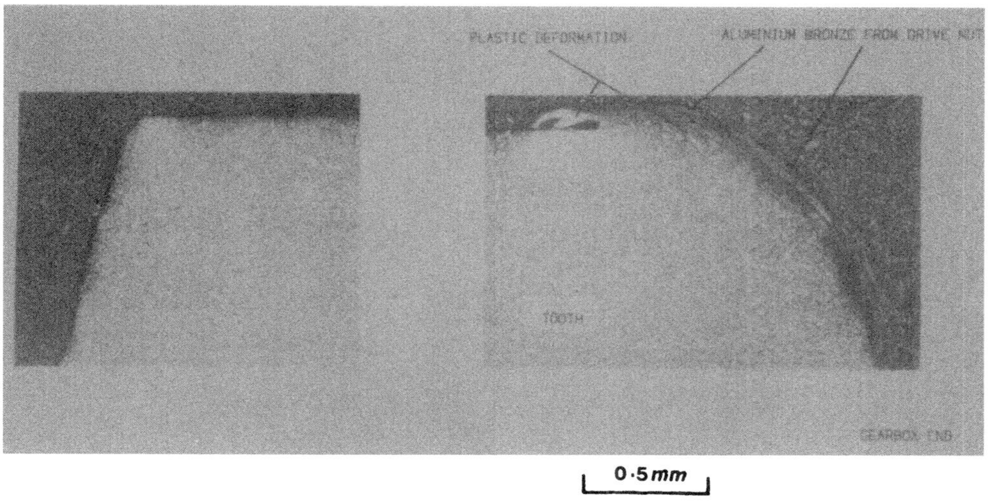

Figure 4: Sections through stem thread

a

b

c

Figure 5: a) and b) Thrust bearings, c) Wear of bronze counterface

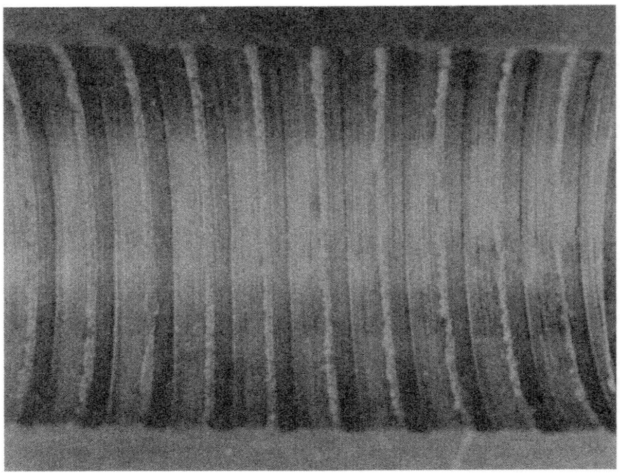

a

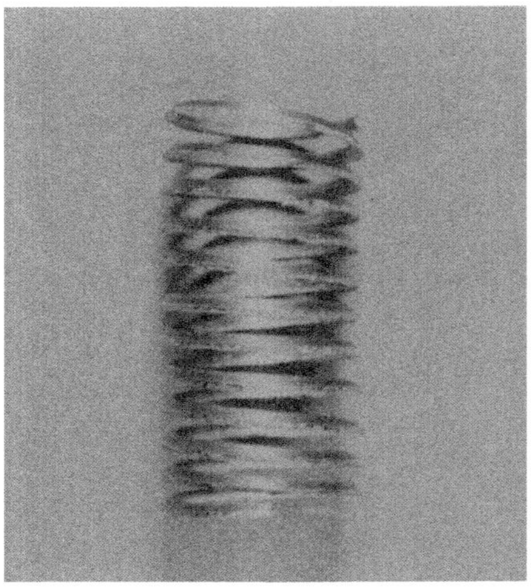

b

Figure 6: a) Failed drive nut, b) Failed drive nut thread, b) Worn drive nut

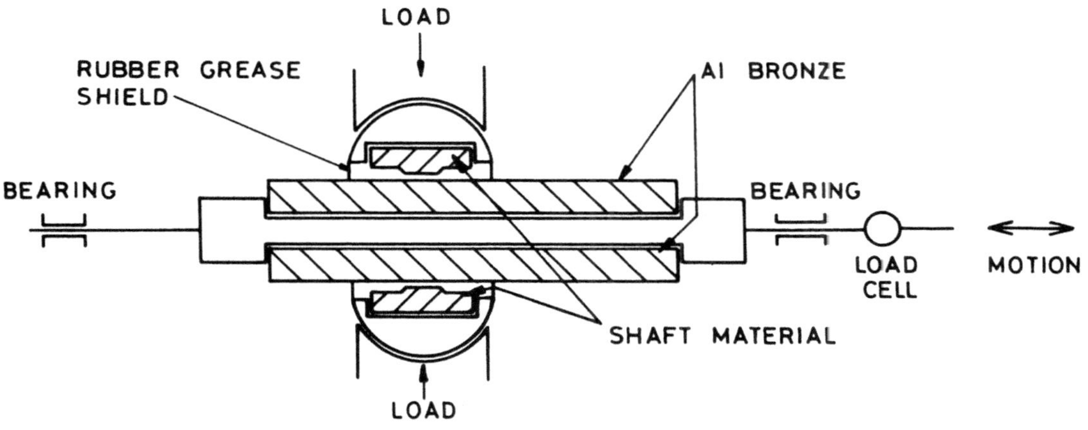

Figure 7: Sliding wear/friction rig

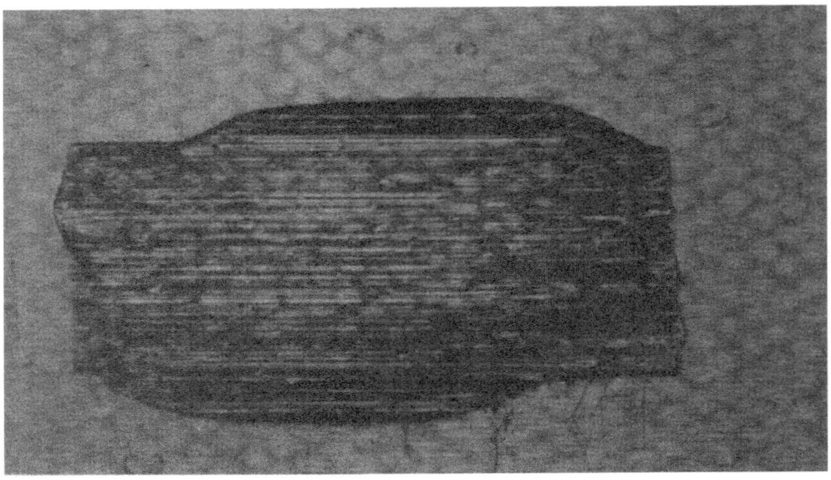

a

b

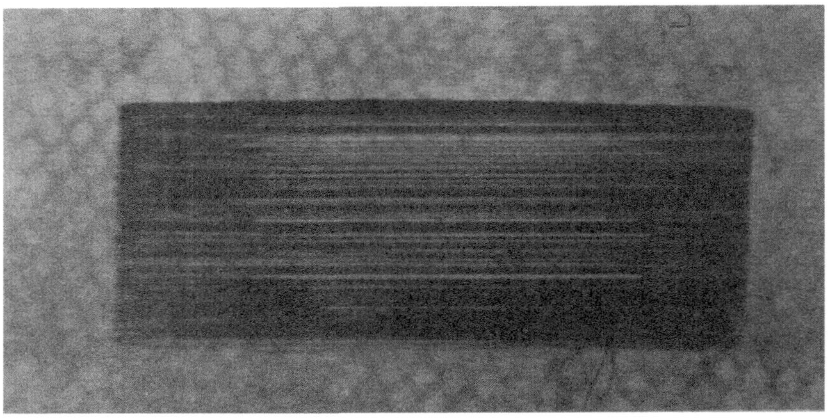

c

Figure 8: Aluminium bronze wear scars, a) Unlubricated, b) Trace lubrication c) Good lubrication

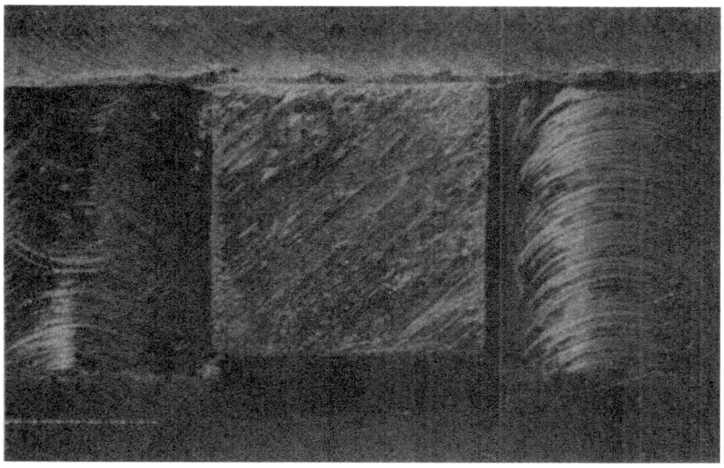

a

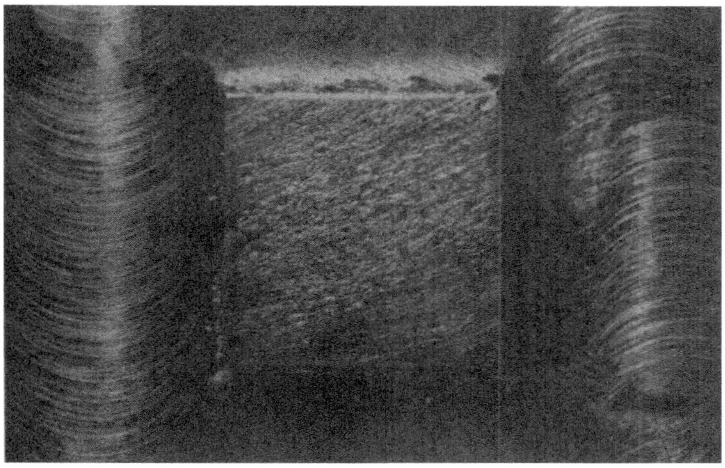

b

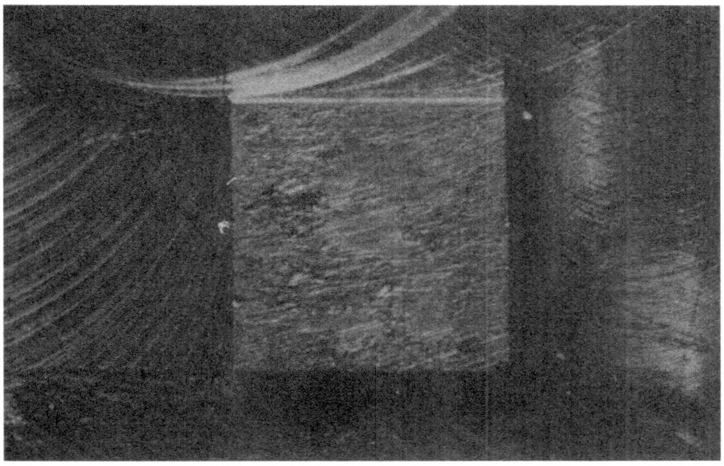

c

Figure 9: Stem material wear surfaces, a) Unlubricated, b) Trace lubrication c) Good lubrication

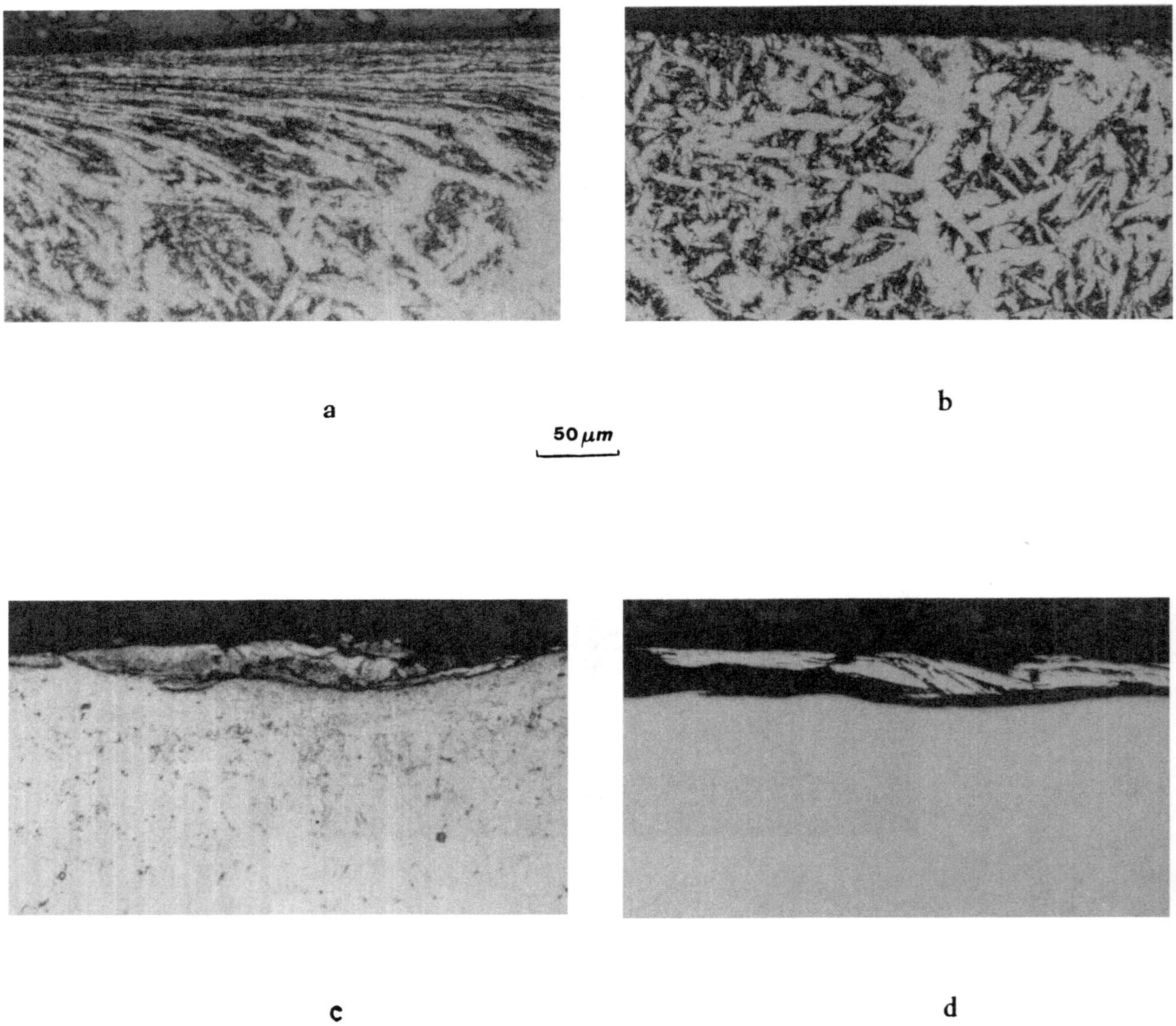

Figure 10: Sections through wear specimens, a) Bronze, unlubricated, b) Bronze, lubricated, c) Steel, unlubricated, d) Steel, lubricated

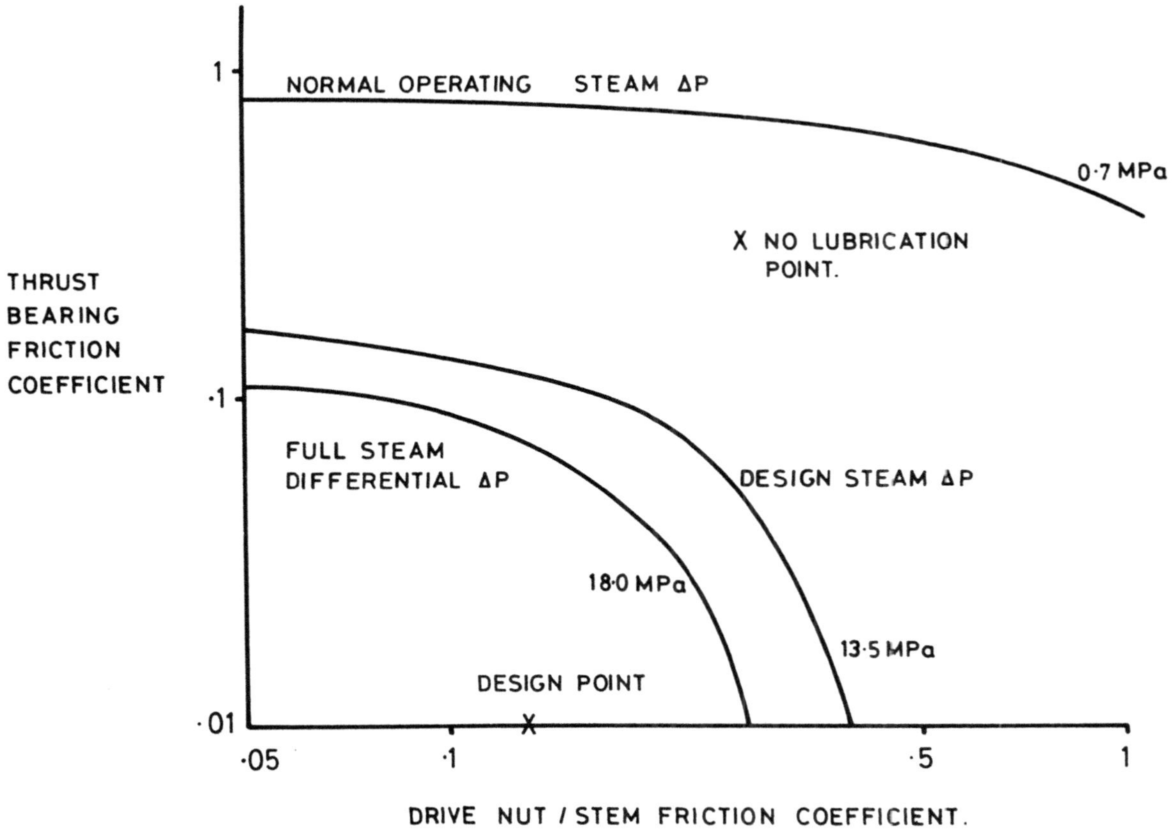

Figure 11: Limiting friction coefficients for valve to operate.

2nd International Conference on

Developments in Valves and Actuators for Fluid Control

Manchester, England: 28-30 March 1988

PAPER A2

TITLE: BELLOWS SEALS FOR ROTARY (1/4 TURN) VALVES

AUTHOR: RICHARD W. CONLEY, P.E.
KEROTEST MANUFACTURING CORPORATION
PITTSBURGH, PENNSYLVANIA USA

SUMMARY:

This paper describes the search for and successful development of an hermetic (zero leakage) bellows stem sealing mechanism for rotary (1/4 turn) valves. An historical perspective on the development of stem seals is offered. Various conceptual designs are examined. Selected designs are evaluated and tested until the objective of a reliable, cost effective seal mechanism with practical applications is achieved. Finally, field testing and in service experience is presented.

1. INTRODUCTION

Valves as we know them have been around for centuries - since the time water was first directed thru a conduit or pipe. It must have been apparent almost from the beginning that once fluid flow is generated, there is an immediate desire to control it so the flow of water, for example, could be started or stopped, reduced or increased, or directed at will so the water could serve its intended purpose with a minimum of effort and waste.

Early conduit designers (or piping engineers) must have placed plugs, or wedges or wafers or disks or rocks or balls or similar control elements into the conduit to achieve the type of control desired. Soon thereafter, he connected a shaft or rod or stem thru the wall of the conduit to the control element so he could change position of the control element from a location external to the conduit. To do this, of course, one had to cut a hole in the wall of the conduit so that this shaft or stem could reach thru to the control element. In its most generic sense then, a valve is a conduit with a control element inside it, a hole thru the conduit and a shaft running thru this hole from the control element to the external operator or controller.

Held in Manchester, England. Organised and sponsored by BHRA, The Fluid Engineering Centre and co-sponsored by the British Valve Manufacturers Association and the Institution of Mechanical Engineers Process Industries Division

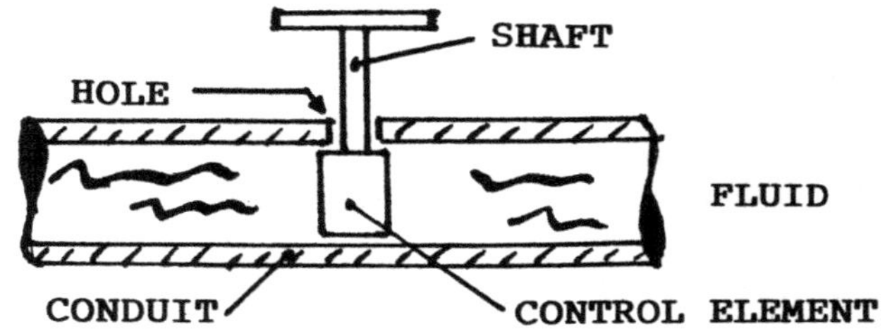

FIGURE 1

It's not hard to imagine the leakage that occured around these early shafts and ancient operators "stuffing" or packing" materials of one type or another into the hole to reduce or try to eliminate this shaft or stem leakage. This ancient practice of stuffing or packing materials into the hole around a shaft or stem must have been so pervasive and effective that this exact practice is still used today to seal most valve stems.

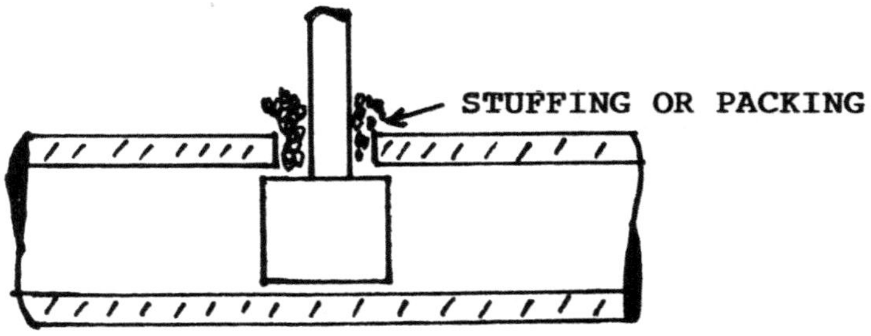

FIGURE 2

In fact, we use the same terminology. We all recognize a stuffing box or packing chamber and stuffing or packing materials and many of us have been trained in the practice of packing a valve. "Packing kits" and "repacking tools" are common elements of our business. The methods and materials we use today are very sophisticated and have resulted from years of effective research and development - but, the practice of "packing" a valve is at least as old as the Romans.

Packings work because they are "packed" or "stuffed" or "forced" into intimate contact with both the stem and the wall of the thru hole, effectively blocking the potential leak paths. And, while the stem is stationery, an effecive seal is generally maintained. However, the very nature of a valve suggests the desire to at least occasionally (often, very frequently) reposition the control element by means of linear, rotary or combination motion of the stem. When the stem is moved, the intimate bond betweeen the stem and the packing is of necessity, broken, and a leak path is created. Leakage, to some degree, always occurs when a stem moves in or thru a packing. This leakage may be so slight as to be undetected by normal instruments or methods, but it always exists and it generally increases as the packing begins to wear. This bond between the packing or stuffing material and the stem may also be broken due to the effects of pressure, temperature, corrosives, age or various combinations thereof. The end result is packings eventually leak at an unacceptable rate and are replaced. The act of repacking or replacing valves after packing leakage becomes unacceptable is still the most common practice employed today.

2. <u>DIAPHRAGM VALVES</u>

The first significant and widespread change in this practice occured in the 1930's when refrigerators were first introduced. These refrigerators depended on freon as the cooling media and it escaped thru conventional valve packings at an alarming rate. Unless packing leakage could be eliminated, the freon refrigerator would have been judged impractical. In response to the need to contain freon in a piping system, the first "packless metal diaphragm" valve was introduced. This valve (still in wide use today) employed a series of metallic hemispherically shaped diaphragms to isolate the fluid media and control element from the stem and operating mechanism. In effect, the "hole thru the conduit" was completely blocked by the diaphragms, rather than stuffed with packing, and the leak path was eliminated.

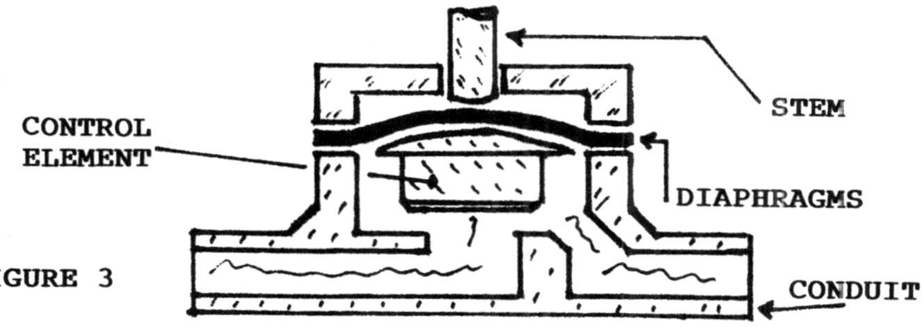

FIGURE 3

Motion of the control element, a globe disk or plug, was controlled by the deformation of the diaphragms which produced a linear motion only.

FIGURE 4

This concept proved highly effective in eliminating stem leakage in freon systems and ultimately many other compressed gas and vacuum services. Years later, this concept would prove itself equally effective in containing radioactive borated water in high pressure, high temperature nuclear power plant operations. The drawback to this concept is that it produces only very small linear motions thus it's not practical to apply to anything but small globe or poppet type valves.

3. LINEAR BELLOWS VALVES

To extend the application range of this new "packless" technology, bellows valves were developed. A bellows, in a sense, is a multiple stack of diaphragms, either welded together at each inside and outside edge, or formed into a series of convolutes from one continuous piece of tubing.

FIGURE 5

This process of stacking diaphragms to form a bellows allowed the application of this technology to be extended to larger and varying types of linear motion gate and globe type valves. The bellows performs exactly the same primary function as the diaphragm. It completely surrounds the stem and provides a pressure tight barrier, effectively covering the conduit stem hole with a very flexible, pressure tight barrier, isolating the control element and fluid from the stem, and eliminating the need to "pack" or "stuff" the stem opening.

FIGURE 6

Bellows gate and globe type valves are now available in a large variety of sizes and materials and their application usage is growing. With bellows valves, packless technology is extended beyond the small diaphragm valves to much larger globe as well as gate type valves.

Until the present time, however, bellows and diaphragm stem sealing technology has been limited to only linear motion gate and globe type valves. It is the intent of this invention to extend zero stem leakage packless technology to the entire class of <u>rotary</u> (commonly 1/4 turn valves) as well as linear motion valves.

4. ROTARY BELLOWS VALVES

4.1 BACKGROUND OF INVENTION

As explained above, conventional "stuffed" or "packed" valves are employed until "unacceptable" leak rate occurs and then the valve is repacked or replaced. Today the definition of an "unacceptable" leak rate is changing rapidly. As chemical processes become more sophisticated, more exotic and often more dangerous chemcials are employed. In these cases, almost any packing leaking is unacceptable because of the potential health risk and/or potential damage to the environment. In addition, efforts to reduce expensive product and/or energy losses (steam for example) are providing economic incentives to eliminate all product or energy losses. Further, efforts to improve chemical product quality and purity demand tighter process control and the elimination of any transient which may result from in-leakage or out-leakage of the system. Thus, the triple driving forces of:

a) **Safety** - to worker and the environment

b) **Economics** - due to potential loss of high cost products or energy and cost of clean-up and insurance

c) **Product Quality** - eliminating all process leak paths

All tend to demand better technology for the elimination of stem packing leakage. Because of the success discussed earlier with linear motion diaphragm and bellows sealed gate and globe valves, these "packless" valves are being employed in more and more process applications. However, most process valving today is of the full ported ball or plug type construction so that there is minimal flow restrictions while the process is in operation, and, quick shutdown or shutoff can be accomplished with a quick, 90 degree or 1/4 turn motion of the valve handle. Therein is the challenge - to adapt diaphragm, bellows or some new sealing technology which has proved quite effective only with linear motion valves to the task of eliminating stem leakage from rotary or as more commonly described, 1/4 turn valves.

4.2 CONCEPT ANALYSIS

Because linear motion diaphragm and bellows stem seals were well known and have proved effective for at least 50 years, it was first thought desireable to convert linear to rotary motion below (inside the valve) the diaphragm or bellows. In this way, the only new technology would revolve around the appropriate mechanical leakages or gears and not in the seal itself.

Several attempts were made to convert short stroke diaphragm deflection into large scale rotary motion. Two problems became apparent:

1. Short diaphragm motion demanded large, complicated gears and linkages to produce rotary motion.

2. When attempting to amplify motion, one lost a corresponding amount of force so that the output forces were only a fraction of the input forces.

Consider, for example, the below illustration:

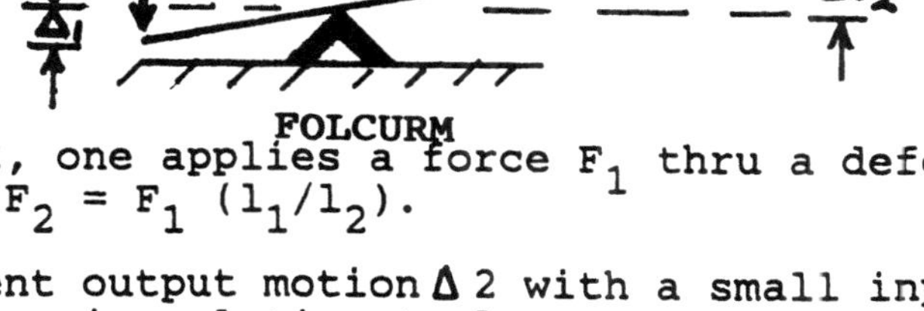

FIGURE 7

FOLCURM

To create motion, $\Delta 2$, one applies a force F_1 thru a deformation $\Delta 1$. The resulting force $F_2 = F_1 (l_1/l_2)$.

To generate sufficient output motion $\Delta 2$ with a small input motion, $\Delta 1$, l_2 must be quite large in relation to l_1.

For example, to amplify the input motion Δ_1 by a factor of 10 at Δ_2, one needs 10 times the output force F_2 at F_1.

$$\text{If } \Delta_2 = 10\,\Delta_1$$
$$F_1 = 10\,F_2$$

This quickly gets very impractical as the desired output motion may not be 10 but rather 100 times the input (diaphragm deflection) motion. Because of this, further attempts to convert short stroke diaphragm motion into rotary motions were abandoned.

Employing a bellows rather than a diaphragm yields signigicantly more input motion available to convert to output motion, which could be then converted to rotary motion. Two problems still persist with this approach.

1. A 10 to 1 or so mechanical advantage is still required. This means the input force must still be considerably higher than the required output force. Because the actual force and torque required to turn rotary valves is high, the required input force is still very, very high. This demands mechanical advantage devices such as screw threads on the input shaft. Then, if a screw thread is used, large rotational input motion can be converted to 90 degrees of output motion. The problem here is that now the valve can no longer be turned 90 degrees to open or close. Many degrees of rotation (3600 degrees or more) are typically required and one of the primary benefits of 1/4 turn valves, its ability to be quickly and safely opened or closed, is lost.

2. The second difficulty with this approach is that an extensive amount of high stressed hardware is now included within the wetted pressure boundary of the valve below the bellows. This seemed undesirable when issues of reliability, chemical compatabiltiy, maintenance, etc. are considered. Thus, for the compound reasons of loss of quick open-close motion, high input forces and extensive hardware now wetted and contained within the primary boundary, this idea was abandoned.

4.3. DISCOVERY

During the ongoing process of examining and experimenting with bellows behavior, a curious observation was made. If one took a bellows and bent it at 90 degrees so that the axis of the top of the bellows is perpendicular to the axis of the base of the bellows such as illustrated below:

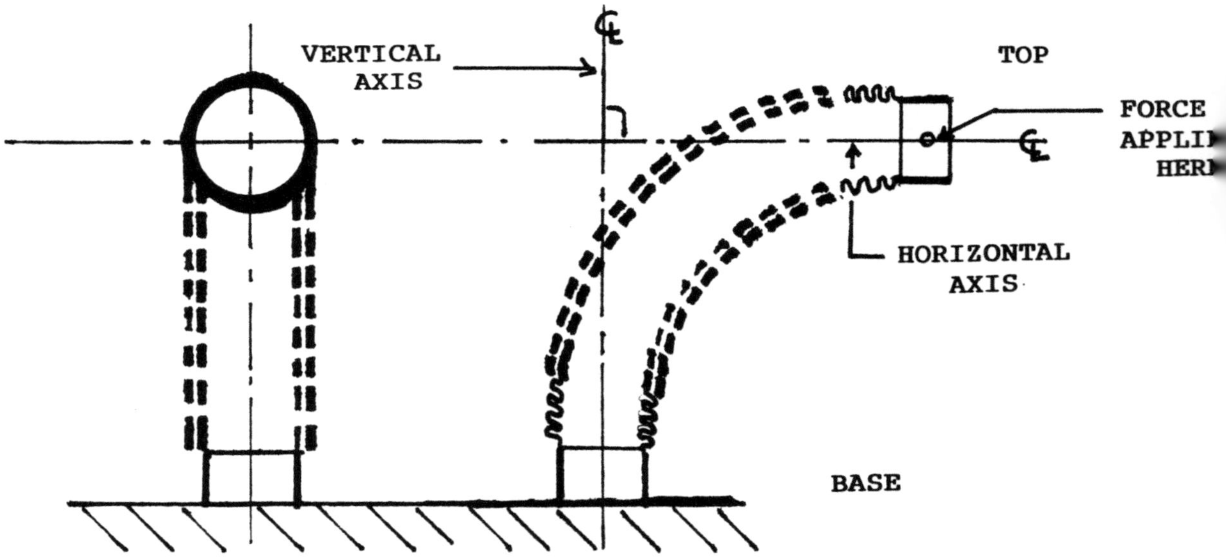

FIGURE 8

Then, one can hold the base stationery and observe the top appearing to rotate about the centerline of the base (℄) as a force is applied at the top end. One can observe this same behavior using a simple flexible drinking straw to represent the bellows, as explained below:

> Consider a flexible straw which does a nice job of illustrating the principle of operation. Hold the long end firmly in your hand (as if it were fixed to the bonnet of a valve) and bend the other end 90 degrees. Then, push on the bent end in the direction which will cause rotation about the fixed end. You will observe the bent end freely rotating as the straw (or bellows) travels in a 90°, 180°, 270°, or 360° arc. At the end of 360 degrees of travel, the straw (bellows) rotating end has returned to its original position without windup or residual torque of any type, all while the fixed end remained fixed. The lines on the straw help you observe the rotation of the free end about the fixed end. The rotation is the result of a force applied at the free end of the straw (bellows) times the moment arm produced when the straw (bellows) is bent at 90 degrees. Observe that no torque is ever applied to the straw (bellows) but simply a force.

4.4 <u>ANALYSIS OF MOTION</u>

A more detailed explanation of the bellows behavior is as follows:

A bellows has unique elastic deformation characteristics which, coupled with the proper design configuration, makes possible its use as a hermetic seal for a rotary valve. The bellows configuration used in the Kerotest design is shown in figure 8. With the bellows bent at 90° degrees and fixed at one end as shown, the bellows can be rotated about the vertical axis. This motion is possible only because there is a cumulative motion of the bellows convolutes that produces a rotation about the <u>bellows</u> axis that varies from zero at the fixed end to a rotation of the free end corresponding one-to-one with the rotation about the vertical axis. The motion illustrated in figure 8 is produced by application of a force at the free end. A torque is not applied at the free end as it is not needed to produce the motion of the bellows.

The explanation of the apparent paradox that one end of the torsionally stiff bellows can rotate with respect to a fixed end without buckling lies in arranging the bellows configuration so that the motion results from a bending action and not from a twisting or torsion action. As illustrated in figure 9, for the same torque, T, and bending moment, M=Fh, the bellows bending deflection, θ, is very large compared to the torsional deformation, $\Delta\theta$. This is the key to understanding the motion of the bellows in the hermetic seal application.

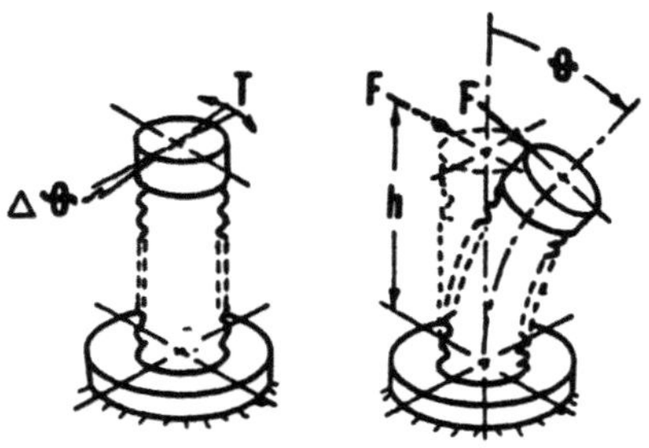

FIGURE 9

To show that bending motions of the bellows alone can produce a rotation of the bellows relative to its axis it is helpful to look at a sequence of pure bending motions as in figure 9.

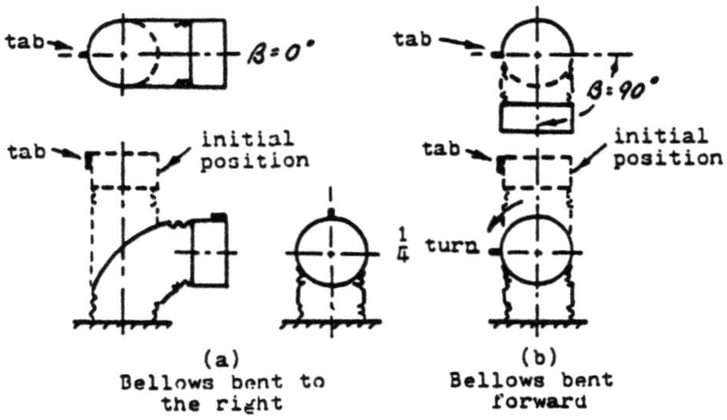

FIGURE 10

A tab is attached to the free end of the bellows to track the orientation of the bellows material as the bellows is first bent down 90 degrees to the right (β = 0), figure 10a, and then 90 degrees forward (β =90), figure 10b; starting each time from the vertical position. This sequence of bending motions results in a final position of the free end equivalent to a clockwise rotation of 90 degrees about the vertical axis and a counterclockwise rotation of the free end about its own axis. A similar sequence of bending motions can be made for any angle β from zero to 360 degrees and careful construction of the diagrams will show that the angular rotation of the free end about its axis is always equal to the rotation of the bellows about the vertical axis. This sequence of motions, bending constantly from the vertical to horizontal axis, just described, is not feasible for a valve seal application.

However, if the bellows is first bent at 90 degrees and a force is applied as in figure 11, bending moments again exist that will produce the same results (motions) as described for the pure bending case illustrated in figure 10.

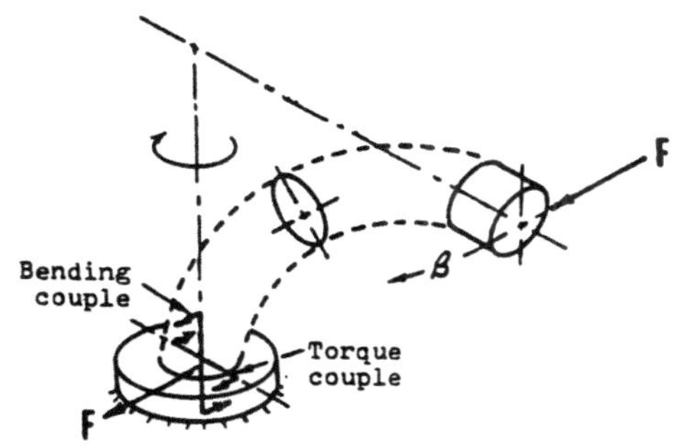

Figure 11

The bending motion is more difficult to visualize but it is clear that a moment exists on each convolute which can be resolved into bending and torsional components. These components vary from a maximum at the fixed end to zero at the free end. The high torsional stiffness of the bellows forces the motion to take place through the weaker bending mode and the net result is an orientation of the bellows material for any angle that is precisely the same as if the position were reached through the sequence of bending actions described above in figure 10.

In the Kerotest hermetic bellows stem seal concept, the stem is enclosed within the bellows and the free end of the bellows is capped. A bearing at the free end and guides placed along the stem mimimize friction as the bellows rotates with respect to the stem. The guides also confine the motion of the bellows to eliminate unwanted deformations and avoid squirm. A schematic sketch of this arrangement is shown in figure 12.

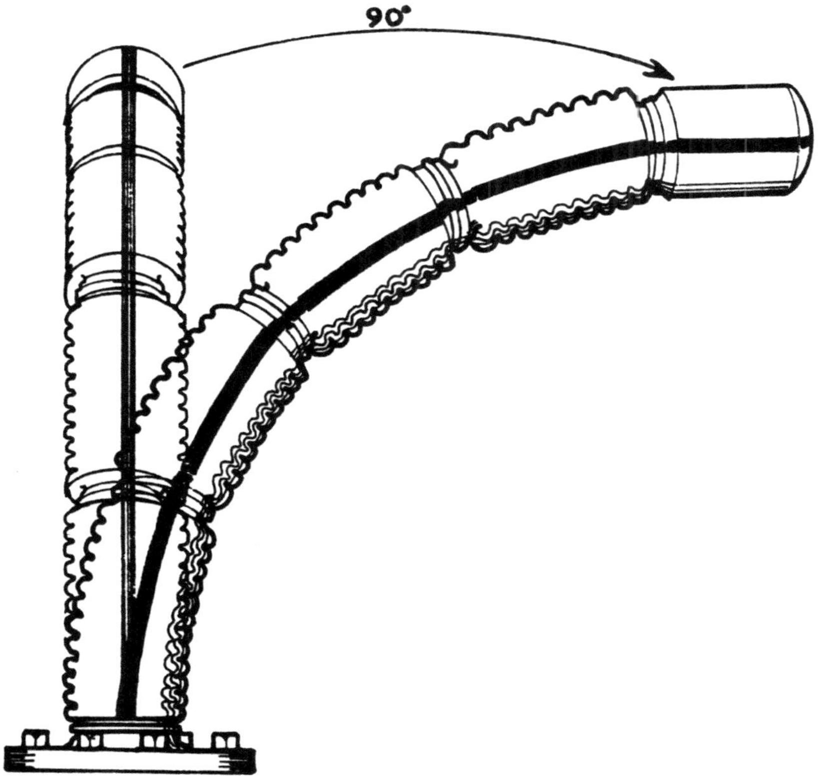

FIGURE 12

(SEE ALSO FIGURE 13)

4.5 HARDWARE DEVELOPMENT

Now that we observe and understand how one can create rotary motion thru a bellows, the next step is to incorporate this concept into practical hardware, suitable for a significant segment of real-world applications.

First, a flange*is welded to the base of the bellows for attachment to the valve (by bolting or welding) and a cap is welded to the top or free end of the bellows. Now the bellows becomes a pressure containing device. Next we come up out of the top of the valve with a stem or stem adapter bent at 90 degrees and install the bellows over the stem. The stem is simply a rigid metal member bent at 90 degrees and completely enclosed or captured within the bellows assembly. Then, any potential leakage up the stem is captured within the bellows and cannot escape to the environment. The valve stem is now part of the internal boundary of the valve - it does not extend into the environment, thus there is no leak path to the environment. The bellows completely captures and encloses the stem, eliminating the need for "stuffing" or packing" around the stem. The result is a "packless" rotary valve. In a sense, it is as if a cap had been installed completely over the valve and welded shut. It's apparent stem leakage would be eliminated with this solution. The problem, of course, is the valve is no longer operable! The Kerotest hermetic bellows stem seal is in effect a flexible, welded cap, always fully enclosing the stem in a leak tight metal housing which is also flexible enough to permit complete rotary operation of the valve, not just thru 90 degrees but to 360 degrees or beyond if desired - without torque or windup in the bellows.

4.6 RESULTS

A flexible metal bellows assembly fully surrounding, enclosing and turning with the stem is the essence of the rotary stem seal device concept. Once the concept was well defined, the task then was to prove it had practical, cost effective applications. It was believed that the following criteria as a minimum had to be achieved:

1. It must be useful on valves up to at least 2" size
2. It must withstand pressures of a ANSI class 150 valve - 225 PSI
3. At least 3000 cycles of full-open-close cycles must be achieved
4. The bellows and associated hardware must be of corrosion resistant materials

Extensive analysis and testing over 4 years in research and development of various bellows materials, thicknesses, lengths, bend ratio, precompression, etc. finally yielded a set of stainless steel and incone assemblies which could satisfy the above criteria. Then, it was decided to extend the application of this technology to larger valves and also to the next higher pressure class - Class 300, i.e. 720 PSI service. Eventually this too was successful and the current embodiment of the design satisfies the following criteria:

1. It is suitable for <u>any</u> rotary ball, plug or butterfly valve
2. It can be adapted for 90 degree, 180 degree, 360 degree or any other degree of rotary motion without limitation
3. It can be applied to typical ball and plug valves thru 6" in size and butterfly valves thru 12" in size
4. It can satisfy the full pressure range for either Class 150-225 PSI or Class 300-720 PSI service
5. All wetted parts are of corrosion resistant materials

*** Refer to figure 13**

Once the primary bellows assembly proved effective, the final job was to wrap it in appropriate auxiliary hardware to survive in its anticipated environment.

4.7 AUXILIARY HARDWARE

There were three essential functions which were to be incorporated into any of the proposed auxillary hardware associated with the bellows stem seal device:

1. It must provide a means of applying a force through the top of free end of the bellows
2. It must protect the bellows from external mechanical or chemical attack
3. It must permit rotation of the bellows assembly relative to the external mechanism

AND 4. It should incorporate a secondary or backup sealing capability in the event of a bellows leak

There are numerous solutions to these problems and we don't believe the optimum arrangment has yet been devised. However, there exists a practical set of hardware which clearly satisfies all these criteria.

Surrounding the bellows assembly is a stainless steel housing. This housing is shaped consistent with the required bellows bend radius and installed length (see figure 13).

It is stainless steel to provide the desired corrosion resistance if it is ever wetted by the process fluid. It has sufficient wall thickness to both achieve the ruggedness and desired strength to resist external mechanical damage and also to satisfy code minimum wall thickness criteria of a Class 300 valve. It is threaded at the top end to allow installation of the handle, to preload the bellows and to form a compression type mechanical seal. It has a secondary "O" ring sealing surface on the lower end and a shoulder to allow installation in the valve.

The housing is installed over the bellows assembly and clamped to the valve. Bearings are installed on both shoulders of the housing to allow free rotation of the housing relative to the clamp and the valve bonnet. A handle is installed via the threads on the upper end of the housing. To operate the valve one simply grips the handle and pulls the housing thru a 90 degree arc to close (or 180 degree or more if desired) or open the valve. When one pulls on the handle, this force is transmitted thru the housing, bearing and bellows cap to the stem installed within the bellows. Then, the stem rotates thru its 90 degree arc, the bellows free end rotates over the stem and within the housing and the housing rotates in the same path as the stem. Torque is created at the lower or output end of the stem due to the force applied times the moment created when the stem was bent. This output torque is directly coupled to the valve rotary element (ball, plug or butterfly) and the element moves to its open, close or intermediate position as desired.

One can see that the product or output of the bellows assembly is <u>torque</u> provided thru a hermetically sealed shaft, to turn a valve rotary element; or, to turn <u>any</u> device within a pressure or vacuum vessel. This device is currently being applied to valve applications but there may be broader applications where it is desired to create high torque rotary motion thru a sealed shaft.

In addition to the manual action described above, this device is also used with conventional valve actuators for remote operations. In this case, the valve handle is replaced (because it is no longer required) by a cap at the top end of the housing. This cap serves the functions of preloading the bellows and the gasket seal so the housing is still an effective secondary sealing chamber. Then, a conventional mount is built around the housing and the actuator installed on the top plate of the mount. The actuator ouput shaft is directly connected to the housing via a mechanical clamp or a welded boss. Then, as the actuator is energized, its output torque turns the housing, the bellows assembly and stem inside and the valve internal rotary element. The exact same actuator which is commonly used with that particular valve is also used when the bellows stem seal device is employed. The additional turning torques required when the bellows stem seal is employed is negligable when compared with the torque required to turn the valve internal rotary element.

The internal annular space between the bellows assembly and the housing can be accessed to detect product if the bellows should ever begin to leak. It can also be purged or a buffer gas added here if desired.

4.7 FIELD TESTING AND APPLICATIONS

Testing of the original bellows seal concept and ongoing improvements have been conducted over the past eight (8) years. Field trial units were installed beginning in 1984 and the units are currently in a wide range of specific applications such as phosgene, hydrogen, benzene, chlorine, amenia, UF_6, high purity fluids and numerous other applications.

The bellows stem seal device is offered for sale as an enhancement to the customer's conventional rotary valves. This device can be adapted to almost any rotary valve within the size (torque requirements determine size limit) and operating pressure limits (720 PSI) of the devices. The customer need not change valves but merely specify his desired valve with the bellows stem seal mechanism.

5.0 CONCLUSION

A bellows stem sealing mechanism has been developed which provides for hermetic sealing of 1/4-turn or rotary valves. It can be adapted to a wide variety of valve types and sizes suitable for a wide range of applications where stem leakage or potential stem leakage is undesirable. This is a quantum step change in the availability of packless technology for valves in sophisticated service. The product is practical, cost effective and qualified for a wide variety of services.

6.0 ACKNOWLEDGEMENT

I would like to acknowledge the assistance of the management and staff of Kerotest who saw the visionary needs of this device and who dedicated very significant resources to the development efforts. I especially acknowledge Mr. Sam Brunetto, Senior Design Engineer of Kerotest, for his tireless efforts in transforming this concept into practical, reliable hardware. Special thanks also to Dr. Campbell Yates, Professor Emeritus and former Head, Department of Mechanical Engineering, University of Pittsburgh, for his technical dissertation on the behavior of the bellows.

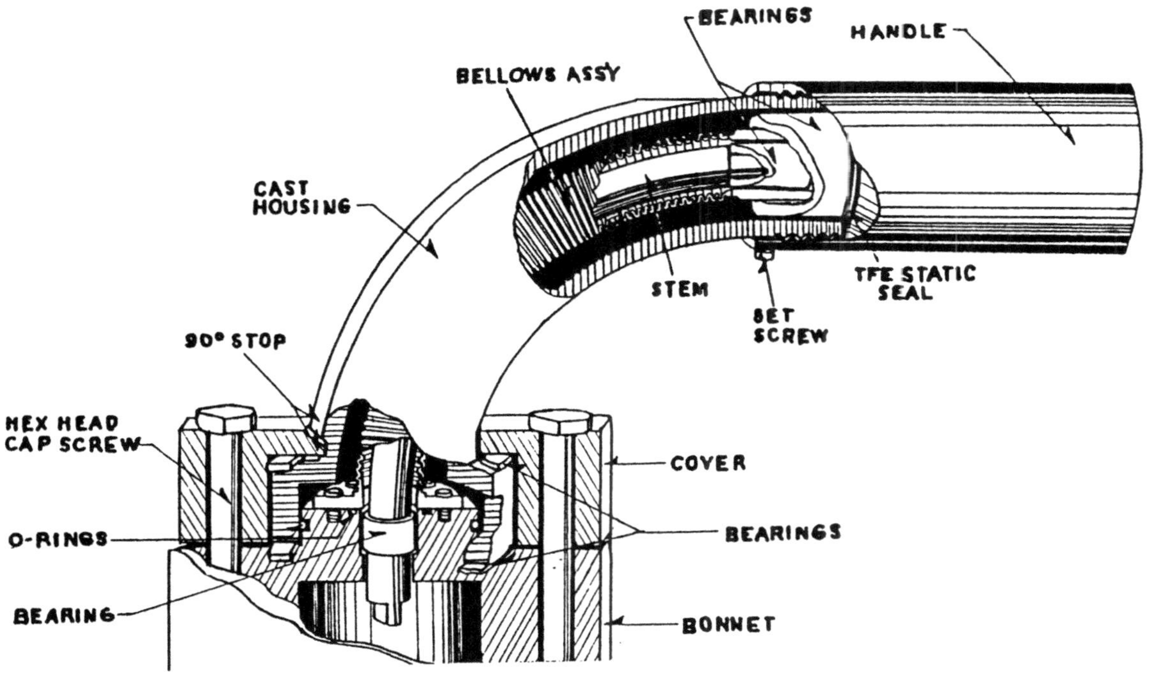

BEARINGS
HANDLE
BELLOWS ASSY
CAST
HOUSING
STEM
SET
SCREW
TFE STATIC
SEAL
90° STOP
HEX HEAD
CAP SCREW
COVER
O-RINGS
BEARINGS
BEARING
BONNET

FIGURE 13

GPSR Compliance

The European Union's (EU) General Product Safety Regulation (GPSR) is a set of rules that requires consumer products to be safe and our obligations to ensure this.

If you have any concerns about our products, you can contact us on ProductSafety@springernature.com

In case Publisher is established outside the EU, the EU authorized representative is:

Springer Nature Customer Service Center GmbH
Europaplatz 3
69115 Heidelberg, Germany

Batch number: 10762211

Printed by Printforce, the Netherlands